# Statphys Kolkata VII



Journal of Physics: Conference Series Volume 297

## Kolkata, India
## 26 – 30 November 2010

**Editors:**

J. K. Bhattacharjee
Jun-Ichi Inoue
Bikas K Chakrabarti
Parongama Sen

ISBN: 978-1-61782-920-8
ISSN: 1742-6588

**Printed from e-media with permission by:**

Curran Associates, Inc.
57 Morehouse Lane
Red Hook, NY 12571

**Some format issues inherent in the e-media version may also appear in this print version.**



Printed by Curran Associates, Inc. (2011)

For permission requests, please contact the Institute of Physics
at the address below.

Institute of Physics
Dirac House, Temple Back
Bristol BS1 6BE UK

Phone: 44 1 17 929 7481
Fax: 44 1 17 920 0979

techtracking@iop.org

**Additional copies of this publication are available from:**

Curran Associates, Inc.
57 Morehouse Lane
Red Hook, NY 12571 USA
Phone: 845-758-0400
Fax: 845-758-2634
Email: curran@proceedings.com
Web: www.proceedings.com

# TABLE OF CONTENTS

**PREFACE**

In the past two decades, a series of international conferences on Statistical Physics, going by the name Statphys Kolkata, have been organized in Kolkata (previously Calcutta) at roughly three-year intervals, the first one being held in 1992–93. The seventh of this series, **Statphys Kolkata VII** (http://www.saha.ac.in/cmp/stat.vii/index.php) was held from 26–30 November 2010. This meeting was organized as part of the Silver Jubilee Celebration of the Satyendra Nath Bose National Centre for Basic Sciences, Kolkata, in collaboration with the Saha Institute of Nuclear Physics, Kolkata.

In Statphys Kolkata VII, a few topics of current interest such as Collective behavior and emergent phenomena, Systems far from equilibrium, Soft matter, and Quantum critical phenomena were given special emphasis, while various other issues of Statistical Physics were also addressed. We were happy to note that the conference attracted a large number of participants, and the talk and poster sessions generated a lot of discussions, arguments and collaborations. The articles appearing in this proceedings are based on the invited talks and selected poster presentations.

We would like to thank the *Journal of Physics Conference Series*, IOP, for publishing the proceedings of the conference, and the referees for their prompt and active support.

The proceedings of the earlier Statphys Kolkata conferences have appeared in Physica A, vol 384 (2007); Physica A, vol 346 (2005); Physica A, vol 318 (2003); Physica A, vol 270 (1999); Physica A, vol 224 (1996); and Physica A, vol 186 (1992).

We would like to take this opportunity to thank all the members of the organizing committee (especially Dr Anjan Kumar Chandra for extensive all-round help), and acknowledge the Centre for Applied Mathematics and Computational Science (CAMCS, Saha Institute of Nuclear Physics, Kolkata) and Satyendra Nath Bose National Centre for Basic Sciences, Kolkata, for their financial support.

Jayanta Kumar Bhattacharjee, Bikas K Chakrabarti, Jun-Ichi Inoue and Parongama Sen
*Convenors and Editors of the proceedings*

STATPHYS-Kolkata VII
Journal of Physics: Conference Series **297** (2011) 012001 doi:10.1088/1742-6596/297/1/012001

# Neuronal avalanches and learning

**Lucilla de Arcangelis**
Department of Information Engineering and CNISM,
Second University of Naples, 81031 Aversa (CE), Italy

E-mail: dearcangelis@na.infn.it

**Abstract.** Networks of living neurons represent one of the most fascinating systems of biology. If the physical and chemical mechanisms at the basis of the functioning of a single neuron are quite well understood, the collective behaviour of a system of many neurons is an extremely intriguing subject. Crucial ingredient of this complex behaviour is the plasticity property of the network, namely the capacity to adapt and evolve depending on the level of activity. This plastic ability is believed, nowadays, to be at the basis of learning and memory in real brains. Spontaneous neuronal activity has recently shown features in common to other complex systems. Experimental data have, in fact, shown that electrical information propagates in a cortex slice via an avalanche mode. These avalanches are characterized by a power law distribution for the size and duration, features found in other problems in the context of the physics of complex systems and successful models have been developed to describe their behaviour. In this contribution we discuss a statistical mechanical model for the complex activity in a neuronal network. The model implements the main physiological properties of living neurons and is able to reproduce recent experimental results. Then, we discuss the learning abilities of this neuronal network. Learning occurs via plastic adaptation of synaptic strengths by a non-uniform negative feedback mechanism. The system is able to learn all the tested rules, in particular the exclusive OR (XOR) and a random rule with three inputs. The learning dynamics exhibits universal features as function of the strength of plastic adaptation. Any rule could be learned provided that the plastic adaptation is sufficiently slow.

## 1. Introduction

Cells in the nervous system can be divided into two categories: neurons and a variety of supporting cells, as neuroglial cells. Neurons are specialized for electrical signaling over long distances, whereas supporting cells are not capable of electrical signaling, despite having an important role in signal transmission. The human brain is estimated to contain 100 billion neurons and several times as many supporting cells. The interconnection of nerve cells via synapses forms an intricate network, which is the foundation on which sensory processes, perception and behaviour are built. The activity of neuronal networks consists in one or more action potentials in a single neuron or an ensemble of neurons. The first case is typical of small networks, as some experimental systems *in vitro*, where isolated spikes can be observed. The presence of a number of action potentials in an ensemble of neurons not always is a consequence of an external stimulus. Neuronal systems exhibit an intense spontaneous activity, known since a long time, whose relation with the response to stimulation is not fully understood yet. It is however well established that spontaneous activity cannot be simply reduced to a background noise uncorrelated to the system response. Indeed, experimental results for the cat visual cortex [1] have shown that the intensity of the response to an external stimulus is roughly proportional to the intensity of the spontaneous activity state of the system when the stimulus is applied. The variability in the response provided to the repeated application of the same stimulus is therefore caused by the different levels of ongoing activity. A similar analysis has been performed at the intracellular level on the same system, confirming that the spatio-temporal structure of the spontaneous activity influences the response signal [2].

The typical form of spontaneous activity consists in the almost synchronous emission of action potentials in a large number of neurons, followed by periods of substantial inactivity. These high activity events, named bursts, are observed both during development and in mature systems and can last from a few to several hundreds milliseconds. In 2003 Beggs and Plenz have identified a novel form of spontaneous activity, neuronal avalanches [3, 4]. Coronal slices of rat somatosensory cortex were placed onto a 8x8 multi-electrode array (MEA) and spontaneous activity was induced by bath perfusion with the glutamate receptor agonist NMDA in combination with a dopamine receptor agonist. The intrinsic activity of the system was monitored by measuring the potential at each electrode. This local field potential (LFP) integrates the electrical activity of neurons placed in the region surrounding the electrode: negative peaks in the LFP measure the influx of positive ions and therefore the cumulative membrane potential variation of the neurons in the region. Experimental data show that before 6 days in vitro activity is mainly composed of sparse activations but during the second week simultaneous activations occur in several electrodes. The novel idea was to examine this electrophysiological signal at a finer temporal scale, which was able to evidence a complex spatio-temporal structure. Indeed, activity starting at one electrode may involve more, not necessarily neighbour, electrodes. Binning time in cells of duration $\delta t$, allows to create a spatio-temporal grid reporting the active electrodes in each temporal cell. A neuronal avalanche is therefore defined as a sequence of successively active electrodes between two temporal bins with no activity. The total number of active electrodes, or alternatively the sum of all LFPs, is defined as the size $s$ of an avalanche and the time interval with ongoing activity as its duration $T$.

The striking result is that both size and duration have no characteristic value, i.e. their distributions exhibit a power law behaviour. The analysis at a finer temporal scale is then able to enlighten the non-synchronous character of the bursts. The exponents of these power law distributions depend on the choice for the temporal bin $\delta t$. Indeed larger bins make active electrodes belonging to different avalanches to merge into the same larger event, leading to a smaller exponent. In order to identify the appropriate value of $\delta t$, Beggs and Plenz verified that if this is equal to the average value of the time delay between two successive LFPs at one electrode for each culture, the exponent does not depend any longer on the specific culture. They were then able to identify the universal scaling behaviour

$$\begin{aligned} P(s) \propto s^{-\alpha} \quad \text{with} \quad \alpha = 1.5 \pm 0.1 \\ P(T) \propto T^{-\beta} \quad \text{with} \quad \beta = 2.0 \pm 0.1. \end{aligned} \tag{1}$$

The power law behaviour for the size distribution is followed by an exponential cutoff due to the finite size of the system, whereas for the duration distribution it extends for about one decade and the exponential cutoff sets in at about 10ms.

These results *in vitro* have been confirmed by extended studies *in vivo* on anaesthetized rats during development [5] and awake rhesus monkeys [6]. Synchronous activity in the bursts exhibits the same scaling behaviour found for neuronal avalanches *in vitro* (Eq.1). The investigation on the spontaneous activity has been performed also for dissociated neurons from different networks as rat hippocampal neurons [7], rat embryos [8] or leech ganglia [7]. Under fixed conditions of humidity and temperature, neurons start to develop a network of synaptic connections and, after a variable period *in vitro*, exhibit spontaneous electrical activity. Different behaviours are observed. Only those system exhibiting a medium level of synchronization between random spikes and synchronized bursts exhibit critical behaviour. For those cultures the scaling behaviour is very robust and in agreement with Eq.1. In particular, the emergence of a critical state has been found to be strongly related to the ageing of the system, namely after the first few weeks *in vitro*, where the behaviour of the system is subcritical, some cultures may self-organize and reach the critical state as they mature [8].

## 2. The model

In order to formulate a new model to study neuronal activity, we have introduced [9, 10] within a SOC approach the three most important ingredients, namely threshold firing, neuron refractory period and activity-dependent synaptic plasticity. We consider a lattice of $N$ sites on which each site represents the cell body of a neuron, each bond a synapse. Each neuron is characterized by the potential $v_i$. Connections among neurons are established by assigning to each neuron $i$ a random out-going connectivity degree, $k_{out_i}$. The distribution of the number of out-connections is then chosen in agreement with the experimentally measured properties of the functionality network [11] in human adults. Functional magnetic resonance imaging has indeed shown that this network has universal scale free properties, namely it exhibits a scaling behaviour $n(k_{out}) \propto k_{out}^{-2}$, independent of the different tasks performed by the patient. We adopt this distribution for the number of pre-synaptic terminals of each neuron, over the range of possible values between 2 and 100, as in experimental data. Two neurons are then connected according to a distance dependent probability, $p(r) \propto e^{-r/r_0}$, where $r$ is their spatial distance [12] and $r_0$ a typical edge length. To each synaptic connection we then assign an initial random strength $g_{ij}$, where $g_{ij} \neq g_{ji}$ and an excitatory or inhibitory character, with a fraction $p_{in}$ of inhibitory synapses. Whenever at time $t$ the value of the potential at a site $i$ is above a certain threshold $v_i \geq v_{\max}$, approximately equal to $-55mV$ for the real brain, the neuron fires, i.e. generates an "action potential, distributing charge to its connected neighbours. The total charge released by a neuron is proportional to the number of synaptic connections, $q_i \propto v_i k_{out_i}$. Each connected neuron receives charge in proportion to the strength of the synapse $g_{ij}$

$$v_j(t+1) = v_j(t) \pm \frac{q_i(t)}{k_{in_j}} \frac{g_{ij}(t)}{\sum_k g_{ik}(t)} \tag{2}$$

where $k_{in_j}$ is the in-degree of neuron $j$ and the sum is extended to all out-going connections of $i$. In Eq.2 it is assumed that the received charge is distributed over the surface of the soma of the post-synaptic neuron, proportional to the number of in-going terminals $k_{in_j}$. The plus or minus sign in Eq.2 is for excitatory or inhibitory synapses, respectively. After firing a neuron is set to a zero resting potential and in a refractory state lasting $t_{ref} = 1$ time step, during which it is unable to receive or transmit any charge. We wish to stress that the unit time step in Eq.2 does not correspond to a real time scale, it is simply the time unit for charge propagation from one neuron to the connected ones. In real system this time could vary and be as large as 100 ms for longer firing periods. The practical implication of the refractory period is that the action potential does not propagate back toward the initiation point and therefore is not allowed to reverberate between the cell body and the synapse. The synaptic strengths have initially equal value, whereas the neuron potentials are uniformly distributed random numbers between $v_{\max} - 1$ and $v_{\max}$. Moreover, a small random fraction (10%) of neurons is chosen to be boundary sites, with a potential fixed to zero, playing the role of sinks for the charge. An external stimulus is imposed at a random site and, if the potential reaches the firing threshold, the neuron fires and an avalanche of firing neurons can propagate through the system.

### *2.1. Plastic adaptation*

As soon as a site is at or above threshold $v_{\max}$ at a given time $t$, it fires according to Eq.2, then the conductance of all the bonds, connecting to active neurons and that have carried a current,

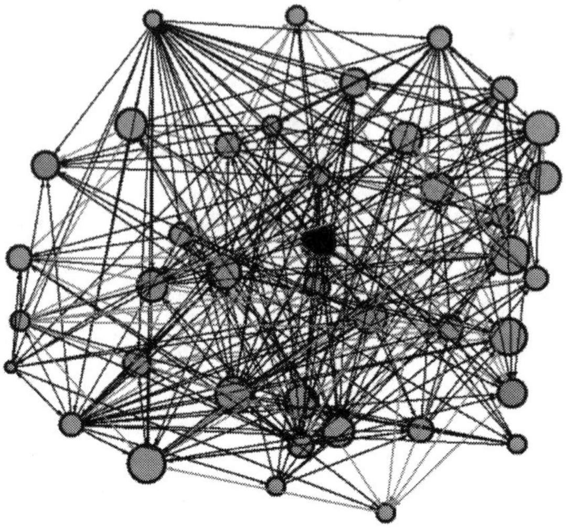

**Figure 1.** One neuronal avalanche involving 15 neurons on the scale-free network with $N = 40$ neurons connected by directed bonds (direction indicated by the arrow at one edge), representing the synapses. The size of each neuron is proportional to the number of in-connections, namely the number of dendrites. The two red neurons are the two input sites, whereas the black neuron is the output. Connections involved in the avalanche propagation are shown in red, whereas inactive connections are black.

is increased in the following way

$$g_{ij}(t+1) = g_{ij}(t) + \delta g_{ij}(t) \tag{3}$$

where $\delta g_{ij}(t) = k\alpha i_{ij}(t)$, with $\alpha$ being a dimensionless parameter and $k$ a unit constant bearing the dimension of an inverse potential. After applying Eq.3 the time variable of our simulation is increased by one unit. Eq.3 describes the mechanism of increase of synaptic strength, tuned by the parameter $\alpha$. This parameter then represents the ensemble of all possible physiological factors influencing synaptic plasticity. Once an avalanche of firings comes to an end, the conductance of all the bonds with non-zero conductance is reduced by the average conductance increase per bond

$$\Delta g = \sum_{ij,t} \delta g_{ij}(t)/N_b \tag{4}$$

where $N_b$ is the number of bonds with non-zero conductance. The quantity $\Delta g$ depends on $\alpha$ and on the response of the brain to a given stimulus. In this way our electrical network "memorizes" the most used paths of discharge by increasing their conductance, whereas the less used synapses atrophy. Once the conductance of a bond is below an assigned small value $\sigma_t$, we remove it, i.e. set it equal to zero, which corresponds to what is known as pruning. These mechanisms correspond to a Hebbian form of activity dependent plasticity, where the conjunction of activity at the presynaptic and postsynaptic neuron modulates the efficiency of the synapse [13]. To insure the stable functioning of neural circuits, both strengthening and weakening rule of Hebbian synapses are necessary to avoid instabilities due to positive feedback [14]. However, differently from the well known Long Term Potentiation (LTP) and Long Term Depression (LTD) mechanisms, in our model the modulation of synaptic strength does not depend on the frequency of synapse activation [15, 16, 17]. It should be also considered that in the living brain synapses exhibiting plasticity are not electrical but chemical. For instance, Hebbian plasticity at excitatory synapses is classically mediated by postsynaptic calcium dependent mechanisms [18]. In our approach the excitability of the postsynaptic neuron is simply modulated by the value of the electrical potential.

## 3. Neuronal avalanches in spontaneous activity

The plasticity procedure is applied for a series of $N_p$ stimuli to adapt the strengths of synapses. In fact, the more the system is active strengthening the used synapses, the more the unused connections will weaken. The plastic adaptation proceeds until few bonds are pruned. The system at this stage constitutes the first approximation to a trained brain, on which measurements are performed. These consist of a new sequence of stimuli, by adding a small quantity to the voltage of a random neuron. We monitor the avalanche activity s function of time.

We measure the size distribution of neuronal avalanches. The size is either the total number of firing neurons, or the to sum of their voltage variations. This distribution exhibits a power law behaviour, with an exponent equal to $1.5 \pm 0.1$, quite stable with respect to parameters. This scaling behaviour is also robust for densities of inhibitory synapses up to 10%, whereas the scaling behaviour is lost for higher densities. Moreover, the distribution of avalanche temporal durations is also a power law with an exponent close to $-2.0$. Both these values show an excellent agreement with experimental data. Moreover, the temporal signal for electrical activity and the power spectrum of the resulting time series have been compared with EEG data [9, 10]. The spectrum exhibits a power law behaviour, $P(f) \sim f^{-0.8}$, with an exponent in good agreement with EEG medical data [19] and physiological signal spectra for other brain controlled activities [20]. This model therefore seems to capture many of the essential ingredients of spontaneous activity, as measured in cortical networks (Eq.1).

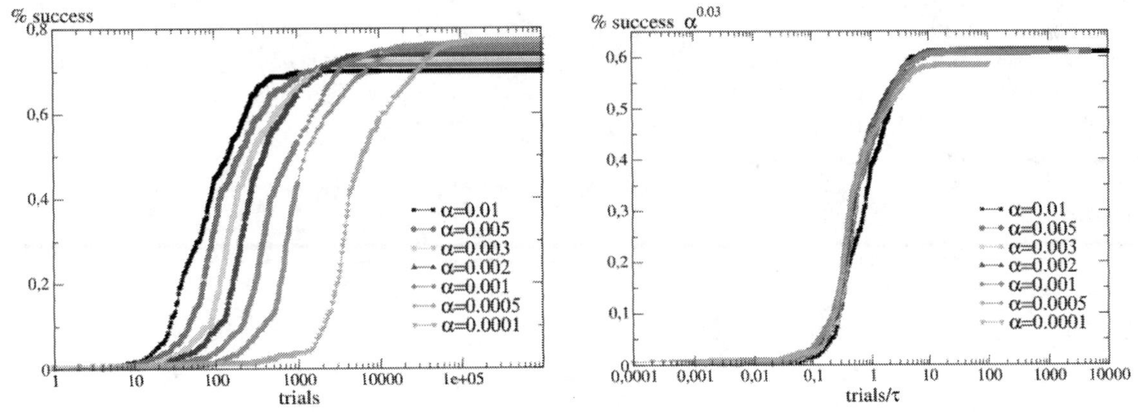


**Figure 2.** *(Left)* Percentage of configurations learning the AND rule as function of the number of learning steps for different plastic adaptation strengths $\alpha$ (decreasing from bottom to top). Data are for 400 realizations of a network with $N = 1000$ neurons, $p_{in} = 0.1$, $r_0 = 15$, $k_{min} = 3$ and $k_d = 5$. *(Right)* Collapse of the curves by rescaling the number of learning steps by the characteristic learning time $\tau = 1/\alpha$ and the percentage of success by $\alpha^{-0.03}$.

## 4. Learning

Next we study the learning performance of this neuronal network acting in a critical state [21]. In order to start activity we identify input neurons at which the rule is applied and the output neuron at which the response is monitored. These nodes are randomly placed inside the network under the condition that they are not boundary sites and they are mutually separated on the network by $k_d$ nodes. $k_d$ represents the chemical distance on the network and plays the role of the number of hidden layers in a perceptron. We test the ability of the network to learn different rules: AND, OR, XOR and a random rule RAN which associates to all possible combinations of binary states at three inputs a random binary output. More precisely, the AND, OR and XOR rules are made of three input-output relations, whereas the RAN rule with three input sites implies a sequence of seven input-output relations. A single learning step requires the application of the entire sequence of states at the input neurons, monitoring the state of the output neuron. For each rule the binary value 1 is identified with the output neuron firing, namely the neuron membrane potential at a value greater or equal to $v_{\max}$ at some time during the activity. Conversely, the binary state 0 at the output neuron corresponds to the physiological state of a real neuron which has been depolarized by incoming ions but fails to reach the firing threshold membrane potential during the entire avalanche propagation. Once the input sites are stimulated, their activity may bring to threshold other neurons and therefore lead to avalanches of firings. We impose no restriction on the number of firing neurons in the propagation and let the avalanche evolve to its end according to Eq.2. If at the end of the avalanche the propagation of charge did not reach the output neuron, we consider that the state of the system was unable to respond to the given stimulus, and as a consequence to learn. We therefore increase uniformly the potential of all neurons by units of a small quantity, $\beta = 0.01$, until the configuration reaches a state where the output neuron is first perturbed. We then compare the state of the output neuron with the desired output.

Plastic adaptation is applied to the system according to a non-uniform negative feedback algorithm. Namely, if the output neuron is in the correct state according to the rule, we keep the value of synaptic strengths. Conversely, if the response is wrong we modify the strengths of those synapses involved in the information propagation by $\pm\alpha/d_k$, where $d_k$ is the chemical distance of the presynaptic neuron from the output neuron. The sign of the adjustment depends on the mistake made by the system: If the output neuron fails to be in a firing state we increase the used synapses by a small additive quantity proportional to $\alpha$. Synaptic strengths are instead decreased by if the expected output 0 is not fulfilled. This adaptation rule intends to mimic the feedback to the wrong answer triggered locally at the output site, for instance by some hormons, and propagating backward towards the input sites.

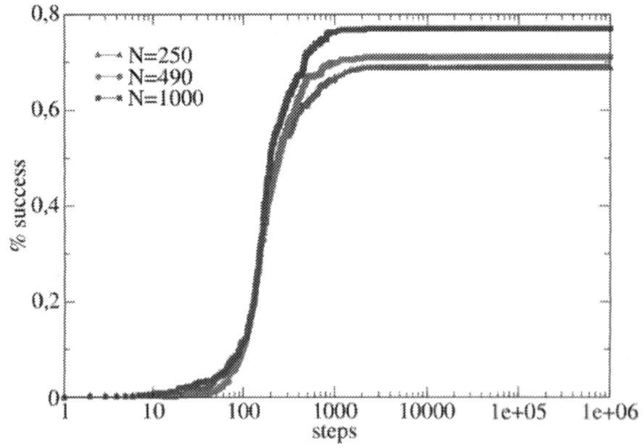


**Figure 3.** Percentage of configurations learning the XOR rule as function of the number of learning steps, for $\alpha = 0.005$, $k_d = 3$, the initial minimum connectivity degree equal to 3 and different numbers of neurons $N$ (from 250 to 1000 bottom to top).

We analyse the ability of the system to learn the different rules. Fig. 2 shows the fraction of configurations learning the AND rule versus the number of learning steps for different values of the plastic adaptation strength $\alpha$. We notice that the larger the value of $\alpha$ the sooner the system starts to learn the rule, however the final percentage of learning configurations is lower. The final rate of success increases as the strength of plastic adaptation decreases. This result is due to the highly non-linear dynamics of the model, where firing activity is an all or none event controlled by the threshold. The result that all rules give a higher percentage of success for weaker plastic adaptation, is in agreement with recent experimental findings on visual perceptual learning, where better performances are measured when minimal changes in the functional network occur as a result of learning [22].

We characterize the learning ability of a system for different rules by the average learning time, i.e. the average number of times a rule must be applied to obtain the right answer, and the asymptotic percentage of learning configurations. This is determined as the percentage of learning configurations at the end of the teaching routine, namely after $10^6$ applications of the rule. The average learning time scales as $\tau \propto 1/\alpha$ for all rules and independently of parameter values. The asymptotic percentage of success increases by decreasing $\alpha$ as a very slow power law, $\propto \alpha^{-0.04\pm0.01}$, where the exponent is the average value over different rules. We check this scaling behaviour by appropriately rescaling the axes in Fig.2. The curves corresponding to different $\alpha$ values indeed all collapse onto a unique scaling function. Similar collapse is observed for the OR, XOR and RAN rules and for different parameters $k_d$ and $p_{in}$. The learning dynamics shows therefore universal properties, independent of the details of the system or the specific task assigned.

Finally we explicitly analyse the dependence of the learning performance and its scaling behaviour on the number of neurons. Indeed, as the system size increases the number of highly connected neurons becomes larger. A well connected system provides better performances, therefore we could expect that the size dependence reflects the same effect. In Fig.3 we show data for a set of parameters and different system sizes. The learning performance indeed improves with the system size since, for the same out degree distribution, the overall level of connectivity improves for larger systems.

## 5. Conclusions

Extensive simulations have been presented for a novel brain model with activity dependent plasticity, implemented on scale free lattices. The ingredients of the model are close to most functional and topological properties of real neuronal networks. The avalanche size and duration distributions show a power law behaviour with exponents $1.5 \pm 0.1$ and $2.0 \pm 0.1$, respectively, compatible the values found experimentally for neuronal avalanches. Next, we investigate the learning ability of this model. The implemented learning dynamics is a cooperative mechanism where all neurons contribute to select the right answer and negative feedback is provided in a non-uniform way. Despite the complexity of the model and the high number of degrees of freedom involved at each step of the iteration, the system can learn successfully even complex rules. The neuronal network has a "universal" learning dynamics, even complex rules can be learned provided that the plastic adaptation is sufficiently slow.

## References

[1] Arieli G A, Sterkin A and Aertsen A 1996 *Science* **273** 1868
[2] Azouz R and Gray C M 1999 *J. Neurosci.* **19** 2209
[3] Beggs J M and Plenz D 2003 *J. Neurosci.* **23** 11167
[4] Beggs J M and Plenz D 2004 *J. Neurosci.* **24** 5216
[5] Gireesh E D and Plenz D 2008 *Proc. Nat. Acad. Sci.* **105**, 7576
[6] Petermann T, Thiagarajan T C, Lebedev M A, Nicolelis M A L, Chialvo D R and Plenz D 2009 *Proc. Nat. Acad. Sci.* **106**, 15921

[7] Mazzoni A, Broccard F D, Garcia-Perez E, Bonifazi P, Ruaro M E and Torre V 2007 PLoS ONE 2:e439
[8] Pasquale V, Massobrio P, Bologna L L, Chiappalone M, and Martinoia S 2008 *Neuroscience* **153**, 1354
[9] de Arcangelis L, Perrone Capano C and Herrmann H J 2006 *Phys. Rev. Lett.* **96**, 028107
[10] Pellegrini G L, de Arcangelis L, Herrmann H J and Perrone Capano C 2007 *Phys. Rev. E* **76**, 016107
[11] Eguiluz V M, Chialvo D R, Cecchi G A, Baliki M and Apkarian A V 2005 *Phys. Rev. Lett.* **94**, 018102
[12] Roerig B and Chen B 2002 *Cerebral Cortex* **12**, 187
[13] Cooper S J 2005 *Neurosci. Biobehav. Rev.* **28**, 851
[14] Desai N S 2003 *J. Physiol. Paris* **97**, 391
[15] Albraight T D et al 2000 *Neuron* Review supplement to vol.**59**
[16] Paulsen O and Sejnowski T J 2000 *Curr. Opin. Neurobiol.* **10** 172
[17] Braunewell K H and Manahan-Vaughan D 2001 *Rev. Neurosci.* **12**, 121
[18] Bi G Q and Poo M M 2001 *Annu. Rev. Neurosci.* **24**, 139
[19] E. Novikov, A. Novikov, D. Shannahoff-Khalsa, B. Schwartz, J. Wright, Phys. Rev. E. **56**, R2387 (1997).
[20] Hausdorff J M et al. 2001 *Physica A* **302**, 138
[21] de Arcangelis L and Herrmann H J 2010 *Proc. Natl. Acad. Sci.* **107**, 3977
[22] Lewis CM, Baldassarre A, Committeri G, Romani G L and Corbetta M 2009 *Proc Natl Acad Sci USA* **106** 17558

Journal of Physics: Conference Series **297** (2011) 012002 doi:10.1088/1742-6596/297/1/012002

# Modules of human micro-RNA co-target network

**Mahashweta Basu[1,†], Nitai P. Bhattacharyya[2], P. K. Mohanty[1]**

[1]Theoretical Condensed Matter Physics Division,
Saha Institute of Nuclear Physics, 1/AF Bidhan Nagar, Kolkata 700064, India.
[2]Crystallography and Molecular Biology Division,
Saha Institute of Nuclear Physics, 1/AF Bidhan Nagar, Kolkata 700064, India.

E-mail: [†]mahashweta.basu@saha.ac.in

**Abstract.** Human micro RNAs (miRNAs) target about 90% of the coding genes and form a complex regulatory network. We study the community structure of the miRNA co-target network considering miRNAs as the nodes which are connected by weighted links. The weight of link that connects a pair of miRNAs denote the total number of common transcripts targeted by that pair. We argue that the network consists of about 74 modules, quite similar to the components (or clusters) obtained earlier [Online J Bioinformatics, 10,280 ], indicating that the components of the miRNA co-target network are self organized in a way to maximize the modularity.

## 1. Introduction

Micro RNAs are a class of small single stranded non-coding RNAs, about 20 to 22 base long, which interfere with the translation of messenger RNAs (mRNAs) by binding to their $3'$ untranslated regions (UTR) [1]. Several computational methods [2] have been developed for predicting the mRNA transcripts which are possible targets of a particular miRNA. For example, 711 nucleotide sequences are predicted as miRNAs [3] of human; their possible targets, (34525 in total) are listed in the mirBASE database [4]. It has been proposed on the basis of theoretical analysis that as large as 90% human genes are targets of miRNA [5]. Regulation of coding genes by miRNAs in combination are also experimentally validated [6].

The abundance of miRNA and their targets provide enormous combinatorial possibilities for regulation. Combinatorial regulation of genes by transcription factor (TF) and miRNAs provides higher complex programs [7]. Recently, taking TFs as important mediators of miRNA-initiated regulatory effects, it was shown [8] that the underlying network is significantly associated with multicellular organismal development, cell development and cell-cell signaling. Combinatorial effect of miRNA modules [9] has been observed in tumor tissues or cell lines. This observation suggests a combinatorial effect of the module associated miRNAs on target gene regulation in selective tumor tissues or cell lines. Synergistic network [10] of miRNAs reveals that miRNA modules associated with diseases are significantly different from modules of miRNAs that does not involve in disease. Possibility of co-regulation of two or more miRNAs in context of gene expressions and relevant biological functions is, however, least explored.

Recently Mookherjee *et. al.* [11] have analyzed the miRNA co-target network (MCN) of *Homo sapiens*, which indicate that several group of miRNAs (so called *clusters*) provide most essential regulations. This topological analysis of miRNA network revealed that about 70 clusters of miRNAs co-target the genes, which are involved in specific pathways. For several clusters, all miRNAs belonging to the cluster are found to be maximally expressed in a specific tissue. Further studies [12], indicate that the clusters are also disease specific. Reorganizing miRNAs into such groups (clusters) helps in identifying cooperative activity of miRNAs. In fact, from these analysis one can predict that, "if one miRNA from a particular cluster is involved in a specific biological pathway or cellular function, the other miRNAs belonging to the same cluster are likely to be involved in the same disease, pathway or function".

Detection of communities, groups, components or clusters have been a focus of recent interest in context of complex networks. Networks like the world wide web [13], the metabolic network [14], the social network [15], protein protein interaction network [16] etc. do possess *community structures*, meaning the vertices tend to divide into groups, with dense connections within the groups and sparse connection existing among the groups. These communities act as the functional units of the network; for example 'ATP synthesis', 'DNA processing', and 'cell cycle control' are well known [17] functional modules of yeast protein-protein interaction network. Evidently, the functional properties of an entire network is quite different from their properties at community level.

In this article, we study the community structures (modules) of miRNA co-target network of human and compare them with the components (clusters) of miRNAs obtained earlier [11]. Since the components of a network are the only disjoint subgraphs, it is expected that the community structures can be better represented by the modules. This is explained schematically in Fig. 1, where the network has 3 components and 6 modules.

In section 2 we briefly review the relevant features of the miRNA co-target network of human and its components (clusters). In section 3 we apply the modularization method introduced by Newmann [18] to analyze this miRNA co-target network and compare the resulting modules with the clusters obtained earlier[11]. Finally, conclusions are given in section 4.

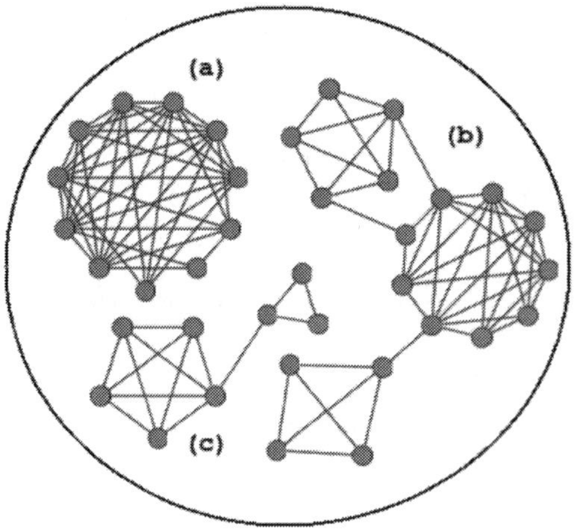


**Figure 1.** Component versus modules : Components (or clusters) are disjoint sub-graphs of a network (they do not have any common link), whereas modules may have relatively fewer connections among them. The above network has 3 components namely (a), (b) and (c) with each component consisting of 1, 3 and 2 modules respectively. Component (b) has three modules consisting of 4, 9 and 5 vertices, similarly component (c) consists of two modules comprising of 3 and 5 vertices.

## 2. Clusters of miRNAs in the co-target network

Let us briefly revisit the main ideas and results of Ref. [11] to understand the construction, topology and components of human miRNA co-target network consisting of 711 miRNAs and their 34525 predicted targets obtained from the miRBase database (*http://microrna.sanger.ac.uk/, version* 10). For convenience, miRNAs are given arbitrary identification number $m = 1, 2, \ldots i, \ldots M$, where $M = 711$. Further, the miRNA co-target network was constructed by considering miRNAs as the nodes and joining every pair of miRNAs having one or more common targets with a link. The total number of co-targeted transcripts $C_{ij}$ of miRNAs $i$ and $j$ is taken to be the weight of the corresponding link. Clearly, the resulting adjacency matrix $C$ is symmetric (with diagonal elements $C_{ii} = 0$).

Mookherjee et. al. [11] have proposed an elegant method for finding the clusters of miRNAs. Since substantially large number of miRNA pairs have only few co-targets, the links between them have small weights, and can be erased to obtain a simplified network. Let $N_q$ be the number of components of the network when all the links having weight less than $q$ are erased. Thus, the network breaks into smaller disjoint subgraphs (components) with rate $\frac{dN_q}{dq}$, which is maximum at $q = q^*$. It was argued that among all the subgraphs of the co-target network obtained at $q^*$, the largest one is the most important; miRNAs belonging to this subgraph are found to down regulate expression of genes involved in several genetic diseases.

To be specific, the human miRNA co-target network breaks into $N_q = 166$ subgraphs at $q = 103$, where the largest subgraph called $\mathcal{G}$ contain 477 miRNAs. To determine how miRNAs are organized within the subgraph $\mathcal{G}$, $q$ is increased further. At $q = 160$ the subgraph $\mathcal{G}$ breaks up into 70 small clusters (the subgraphs having two or more miRNAs) and 147 independent miRNAs. Out of total 70, 18 clusters arise from the seed sequence[1] similarities and 11 clusters are organized into the same genomic region (5 inter-genomic clusters also show seed sequence similarities). Most of the clusters are found to be either pathways, tissues and/or diseases specific.

In the following section we aim at investigating the modular structure of miRNA co-target network. Figure 1 schematically describes, why a network is better represented by its modules than its components (disjoint subgraphs).

## 3. Modular structure of miRNA network

The identification of community structure is one of the many challenging problems in various scientific field. A large variety of community detection techniques have been developed based on centrality measures, link density, percolation theory etc. Recently, Newman *et. al.*[18] proposed a method of finding community structure of a network based on maximization of the *modularity*. This method is further generalized to include weighted networks[19].

The most obvious way of finding groups in a network is to minimize the number of edges connecting the groups. Simply rearranging the network, such that only few edges exist between the communities, is not enough. Rather one must rearrange it in a way that communities are connected with fewer than expected edges. One can associate a score called modularity $Q$ [20] for each possible partition of a network. $Q$ is defined up to a multiplicative factor as the number of edges present within the groups minus the expected number in an equivalent random network. Since positive values of $Q$ indicates possible presence of community structure, one need to look for a partition for which the modularity is preferably large and positive.

A good partition of a network can be obtained by maximizing the modularity index $Q$ defined as follows. Let us consider a network with $n$ vertices labeled by $i = 1, \ldots n$, and $m$ links. The corresponding adjacency matrix is $A_{ij}$. Let the degree of each vertex $i$ is $k_i = \sum_j A_{ij}$, thus $m = \frac{1}{2}\sum_i k_i$. If the network is to be partitioned into two groups, one associates a quantity $s_i$

[1] The nucleotides 2 - 7 of the miRNAs are called seed sequences.

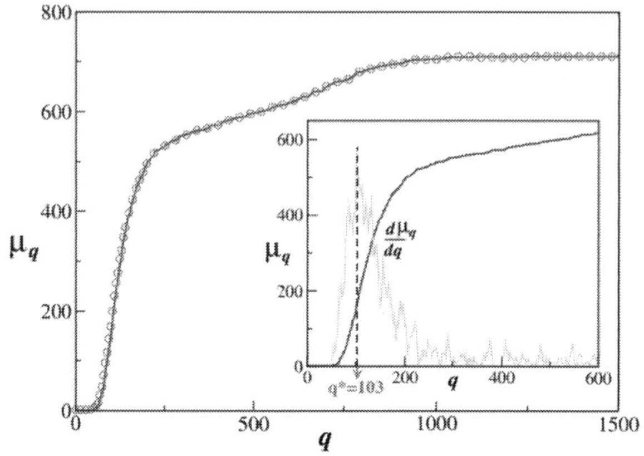


**Figure 2.** When all the links having weight less than $q$ are erased, the miRNA co-target network breaks into $N_q$ distinct components comprising of $\mu_q$ modules. The main figure compares $\mu_q$ (solid line) with the $N_q$ (circles). The inset shows that the $\frac{d\mu_q}{dq}$ is maximum at $q = 103$, which is same as the value obtained from $\frac{dN_q}{dq}$ earlier [11].

which takes a value $s_i = +1(-1)$ if vertex $i$ belongs to group 1 (group 2).

Correspondingly the modularity is given by

$$Q = \frac{1}{4m} \sum_{ij} \left( A_{ij} - \frac{k_i k_j}{2m} \right) (1 + s_i s_j), \tag{1}$$

where $k_i k_j / 2m$ is the expected number of links between $i$ and $j$, if edges were placed at random. The term $(1 + s_i s_j)$ is 0 (1) if vertices $i$ and $j$ belong to different (same) group; this assures that $Q$ is maximum when two groups are connected by smaller than expected number of links. In the following we apply this procedure to obtain modular structure of MCN.

MCN is a undirected weighted network, where the weight of the link $C_{ij}$ corresponds to the number of transcripts being co-targeted by the concerned pair of miRNAs $i$ and $j$. The diagonal elements are taken $C_{ii} = 0$, as usual. It has been pointed out in Ref. [11] that the weights vary widely between 1 to 1253, indicating that most of the links with small weights can be erased to obtain a simpler network. However, the connectivity of the network changes when links having weight less than a predefined value $q$ are erased. Taking the adjacency matrix $C^q$, defines as

$$C^q_{ij} = \begin{cases} 1 & \text{if } C_{ij} > q \\ 0 & \text{otherwise,} \end{cases} \tag{2}$$

Mookherjee *et. al.* [11] have calculated the number of components $N_q$ by varying $q$. Since, this adjacency matrix $C^q$ is unweighted (as it keeps the information of connectivity ignoring the actual weights) one can apply the idea of modularity maximization[18] to detect the communities or modules present there. Let $\mu_q$ be the total number of modules of $C^q$. Clearly $\mu_q \geq N_q$, as each component can either have one module, *i.e.* itself, or it can break into two or more modules. In Fig. 2 we have compared $\mu_q$, obtained from modularization methods, with the components $N_q$ obtained earlier. It is evident that $\mu_q \simeq N_q$; a negligible small positive difference is not visible in the figure. This brings us to conclude that the components of the network are self organized in a way that modularity (given by Eq. (1)) is maximized.

We also find that $\frac{d\mu_q}{dq}$ is maximum at $q = 103$, which is the value obtained from $\frac{dN_q}{dq}$ earlier [11]. Now let us have a closer look at the size of the components and that of the modules

 doi:10.1088/1742-6596/297/1/012002

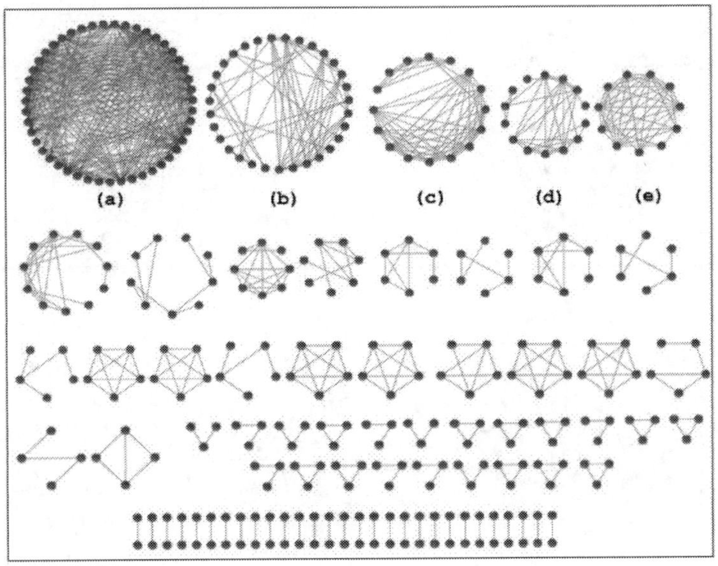


**Figure 3.** Components of $\mathcal{G}$ (which has 477 miRNAs) : Only the clusters (components of size larger than 1) are shown. Each of the large clusters with size $(7, 8, 9, 14, 16, 31, 47)$ appear once. The other 63 clusters comprises of $(2; 29)$, $(3; 21)$, $(4; 2)$, $(5; 7)$, $(6; 2)$,$(11; 2)$, where $(m; n)$ denotes that cluster of size $m$ appears $n$ times.

| Size | 2 | 3 | 4 | 5 | 6 | 7 | 8 | 9 | 11 | 12 | 13 | 14 | 17 | 26 | 37 | 57 | 98 | 101 | 477 |
|---|---|---|---|---|---|---|---|---|---|---|---|---|---|---|---|---|---|---|---|
| No. of clusters | 24 | 7 | 1 | 2 | 0 | 3 | 0 | 0 | 0 | 0 | 0 | 0 | 0 | 0 | 0 | 0 | 0 | 0 | 1 |
| No. of modules | 26 | 8 | 2 | 3 | 1 | 4 | 1 | 2 | 2 | 2 | 2 | 1 | 1 | 1 | 1 | 1 | 1 | 1 | 0 |

**Table 1.** Comparison of number of modules of the miRNA network with the number of components, at $q = q* = 103$.

obtained at $q* = 103$. This is listed in Table-I. Note, that the modularity maximization algorithm organizes the network into several small modules and few moderate size modules as $26, 37, 57, 98, 101$. Whereas in terms of components the network breaks into few clusters of small sizes (*e.g.* $2, 3, 4, 5, 7$) along with a *giant* cluster ($\mathcal{G}$) of size 477 [11]. Evidently at $q* = 103$, MCN has *one* distinctly large component containing 477 miRNAs, compared to the moderate size modules those appear with competitive sizes (101 and 98). The largest component must have been broken into these smaller modules. Thus, as far as 'identifying a large set of relevant miRNAs' (one like $\mathcal{G}$) is concerned, one can reliably consider the component $\mathcal{G}$ as the optimal set of miRNAs, which co-regulate the gene expressions. Further, to understand how miRNAs are organized within $\mathcal{G}$, we calculate its modules by taking $q = 160$, which is the same value of $q$ used in [11], to obtains the clusters (in total 70). All the components of $\mathcal{G}$ having two or more miRNAs (referred to as clusters), are shown in Fig. 3 in decreasing order of their sizes. The first five, named as (a) to (e) have $47, 31, 16, 14$ and 9 miRNAs respectively.

It would be interesting to look at the community structure of $\mathcal{G}$ at this value of $q = 160$. Using the modularization algorithm [18], we find that $\mathcal{G}$ contains 72 modules (total 330 miRNAs) and 147 single miRNAs. Whereas in terms of components, $\mathcal{G}$ had 70 clusters (total 330 miRNAs) and 147 independent miRNAs [11]. The detailed study of modules reveals that only two of the 70 clusters are broken into smaller modules : cluster (a) in Fig. 3 with 47 miRNA, has two modules $a_1$ and $a_2$ of size 34 and 13 respectively, cluster (b) with 31 miRNAs, breaks into two modules $b_1$ (22 miRNAs) $b_2$ (9 miRNAs). Such modular structures of (a) and (b) were not apparent in Fig. 3; we redraw these graphs keeping all miRNAs in same module close to each other. The

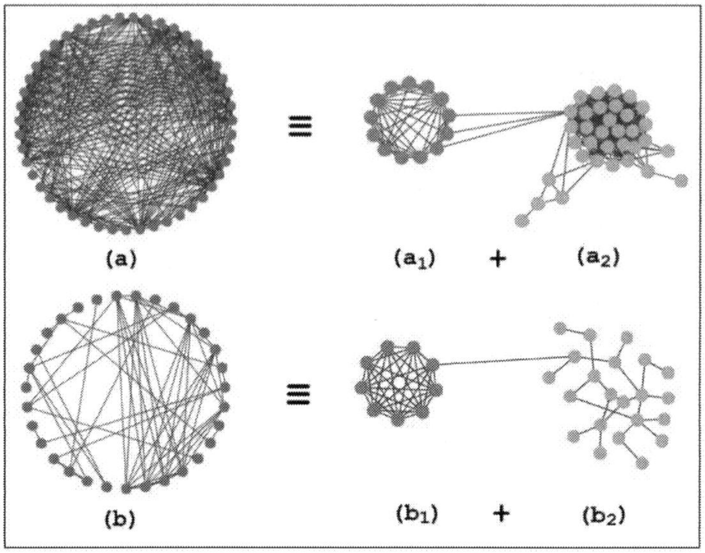


**Figure 4.** The cluster (a) and (b) of Fig. 3 are redrawn to visualize the the community structure obtained through modularization. Clearly (a) which contains 47 miRNAs has two modules $a_1$ (34 miRNAs) and $a_2$ (13 miRNAs), and (b) has two modules $b_1$ (22 miRNAs) and $b_2$ (9 miRNAs).

resulting graph (Fig. 4 (a) and (b)) clearly show the existence of modular structures.

In summary, the community structure in these networks are very similar to the components (or clusters) obtained earlier in [11]. Only few large components show further small sub-structures, indicating that the existing components of MCN are already optimally modularized. Implications of these results will be discussed in section 4.

Few comments are in order. It is quite evident that cluster (c) in Fig. 3, containing 16 miRNAs, might have sub-structures of size 11 and 5, which could not be obtained when the modularization algorithm is applied to the un-weighted graph $\mathcal{G}$ containing 477 miRNAs. In this analysis, the actual weights were ignored, *i.e.*, all links having weight more than 160 are considered identical irrespective of their actual weight. When we keep these weights and use the modified version of the algorithm [19], that works for weighted networks, the cluster (c) shows the predicted substructures. In addition, some other modules, such as $a_2$ which has 34 miRNAs also show further sub-structures of size 30 and 4. These four nodes, turns out to be those shown in the left side of $a_2$ in Fig. 4.

It appears that Newman's algorithm, both for un-weighted and weighted network, provides only the sub-structures of large components. This is because, modularity of the network is inversely proportional to the total number of links ( see Eq. (1)). Thus, the total modularity of a network with many components is not substantially altered by re-structuring the small components into smaller sub-structures. It is only, the re-structuring of larger components which can change the modularity appreciably. To overcome this difficulty, one must find modular structures of individual components, instead of looking at the community structure of the whole network.

## 4. Conclusion

To our surprise, the community structure of human miRNA co-target networks is very similar to the existing components or clusters. Only few large components show smaller sub-structures.

Most of the components do not show any further substructures, indicating that the miRNA co-target network inherently consists of optimally modularized structures. It is quite possible that, during the evolution of miRNAs, first the the modular structures are formed, optimized and then they join with other modules to provide essential regulation for complex life structures. Further study in these directions is required to verify such hypothesis.

[1] Chen K and Rajewsky N 2007 *Nature Reviews - Genetics* **8** 93.
[2] Majoros W H and Ohler U 2007 *BMC Genomics* **8** 152; Grimson A, Farh K K H, Johnston W K, Engele P G, Lim L P and Bartel D P 2007 *Molecular Cell* **27**, 91105.
[3] Jones S G, Saini H K, Dongen S V and Enright A J 2008 *Nucl. Acids Res.* **36** D154.
[4] miRBase : *http://microrna.sanger.ac.uk/, version 10.*
[5] Miranda K C, Huynh T, Tay Y, Ang Y S, Tam W L, Thomson A M, Lim B and Rigoutsos I 2006 *Cell* **126** 1203.
[6] Krek A, Grun A D, Poy M N, Wolf R, Rosenberg L, Epstein E J, MacMenamin P, Piedade I, Gunsalus K C, Stoffel M, *et. al.* 2005 *Nature Genetics* **37** 495.
[7] Zhou Y, Ferguson J, Chang J T and Kluger Y 2007 *BMC Genomics* **8** 396.
[8] Tu K, Yu H, Hua Y J, Li Y Y, Liu L, Xie L and Li Y X 2009 *Nucl. Acids Res.* **37** 5969.
[9] Bandyopadhyay S, Mitral R, Maulik U and Zhang M Q 2010 *Silence* **1**:6.
[10] Xu J *et. al.* 2010 *Nucl. Acids Res.* **39**(3) 825.
[11] Mookherjee S, Sinha M, Mukhopadhyay S, Bhattacharyya N P and Mohanty P K 2009 *Online Journal of Bioinformatics* **10**,(2) 280.
[12] Mookherjee S, Sinha M, Mukhopadhyay S, Bhattacharyya N P and Mohanty P K *arXiv:0901.4211* .
[13] Albert R, Jeong H and Barabási A L 1999 *Nature* **401** 130.
[14] Jeong H, Tombor B, Albert R, Oltvai Z N and Barabási A L 2000 *Nature* **407**, 651.
[15] Barabási A L 2005 *Science* **308** 639; Barabási A L, Jeong H, Ravasz R, Néda Z, Vicsek T and Schubert A 2002 *Physica* **A 311** 590.
[16] Jeong H, Mason S, Barabási A L and Oltvai Z N 2001 *Nature* **411** 41; Lim J, Hao T, Shaw C, Patel A J, Szabó G, Rual J F, Fisk C J, Li N, Smolyar A, Hill D E, Barabási A L, Vidal M and Zoghbi H Y 2006 *Cell* **125** 801.
[17] Zhou H and Lipowsky R 2004 *Lecture Notes Comput. Sci.* **3038** 1062 (*Preprint http://power.itp.ac.cn/ zhouhj/works/publications.html*).
[18] Newman M E J 2006 *PNAS* **103** 8577.
[19] Clauset A, Newman M E J and Moore C 2004 *Phys. Rev.* **E 70** 066111.
[20] Newman M E J and Girvan M 2004 *Phys. Rev.* **E 69** 026113.

# Opinion dynamics model with domain size dependent dynamics: novel features and new universality class

**Soham Biswas[1], Parongama Sen[1], Purusattam Ray[2]**

[1]Department of Physics, University of Calcutta, 92 Acharya Prafulla Chandra Road, Kolkata 700009, India

[2]Institute of Mathematical Sciences, CIT Campus Taramani, Chennai 600 113, India.

E-mail: soham.physics@gmail.com, parongama@gmail.com, ray@imsc.res.in

**Abstract.**

A model for opinion dynamics (Model I) has been recently introduced in which the binary opinions of the individuals are determined according to the size of their neighboring domains (population having the same opinion). The coarsening dynamics of the equivalent Ising model shows power law behavior and has been found to belong to a new universality class with the dynamic exponent $z = 1.0 \pm 0.01$ and persistence exponent $\theta \simeq 0.235$ in one dimension. The critical behavior has been found to be robust for a large variety of annealed disorder that has been studied. Further, by mapping Model I to a system of random walkers in one dimension with a tendency to walk towards their nearest neighbour with probability $\epsilon$, we find that for any $\epsilon > 0.5$, the Model I dynamical behaviour is prevalent at long times.

## 1. Introduction

Sociophysics has emerged as one of the important areas of research during recent times. The concepts of statistical physics find application to many situations that occur in a social system with the assumption that individual free will or feelings do not take crucial role in these situations [1, 2]. One of the major issues that has attracted lots of attention is how opinions evolve in a social system. Starting from random initial opinions, dynamics often leads to a consensus which means a major fraction of the population support a certain cause, for example a motion or a candidate in an election etc.

Simulating human behaviour by models effectively implies quantifying the outcome of the behavior by suitable variables having continuous or discrete values. Different dynamical rules are proposed for the evolution of these variables, depending on how these variables change with time following social interactions. Thus, a social system can be treated like a physical system. For example, in case of opinion dynamics, if the opinions have only discrete binary values, the social system can be regarded as a magnetic system of Ising spins.

In this context, Schelling model [3], proposed in 1971, seems to be the very first model of opinion dynamics. Since then, a number of models describing the formation of opinions in a social system have been proposed [4]. While on one hand these models attempt an understanding of how a society behaves and social viewpoints evolve, on the other hand, these provide rich complex dynamical physical systems suitable for theoretical studies.

Dynamics of complex systems has become a subject of extensive research from several aspects. For many such systems, e.g., traffic or agent based models, one cannot define a conventional

Hamiltonian or energy function. The only method by which one can study the steady state behaviour of such systems is by looking at the long time dynamics. Nonequilibrium dynamics involves the evolution of a system from a completely random initial configuration and associated with this evolution are several phenomena of interest like domain growth or persistence that have been studied, for example, in spin systems. Since in many sociophysics model, one can have variables analogous to spin variables, these phenomena can be readily studied here. An important objective is to identify dynamical universality classes by estimating the relevant dynamical exponents.

Another point of interest in studying dynamics is that many systems may have identical equilibrium behaviour but behave differently as far as dynamics is concerned. For example, Ising spin dynamics with or without conservation belong to different dynamic universality class although their equilibrium behaviour is identical.

Apart from the dynamical behaviour, different kinds of phase transitions have also been observed in these models by introducing suitable parameters. One such phase transition can be from a homogeneous society where everyone has the same opinion to a heterogeneous one with mixed opinions [5].

Change in the opinion of an individual takes place in different ways in different models. For example in the Voter model [6], an individual simply follows the opinion of a randomly chosen neighbour while in the Sznajd model [7], the opinion of one or more individuals are changed following more complicated rules.

In this article, we review the dynamical studies in a recently introduced model [8] (to be referred to as Model I henceforth) and its variants in which a new rule of updating (discussed in detail in section 2) is introduced. In one dimension, Model I can be visualized as an Ising spin chain (the binary opinions are represented by Ising spins) with a Glauber-like dynamics, where a spin only at the domain boundary can flip. In this new model, the state of the spin is determined by the state of the neighboring domain larger in size (detailed description is given in the next section). Model I shows strikingly different dynamical behavior compared to known models of opinion dynamics or spin dynamics. Several observables show power law decay and the exponents strongly suggest a new universality class.

The introduction of disorder in various forms have also been considered which shows that the Model I dynamical behaviour is not affected by annealed disorder. We also report (in section 3) new results when a mapping of this model is made to an equivalent reaction diffusion model where the walkers move towards their nearest neighbour with probability $\epsilon$. Once again Model I dynamical behaviour is seen to exist at long times for any $\epsilon > 0.5$ (which means a bias towards the nearest neighbour) showing the extreme robustness of the model.

## 2. Description of the Model I

In a model of opinion dynamics, the key feature is the interaction of the individuals. Usually, in all the models, it is assumed that an individual is influenced by its nearest neighbours. Model I is a one dimensional model of binary opinion in which the dynamics is dependent on the *size* of the neighbouring domains as well. Here an individual changes his/her opinion in two situations: first when the two neighbouring domains have opposite polarity, and in this case the individual simply follows the opinion of the neighbouring domain with the larger size. This case may arise only when the individual is at the boundary of the two domains. An individual also changes his/her opinion when both the neighbouring domains have an opinion which opposes his/her original opinion, i.e., the individual is sandwiched between two domains of same polarity. It may be noted that for the second case, the size of the neighbouring domains is irrelevant. When the two neighbouring domains are of the same size but have opposite polarity, the individual will change his/her orientation with fifty percent probability.

The binary opinions can be represented by a system of Ising spins where the up and down states correspond to the two possible opinions. The two rules followed in the dynamical evolution in the equivalent spin model are shown schematically in Fig. 1 as case I and II. In the first case the spins at the boundary between two domains will choose the state of the left side domain (as it is larger in size). For the second case the down spin flanked by two neighbouring up spins will flip.

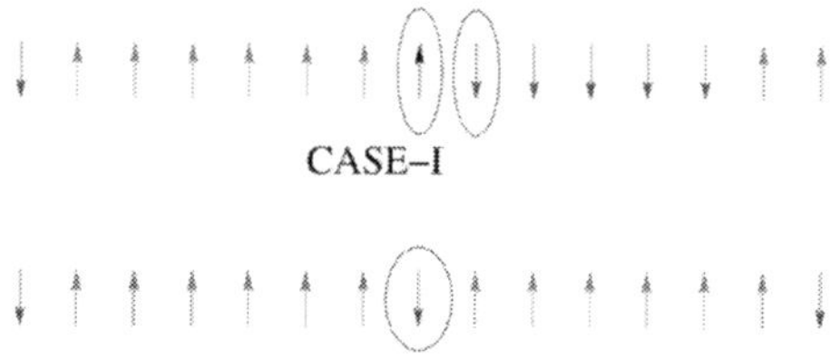


**Figure 1.** Dynamical rules for Model I: in both cases the encircled spins may change state; in case I, the boundary spins will follow the opinion of the left domain of up spins which will grow. For case II, the down spin between the two up spins will flip irrespective of the size of the neigbouring domains.

The main idea in Model I is that the size of a domain represents a quantity analogous to 'social pressure' which is expected to be proportional to the number of people supporting a cause. An individual, sitting at the domain boundary, is most exposed to the competition between opposing pressures and gives in to the larger one. This is what happens in case I shown in Fig.1. The interaction in case II on the other hand implies that it is difficult to stick to one's opinion if the entire (immediate) neighbourhood opposes it.

Defining the dynamics in this way, one immediately notices that case II corresponds to what would happen for spins in a nearest neighbour Ferromagnetic Ising Model (FIM) in which the dynamics at zero temperature is simply an energy minimization scheme. However, the boundary spin in the FIM behaves differently in case I; it may or may not flip as the energy remains same. In the present model, the dynamics is deterministic even for the boundary spins (barring the rare instance when the two neighbourhoods have the same size in which case the spin flips with fifty percent probability).

In this model, the important condition of changing one's opinion is the size of the neighbouring domains which is not fixed either in time or space. This is the unique feature of this model. In the most familiar models of opinion dynamics like the Sznajd model [7] or the voter model [6], one takes the effect of nearest neighbours within a given radius and even in the case of models defined on networks [9], the influencing neighbours may be nonlocal but always fixed in identity.

In the equivalent spin model, if $L_+$ is the number of up spins and $L_-$ is the number down spins, the order parameter is defined as $m = |L_+ - L_-|/L$. This is identical to the (absolute value of) magnetization. Starting from a random initial configuration, the dynamics in Model I showed that it leads to a final state with $m = 1$, i.e. a homogeneous state where all spins have the same value (either +1 or -1). It is not difficult to understand this result; in absence of any fluctuation, the dominating neighbourhood (domain) simply grows in size ultimately spanning the entire system.

In the spin picture, the dynamics can be described in terms of the movement of the domain walls and as the dynamics progresses, number of domain walls goes on decreasing. Monte Carlo simulations showed that the domain dynamics and the dynamics of the order parameter obey conventional power law variations: the fraction of domain walls $D(t) \propto t^{-1/z}$ with $z = 1.00 \pm 0.01$ and order parameter $m(t) \propto t^{1/2z}$ (Fig. 2).

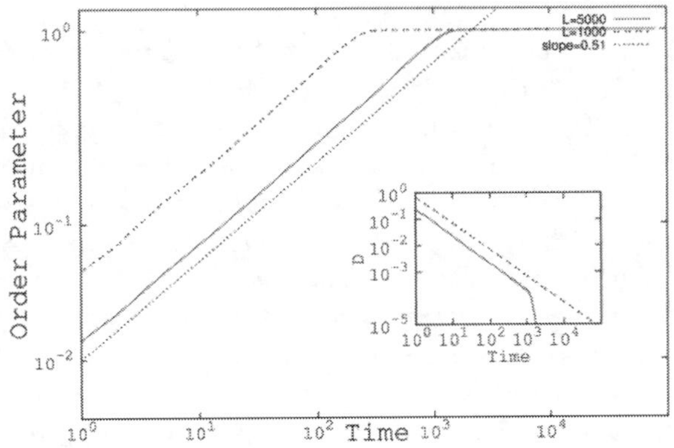


**Figure 2.** Variation of the order parameter $m$ with time for two different system sizes along with a straight line (slope 0.51) shown in a log-log plot. Inset shows the decay of fraction of domain wall $D$ with time.

The persistence measure (i.e., probability that a spin has not flipped till time $t$) [10] showed the familiar power law behaviour: $P(t) \propto t^{-\theta}$, where $\theta$ is the persistence exponent. For finite system of size $L$, $P(t, L)$ is known to behave as [11, 12]

$$P(t, L) \propto t^{-\theta} f(L/t^{1/z}), \tag{1}$$

and at large times, the persistence probability saturates at a value $\propto L^{-\alpha}$. Therefore, for $x << 1$, $f(x) \propto x^{-\alpha}$ with $\alpha = z\theta$. For large $x$, $f(x)$ is a constant. Estimate for both $z$ and $\theta$ using the above scaling form showed that $\theta = 0.235 \pm 0.003$, and a $z$ value $1.04 \pm 0.01$ (Fig. 3).

Model I therefore showed a novel dynamical behaviour with values of $z$ and $\theta$ quite different from those of the one dimensional Ising model [13] and other opinion/voter dynamics models [14, 15, 16]. Specifically in the Ising model, $z = 2$ and $\theta = 0.375$ and for the Sznajd model the persistence exponent is equal to that of the Ising model.

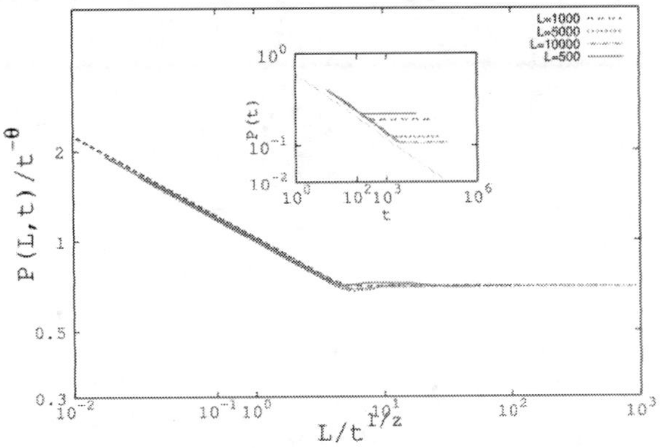


**Figure 3.** The collapse of scaled persistence probability versus scaled time using $\theta = 0.235$ and $z = 1.04$ is shown for different system sizes. Inset shows the unscaled data.

### 2.1. *Model I with a cutoff*

As a variation of the Model I, one can introduce a cutoff [17] denoted by a parameter $p$, such that the maximum size of the neighbouring domains an individual can sense is given by $R = pL/2$ in a one dimensional system of $L$ agents with periodic boundary condition. $p = 1$ corresponds to the original Model I. A finite cutoff (i.e., $p < 1$) puts a restriction on the domain sizes which may correspond to geographical, political, cultural boundaries etc. The case with uniform cutoff signifies that all the individuals have same kind of restriction. The case with random cutoffs, which is perhaps closer to reality, has also been studied briefly. Once again one can represent the system by Ising spins.

Whenever $R$ is kept finite, that is, $R$ does not scale with system sizes (implying $p \to 0$ in the thermodynamic limit), the dynamics leads to the equilibrium configuration of all spins up/down and the dynamic exponents also turn out to be identical to those corresponding to the nearest neighbour Ising values (i.e., $\theta = 0.375$ and $z = 2$).

The results with a nonzero $p$ showed that there is a crossover phenomena in dynamics. For an initial time $t_1 = pL/2$, the dynamics remains Model I-like as the cutoff does not affect the dynamics till this time. Beyond this time, the dynamics becomes diffusive. Such a crossover is, however, not apparent from the log-log plots of the relevant dynamical quantities (Fig. 4 ). This is because the crossover occurs between two different types of phenomena. The first one is pure coarsening in which domain walls prefer to move towards their nearest neighbours as in Model I and one gets the expected power-law behaviour. There is a correlation length which is growing; as long as this length is smaller than $R$ one will observe ballistic growth (as in Model I). When the correlation length is larger than $R$ one observes domain diffusion. The diffusive behavior is not evident in the coarsening process because ballistic dynamics "leaves" the system into a non-typical configuration which is evidently far from those on diffusion paths. At $t_1$, some special configurations are generated in which only a few domain walls remain, the density of domain walls going to zero in the thermodynamic limit. Therefore beyond $t_1$, the second phenomenon involves pure diffusion of these domain walls which remain non-interacting up to large times. Only when the distance between two domain walls become $\leq R$ the Model I like dynamics prevails again before the two are annihilated. This takes place over a much shorter time scale such that the physical quantities attain saturation values ($m = 1$ and $D = 0$) very fast in the last part of the dynamics (Fig. 4 ).

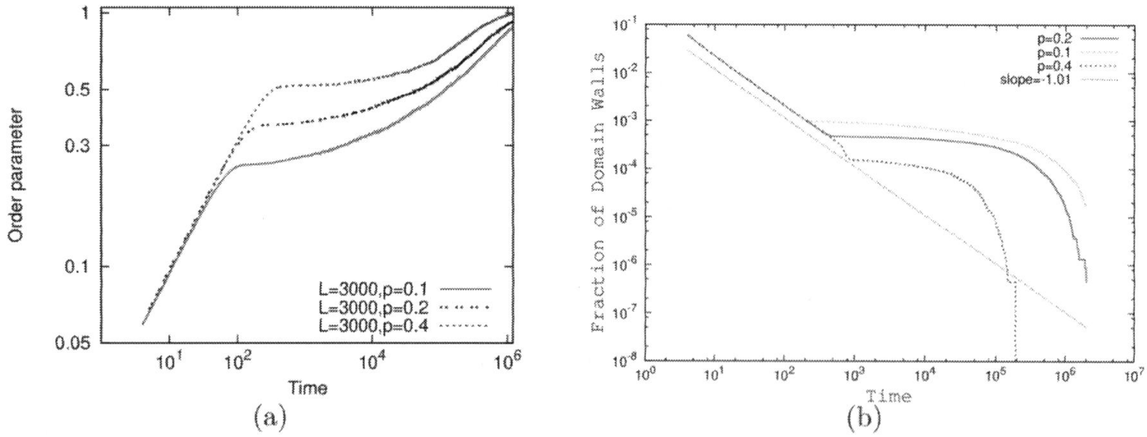


**Figure 4.** Variation of the (a) order parameter and the (b) fraction of domain walls with time for different finite cutoff $p = 0.1, 0.2, 0.4$ ($L = 3000$).

The only dynamic exponent in the diffusive regime is the diffusion exponent $z = 2$, which is distinct from the growth exponent $z = 1$. So the two dynamic exponents not only differ in magnitude, but they are also connected to distinct phenomena. Using both numerical data and some analytical arguments, it could be shown that the behaviour at $t > t_1$ is indeed diffusive. The saturation time to reach equilibrium was found to be

$$t_{sat} = apL + b(1-p)^3L^2, \tag{2}$$

which also showed that for $L \to \infty$, $z = 2$.

The persistence behaviour, however, does not show any power law behaviour corresponding to the diffusive behaviour, i.e., it does not decay like $t^{-0.375}$ as in a usual reaction diffusion system. This is because of the special configurations in which the system is left at time $t_1$. However, the long time behaviour showed that the persistence probability saturates in time and decays with the system size as $L^{-\alpha}$ with $\alpha = 0.235$, which is the Model I value. In fact one can obtain a collapse for the persistence data using $\alpha = 0.235$ in both the regions $t < t_1$ (with $z = 1$) and $t > t_1$ (with $z = 2$) for any $p \neq 0$. Thus the crossover phenomenon occurs with a novel characteristic behaviour.

### 2.2. *Effect of disorder*

The Model I described so far has no disorder, which can be introduced in several possible ways. We discuss three cases in the following.

*2.2.1. Effect of rigidity parameter:* Since every individual is not expected to succumb to social pressure, Model I can be modified by introducing a parameter $\rho$ called rigidity coefficient which denotes the probability that people are completely rigid and never change their opinions [8]. This means there are $\rho N$ rigid individuals (chosen randomly at time $t = 0$), who retain their initial state throughout the time evolution. Thus the disorder is quenched in nature. Such rigid individuals had been considered earlier in [18].

The limit $\rho = 1$ corresponds to the unrealistic noninteracting case when no time evolution takes place; $\rho = 1$ is in fact a trivial fixed point. For other values of $\rho$, the system evolves to a equilibrium state.

The time evolution changes drastically in nature (compared to Model I) with the introduction of $\rho$. All the dynamical variables like order parameter, fraction of domain wall and persistence attain a saturation value at a rate which increases with $\rho$. Power law variation with time can only be observed for $\rho < 0.01$ with the exponent values same as those for $\rho = 0$. The saturation or equilibrium values on the other hand show the following behaviour:

$$\begin{aligned} m_s &\propto N^{-\alpha_1}\rho^{-\beta_1} \\ D_s &\propto \rho^{-\beta_2} \\ P_s &= a + b\rho^{-\beta_3} \end{aligned} \tag{3}$$

where in the last equation $a$ is a constant $\simeq 0.06$ independent of $\rho$. The values of the exponents are $\alpha_1 = 0.500 \pm 0.002$, $\beta_1 = 0.513 \pm 0.010$, $\beta_2 = 0.96 \pm 0.01$ and $\beta_3 = 0.430 \pm 0.01$.

It can be naively assumed that the $N\rho$ rigid individuals will dominantly appear at the domain boundaries such that in the first order approximation (for a fixed population), $D \propto 1/\rho$. This would give $m \propto 1/\sqrt{\rho}$ indicating $\beta_1 = 0.5$ and $\beta_2 = 1$. The numerically obtained values are in fact quite close to these estimates.

The results obtained above can be explained in the following way: with $\rho \neq 0$, the domains cannot grow freely and domains with both kinds of opinions survive making the equilibrium $m_s$ less than unity. Thus the society becomes heterogeneous for any $\rho > 0$ when people do not follow the same opinion any longer. The variation of $m_s$ with $N$ shows that $m_s \to 0$ in the thermodynamic limit for $\rho > 0$. Thus not only does the society become heterogeneous at the onset of $\rho$, it goes to a completely disordered state analogous to the paramagnetic state in magnetic systems. Thus one may conclude that a phase transition from a ordered state with $m = 1$ to a disordered state ($m = 0$) takes place for $\rho = 0^+$.

Since the role of $\rho$ is similar to domain wall pinning, one can introduce a depinning probability factor $\mu$ which in this system represents the probability for rigid individuals to become non-rigid during each Monte Carlo step. $\mu$ relaxes the rigidity criterion in an annealed manner in the sense that the identity of the individuals who become non-rigid is not fixed (in time). If $\mu = 1$, one gets back Model I whatever be the value of $\rho$, and therefore $\mu = 1$ signifies a line of (Model I) fixed points, where the dynamics leads the system to a homogeneous state.

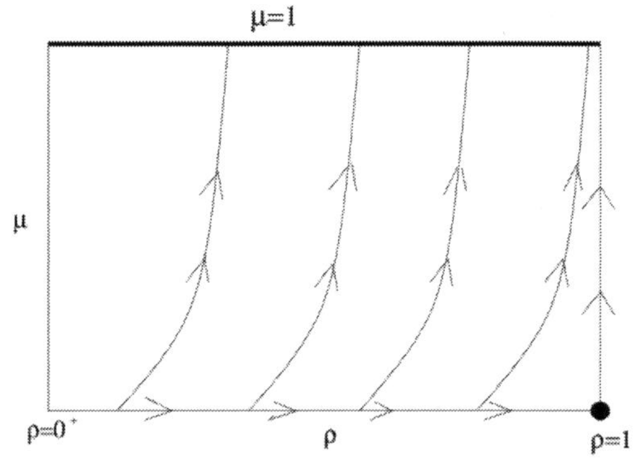


**Figure 5.** The flow lines in the $\rho - \mu$ plane: Any non-zero value of $\rho$ with $\mu = 0$ drives the system to the disordered fixed point $\rho = 1$. Any nonzero value of $\mu$ drives it to the ordered state ($\mu = 1$, which is a line of fixed points) for all values of $\rho$.

With the introduction of $\mu$, one has effectively a lesser fraction $\rho'$ of rigid people in the society, where

$$\rho' = \rho(1 - \mu). \tag{4}$$

The difference from the previous model with quenched rigidity is, of course, that this effective fraction of rigid individuals is not fixed in identity (over time). Thus when $\rho \neq 0, \mu \neq 0$, we have a system in which there are both quenched and annealed disorder. It is observed that for any nonzero value of $\mu$, the system once again evolves to a homogeneous state ($m = 1$) for all values of $\rho$. Moreover, the dynamic behaviour is also same as Model I with the exponent $z$ and $\theta$ having identical values. This shows that the nature of randomness is crucial as one cannot simply replace a system with parameters $\{\rho \neq 0, \mu \neq 0\}$ by one with only quenched randomness $\{\rho' \neq 0,\ \mu' = 0\}$ as in the latter case one would end up with a heterogeneous society. We therefore conclude that the annealed disorder wins over the quenched disorder; $\mu$ effectively drives the system to the $\mu = 1$ fixed point for any value of $\rho$. This is shown schematically in a flow diagram (Fig. 5). It is worth remarking that it looks very similar to the flow diagram of the one dimensional Ising model with nearest neighbour interactions in a longitudinal field and finite temperature.

*2.2.2. Effect of thermal-noise like disorder :* In thermodynamical systems, the effect of thermal disorder is a highly important issue. In Model I which describes a social system, a disorder analogous to thermal noise has been introduced [19].

Journal of Physics: Conference Series **297** (2011) 012003 doi:10.1088/1742-6596/297/1/012003

Let $d_{up}$ and $d_{down}$ be the sizes of the two neighbouring domains of type up and down of a spin at the domain boundary (excluding itself). In Model I, probability $P(up)$ that the said spin is up is 1 if $d_{up} > d_{down}$, 0.5 if $d_{up} = d_{down}$ and zero otherwise. In the simplest possible way to introduce disorder, one may take the probability of a boundary spin to be up as $P(up) = d_{up}/(d_{up} + d_{down})$. However, there is no parameter controlling the stochasticity here and moreover, the results are identical to the original Model I and therefore this kind of stochasticity is not of much importance.

In order to introduce a noise like parameter which can be tuned, the probability that a spin at the domain boundary is up was taken to be

$$P(up) \propto e^{\beta(d_{up}-d_{down})}, \tag{5}$$

and it is down with probability

$$P(down) \propto e^{\beta(d_{down}-d_{up})}. \tag{6}$$

The normalized probabilities are therefore $P(up) = \exp \beta\Delta/(\exp(\beta\Delta) + \exp(-\beta\Delta))$ and $P(down) = 1 - P(up)$, where $\Delta = (d_{up} - d_{down})$.

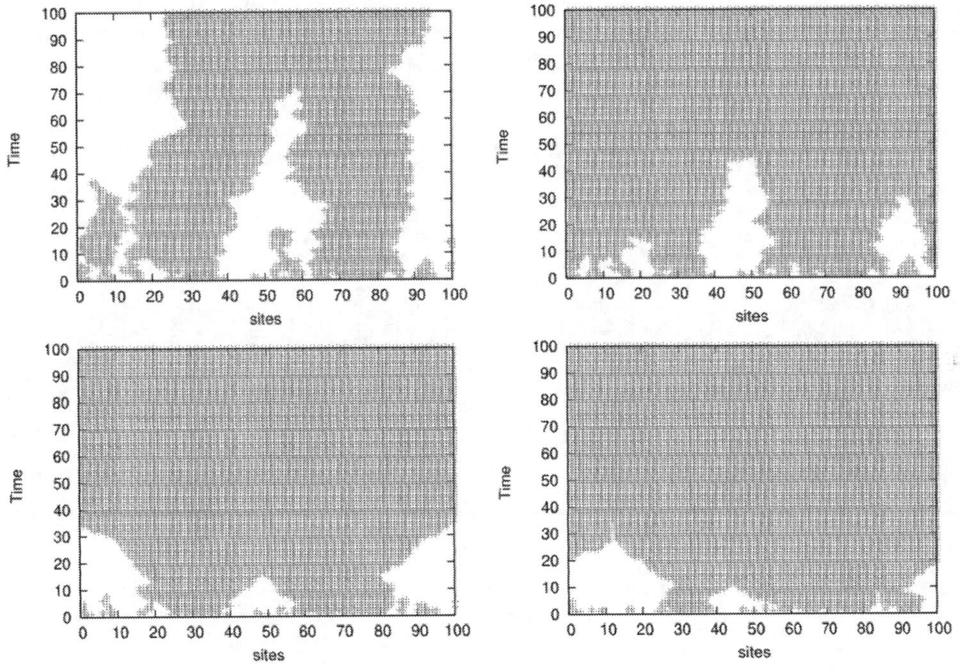


**Figure 6.** Snapshots of the system in time for different values of $\beta = 0.0, 0.005, 0.1$, and $\beta \to \infty$ (top to bottom) showing that even for very small non-zero $\beta$, the system equilibrates towards a homogeneous state much faster compared to the $\beta = 0$ case. These snapshots are for a $L = 100$ system.

Obviously, $\beta \to \infty$ corresponds to Model I while letting $\beta = 0$ we have equal probabilities of the up and down states, making it equivalent to the zero temperature dynamics of the nearest neighbour Ising model. Since the equilibrium states for the extreme values $\beta \to \infty$ and $\beta = 0$ are homogeneous (all up or all down states), it is expected that for all values of $\beta$ they will remain so as is indeed the case.

Journal of Physics: Conference Series **297** (2011) 012003 doi:10.1088/1742-6596/297/1/012003

It is useful to show the snapshots of the evolution of the system over time for different $\beta$ (Fig.6): to be noted is the fact that for any non-zero $\beta$ however small, the system equilibrates very fast compared to the Ising limit $\beta = 0$.

The detailed dynamical studies in fact showed that the system belongs to the Model I dynamical class for any finite $\beta$ and a dynamic transition takes place exactly at $\beta = 0$. Certain subtleties in this model need to be mentioned: with respect to Model I, $\beta = 0$ is the maximum noise and its inverse may be thought of an effective temperature. On the other hand, from the Ising model viewpoint, $\beta$ plays the role of noise. However it is not equivalent to thermal fluctuations which can affect the state of any spin. With $\beta$, flipping of spins can still occur at the domain boundaries only. Hence, even with this noise, the equilibrium behaviour is not disturbed for any value of $\beta$ (even for $\beta \to \infty$ which corresponds to Model I) while in contrast, any non-zero temperature can destroy the order of a one dimensional Ising model.

*2.2.3. Model with random cutoffs :* Previously we have discussed the case when a uniform cutoff is introduced to the system. One can introduce randomness in the cutoff parameter $p$ varying from 0 to 1, (chosen randomly from a uniform distribution) and associated with each individual a different value of $p$ [17]. The randomness is quenched as the value of $p$ assumed by any individual is fixed for all times. In this case, the variation of the magnetization, domain walls and persistence show power law scalings with exponents corresponding to Model I only for an initial range of time; however, calculation of $z$ from the saturation times gives $z = 1$. The later time behaviour appears to deviate from the Model I behaviour and further studies are required to analyses the exact dynamical behaviour.

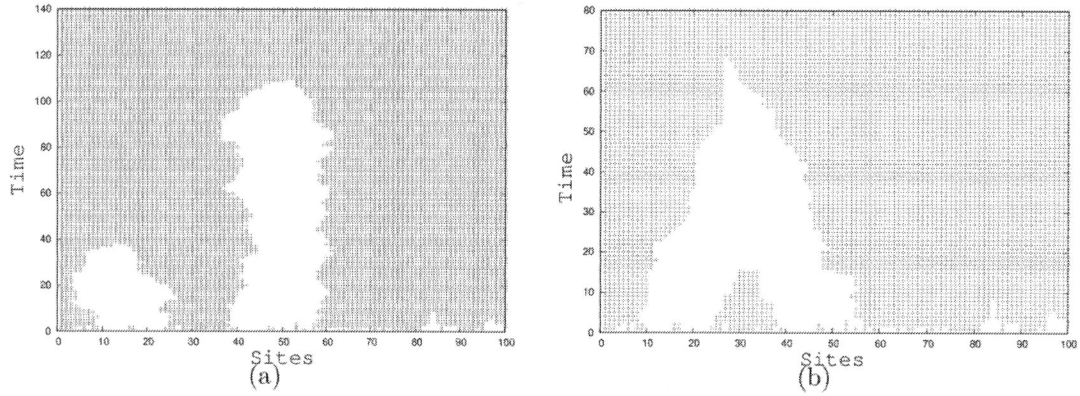


**Figure 7.** Snapshots of the system in time for (a) random cutoff parameter $p$ varying from 0 to 1, and (b) uniform cutoff parameter $p$. These snapshots are for a $L = 100$ system.

Even snapshots of the system (when compared to the uniform cutoff or infinite cutoff cases) do not help in understanding the phenomenon much (Fig 7).

## 3. Mapping of the opinion dynamics model to reaction diffusion system

The opinion dynamics model discussed so far are models where the dynamics is described in terms of the Ising spins that mimic the binary opinions an individual can have. In this model, a spin deep inside a domain does not flip. The dynamics is governed by the flipping of the spins only at the domain walls. The dynamics, in this respect, is reminiscent of the zero temperature Glauber dynamics of the kinetic Ising model. The motions of the domain walls can be viewed as the motions of the particles $A$ with the reaction $A+A \to \emptyset$. This means the particles are walkers and when two particles come on top of each other they are annihilated. The annihilation reaction ensures domain coalescence and coarsening. Unlike that in Glauber Ising model, the walkers $A$ corresponding to Model I do not perform random walks. These walkers move *ballistically towards their nearest neighbours*. This bias, as we have seen before, gives rise to a new universality class than that of conventional reaction diffusion system [20].

We have also studied $A + A \to \emptyset$ model with the particles $A$ performing random walk with a bias $\epsilon$ towards their nearest neighbors. We have taken $\epsilon$ as the probability that a walker walks towards its nearest neighbour. Clearly, $\epsilon = 0.5$ corresponds to usual reaction diffusion system with the particles performing random walk. On the other hand, $\epsilon = 1$ is equivalent to our Model I as has been described above. We have studied the dynamics of the reaction diffusion system for different values of $\epsilon$ in the range $[0^+, 1.0]$.

In reaction diffusion systems, the growth of domains is given by the number of surviving walkers. Persistence $P(t)$ in these systems is defined as the fraction of sites unvisited by any of the walkers $A$ till time $t$. Figures 8 and 9 show the decay of persistence and fraction of walkers with time for different values of $\epsilon > 0.5$.

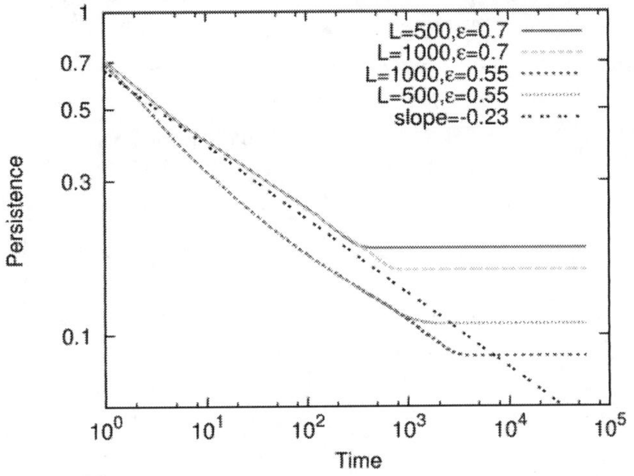


**Figure 8.** Decay of persistence with time for $\epsilon = 0.7$ and $\epsilon = 0.55$

We find that for $\epsilon > 0.5$, the Model I behaviour is observed, namely: $z \simeq 1.0$ and $\theta \simeq 0.235$, with some possible correction to the scaling which becomes weaker as $\epsilon$ is increased. For example, there is a logarithmic correction to scaling for the decay of the fraction of domain walls which takes the form $t^{-1}(1 + \alpha(\epsilon)\log(t))$ where $\alpha(\epsilon) \to 0$ as $\epsilon \to 1$. One can compare the above model with the cases discussed in section 2.2.3, where the introduction of stochastic dynamics also occurred with a bias towards the larger domain. In case of thermal disorder, the parameter comparable to $\epsilon$ is $\beta$. In the present case, the exact sizes of the domains do not matter (which is important for the case with $\beta$) but the results are consistent in the sense that any bias towards the larger domain (or nearest walker) makes the system behave like Model I.

In this model, we have also studied the $\epsilon < 0.5$ region where the opposite happens, the walker has a bias towards the further neighbour. Obviously domain annihilations take place very slowly now, even slower than $1/\log(t)$ and the dynamics continues for very long times. Consequently, the persistence probability no longer shows a power law variation now but falls exponentially to zero (Fig 10).

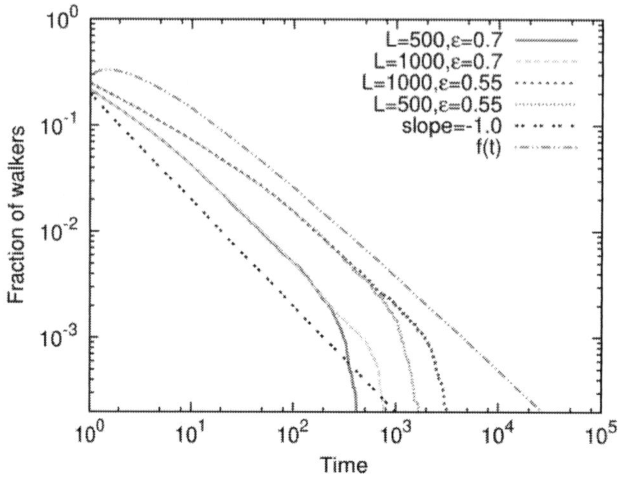


**Figure 9.** Decay of number of walkers with time for $\epsilon = 0.7$ and $\epsilon = 0.55$. There is a logarithmic correction to scaling for the value of $\epsilon = 0.55$. The form of $f(t)$ is $t^{-1}(1+\alpha\log(t))$ with $\alpha = 3.92$

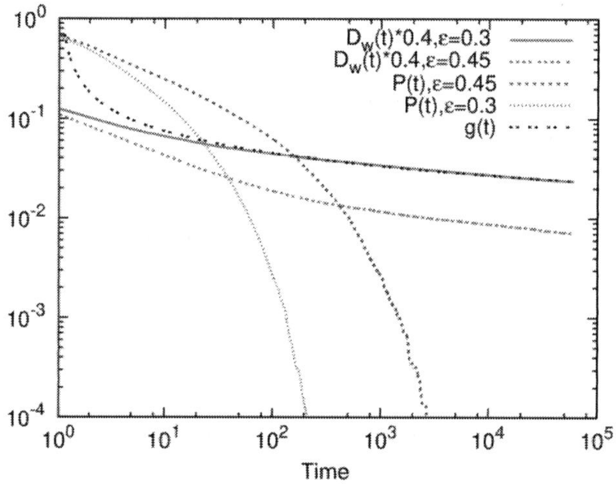


**Figure 10.** Decay of persistence and number of walkers with time for $\epsilon = 0.3$ and $\epsilon = 0.45$. The form of $g(t)$ is $a(1/\log(t))$ where $a$ is any constant.

## 4. Summary and Discussions

We have reviewed the dynamics of a recently proposed opinion dynamics model (Model I and its variations) [8]. Model I in one dimension belongs to a new dynamical universality class with novel dynamical features not encountered in any previous models of dynamic spin system or opinion dynamics. In the corresponding reaction diffusion system $A + A \to \emptyset$, we have introduced a probability $\epsilon$ of random walkers $A$ moving towards their nearest neighbors. $\epsilon = 0.5$ corresponds to the particles $A$ performing unbiased random walks and the system belongs to the dynamical universality class of zero temperature Glauber Ising model. We find that for $\epsilon > 0.5$, the system still shows power law behavior of domain growth and persistence but with a universality class of that of Model I. For $\epsilon < 0.5$, the domain grows logarithmically and the persistence decays exponentially in time.

We have discussed quite a few variations on Model I with quenched and annealed disorders

Journal of Physics: Conference Series **297** (2011) 012003 doi:10.1088/1742-6596/297/1/012003

and cutoffs in the interaction range. In presence of quenched disorders, like the presence of few rigid spins which do not flip, the dynamics changes drastically. The dynamics ends up in heterogeneous equilibrium phases, which is fully disordered in the sense that no consensus can be reached as the order parameter goes to zero in the thermodynamic limit, with identical behaviour irrespective of the percentage of the rigid spins in the system. However, the dynamics shows extreme robustness against annealed disorder. We have also discussed the crossover phenomena in the dynamics in presence of cutoffs in the range over which a spin recognizes its neighboring domains.

All the models discussed in the present article can easily be extended to higher dimensions and its universality class determined. Phase transitions occurring at non-extreme values of suitably defined parameters may also be expected in higher dimensions.

Acknowledgements: SB and PS acknowledge financial support from DST project SR/S2/CMP-56/2007. Partial computational help has been provided by UPE project. SB acknowledges the hospitality of IMSc where part of the work was done. PR acknowledges the Department of Physics, University of Calcutta for hospitality.

## References

[1] *Econophysics and Sociophysics: Trends and Perspectives*, Eds. Chakrabarti B K, Chakrabarti A and Chatterjee A, Wiley VCH, Berlin (2006).
[2] Castellano C, Fortunato S and Loreto V (2009) *Rev. Mod. Phys.* **81** 591.
[3] Schelling T C (1971) *J. Math. Sociol.* **1**: 143
[4] Stauffer D, in *Encyclopedia of Complexity and Systems Science* edited by Meyers R A (Springer, New York, 2009);
[5] Baronchelli A, Dall'Asta L, Barrat A, and Loreto V, *Phys. Rev. E* **76**, 051102 (2007);
Castellano C, Marsili M and Vespignani A, *Phys. Rev. Lett.* **85** 3536 (2000).
[6] Liggett T M *Interacting Particle Systems: Contact, Voter and Exclusion Processes* (Springer-Verlag Berlin 1999).
[7] Sznajd-Weron K and Sznajd J, *Int.J.Mod.Phys C* **11** 1157 (2000).
[8] Biswas S and Sen P, *Phys. Rev. E* **80**, 027101 (2009).
[9] Castellano C, Vilone D and Vespignani A, *Europhys. Lett.* **63** 153 (2003).
[10] Derrida B, Bray A J and Godreche C, *J.Phys. A* **27** L357 (1994).
[11] Manoj G and Ray P, *Phys. Rev. E* **62** 7755 (2000);
Manoj G and Ray P, *J. Phys A* **33** 5489 (2000).
[12] Biswas S, Chandra A K and Sen P, *Phys. Rev.* E **78**, 041119 (2008)
[13] Derrida B in *Lecture notes in Physics* **461** 165 (Springer, 1995);
Derrida B, Hakim V, Pasquier V, *Phys. Rev. Lett.* **75** 751 (1995)
[14] Stauffer D and de Oliveira P M C, *Eur. Phys. J B* **30** 587 (2002.)
[15] Sanchez J R, arXiv:cond-mat/0408518v1
[16] Shukla P, *J. Phys. A Math. Gen.* **38** 5441 (2005).
[17] Biswas S and Sen P, *Journal of Physics A* **44** 145003 (2011).
[18] Galam S, *Physica A* **381** 366 (2007)
[19] Sen P, *Phys. Rev. E* **81**, 032103 (2010)
[20] Redner S, in *Nonequilibrium Statistical Mechanics in One Dimension* edited by Privman V (Cambridge University Press, Cambridge, 1997)

Journal of Physics: Conference Series **297** (2011) 012004 doi:10.1088/1742-6596/297/1/012004

# Phase transitions and non-equilibrium relaxation in kinetic models of opinion formation

**Soumyajyoti Biswas**[1], **Anjan Kumar Chandra**[1], **Arnab Chatterjee**[2,3], **Bikas K. Chakrabarti**[1,4,5]

[1]Theoretical Condensed Matter Physics Division, Saha Institute of Nuclear Physics, 1/AF Bidhannagar, Kolkata 700064, India.
[2]CMSP Section, The Abdus Salam International Centre for Theoretical Physics, Strada Costiera 11, Trieste I-34014 Italy.
[3]Centre de Physique Théorique (CNRS UMR 6207), Université de la Méditerranée Aix Marseille II, Luminy, 13288 Marseille cedex 9, France.
[4]Centre for Applied Mathematics and Computational Science, Saha Institute of Nuclear Physics, 1/AF Bidhannagar, Kolkata 700064. India.
[5]Economic Research Unit, Indian Statistical Institute, 203 Barrackpore Trunk Road, Kolkata 700108, India.

E-mail: `soumyajyoti.biswas@saha.ac.in`, `anjan.chandra@saha.ac.in`, `arnab.chatterjee@cpt.univ-mrs.fr`, `bikask.chakrabarti@saha.ac.in`

**Abstract.** We review in details some recently proposed kinetic models of opinion dynamics. We discuss several variants including a generalised model. We provide mean field estimates for the critical points, which are numerically supported with reasonable accuracy. Using non-equilibrium relaxation techniques, we also investigate the nature of phase transitions observed in these models. We also study the nature of correlations as the critical points are approached.

## 1. Introduction

Application of statistical physics to understand social dynamics is an interesting and very active area of research [1, 2, 3, 4, 5]. The key question one asks is how a set of interacting individuals choose between different options (vote, language, culture, opinions etc), leading to a state of 'consensus' in one such option, or a state of coexistence of many of them. Opinion dynamics is one of the most important aspects of a society. It is a collective dynamical phenomena. Numerous models have been introduced so far to study the dynamics that leads to different opinion states and the processes that determine transitions between such states. Voter model which has a binary opinion variable with the opinion alignment proceeding by a random choice of neighbours [6] and the Sznajd discrete opinion formation model where more than just a pair of spins is associated with the decision making procedure [7] are two such examples. In some other models more than two opinions and also opinion as a continuous variable has been considered [8, 9, 10, 11]. Models where the range of interactions are more than nearest neighbors but finite and even time dependent (see Ref [12] and references therein) are also studied. 'Opinions' are subject to changes due to binary or group interactions, global feedback and even external factors. The usual interest in these studies lies in the distinct steady state properties: a phase characterized by individuals with widely different opinions and another phase with a measured fraction of individuals with similar opinions.

We focus our attention to a specific class of models proposed recently [13, 14], having apparent similarity with kinetic models of wealth exchange [15]. The opinions of individuals are continuous variables in $[-1, 1]$ which change due to binary interactions. The only parameter in these models is 'conviction', which is a measure of how much an individual sticks to his/her previous opinion while interacting with another. The system of such individuals, or the 'society', reaches a state of 'consensus' if this parameter stays above a threshold. One can also generalise this model [16] by introducing another parameter modeling the 'influence' of the other individual. This study models the fact that the ability to influence need not be identical to one's conviction. When 'influence' and 'conviction' are identical, one gets the original model [13, 14]. The case of 'consensus' is an ordered phase (full or partial) while another 'disordered' phase exists, characterized by null (or very small) value of individual opinions.

In this article, we will first review the original model and its generalisation. Then we will present some interesting features of the above model [13, 14], and propose some simplifications and variants. In particular, using Non-Equilibrium Relaxation (NER) [17], we study the phase transition observed in these models. The associated critical exponents are obtained. Further, we propose a simpler variant of the original model to capture the minimum ingredients required to obtain the phase transition. Also the effect of global feedback is studied. The paper is organized as follows: In Sec. 2 and Sec. 3 we will discuss the original model and two of its variants. In Sec. 4, we discuss the general results, and the non-equilibrium relaxation in Sec. 5. Next, we discuss a generalized model in Sec. 6, a map in Sec. 7 and also introduce a new model in Sec. 8. We conclude with a summary and discussions of our main results in Sec. 9.

## 2. The LCCC model

The basic idea of this model originated from a multi-agent statistical model of closed economy [15] where $N$ agents exchange a quantity $w$ defined as wealth. Initially each agent begin with a certain amount of wealth $w_i$, $i = 1, 2, \ldots, N$ such that $w_i > 0$ and the total wealth $W = \sum_i w_i$ is conserved. The system evolves with a prescribed trading rule where agents interact with each other through a pairwise interaction characterised by a "saving" parameter $\lambda$, with $0 \leq \lambda \leq 1$. The dynamics (CC model) is as follows :

$$\begin{aligned} w_i' &= \lambda w_i + \epsilon(1-\lambda)(w_i + w_j) \\ w_j' &= \lambda w_j + (1-\epsilon)(1-\lambda)(w_i + w_j) \end{aligned} \tag{1}$$

where $\epsilon$ is a stochastic variable that changes with time and as $w$ is a conserved quantity, for each transaction $w_i' + w_j' = w_i + w_j$, where $w_i'$ and $w_j'$ are the agent wealth after the transaction. The functional form for steady state distribution $f(w)$ is seen to be close to the $\Gamma$ distribution [18].

Following this, Lallouache et. al. [13, 14] proposed a minimal multiagent model for the collective dynamics of opinion formation. Let $o_i(t) \in [-1, +1]$ be the opinion of an individual $i$ at time $t$. In a system of $N$ individuals, opinions change out of binary interactions:

$$\begin{aligned} o_i(t+1) &= \lambda[o_i(t) + \epsilon o_j(t)] \\ o_j(t+1) &= \lambda[o_j(t) + \epsilon' o_i(t)] \end{aligned} \tag{2}$$

where $\epsilon$, $\epsilon'$ are drawn randomly from uniform distributions in $[0, 1]$. Here, $\lambda$ is a parameter, which is interpreted as 'conviction'. The above model [13, 14] (LCCC model hereafter) considers a society where everyone has the same value of conviction $\lambda$. It is important to note that there are no conservation laws here. The opinions are bounded, i.e., $-1 \leq o_i(t) \leq 1$. The ordering in the system is measured by a quantity (order parameter) $O = |\sum_i o_i|/N$. Another important quantity is the so called 'condensation fraction' $p$, which is the fraction of the agent having $o_i = \pm 1$. Numerical simulations show that the multiagent system (dynamics given by Eqn. (2)) goes into either of the two possible phases: for any $\lambda \leq \lambda_c$, $o_i = 0\ \forall i$, while for $\lambda > \lambda_c$, $O > 0$ and $O \to 1$ as $\lambda \to 1$, with $\lambda_c \approx 2/3$. $\lambda_c$ is the critical point of the phase transition. The relaxation time, defined as the time to reach a stationary value of $O$ in time, diverges as $\tau \sim |\lambda - \lambda_c|^{-z}$ when $\lambda \to \lambda_c$ on either side. A similar behavior is also observed for $p$. Although the values of the exponents differ ($z \approx 1.0 \pm 0.1$ and $z \approx 0.7 \pm 0.1$ with $O$ and $p$ respectively, reported in [14]), the critical points are same. The order parameter exponent $\beta$ is defined (for any order parameter $X$) as:

$$X \sim (\lambda - \lambda_c)^{\beta}. \tag{3}$$

For $O$ its value is $0.10 \pm 0.01$. Now if one also attempts to fit the growth of $p$ in a similar form, one finds its value to be $0.95 \pm 0.02$ .

A mean field calculation can be proposed for the fixed point $o^*$:

$$o^*[1 - \lambda(1 + \langle\epsilon\rangle)] = 0, \tag{4}$$

from which it is easy to show that the critical point is $\lambda_c = 1/(1 + \langle\epsilon\rangle)$ ($\langle\ldots\rangle$ refers to average). For uniform random distribution of $\epsilon$, $\langle\epsilon\rangle = 1/2$ and hence, $\lambda_c = 2/3$. It is important to note that this mean-field treatment does not incorporate the cut-offs at $\pm 1$ and yet gives the correct critical point. We also note that the underlying topology ($1d$, $2d$ or infinite range) has barely any effect on the critical point.

It is appropriate to mention here that a map version of the model [13, 14] has also been proposed:

$$o(t+1) = \lambda(1 + \epsilon(t))o(t) \tag{5}$$

where $\epsilon(t)$ is drawn randomly from a uniform distribution in $[0, 1]$, and $o(t)$ is bounded in $[-1, +1]$ as before. The critical value $\lambda_c$ can be analytically shown to be $\exp[-(2\ln 2 - 1)] \approx 0.6796$ [14].

## 3. Two more models

### *3.1. A simpler model*

A simpler model (model C hereafter) can be proposed:

$$\begin{aligned} o_i(t+1) &= \lambda o_i(t) + \epsilon o_j(t) \\ o_j(t+1) &= \lambda o_j(t) + \epsilon' o_i(t) \end{aligned} \tag{6}$$

where $\epsilon$, $\epsilon'$ are drawn randomly from an uniform distribution in $[0, 1]$ as earlier. Here, $\lambda$ is a parameter as in LCCC model, which is interpreted as 'conviction'. An individual $i$ upon meeting another individual $j$, retains his own opinion proportional to his conviction but also picks up a random influence of that of the other. Numerical simulation reveals that the system goes into either of two possible phases: for any $\lambda \leq \lambda_c$, $O_i = 0\ \forall i$, a *symmetric* phase, while for $\lambda > \lambda_c$, $O > 0$ and goes to 1 as $\lambda \to 1$, a *symmetry broken* phase, with $\lambda_c \approx 1/2$. $\lambda_c$ is the critical point of the phase transition (Fig. 1).

A mean field calculation similar to the one discussed earlier, can be proposed for the fixed point $o^*$:

$$o^*(1 - \lambda - \langle\epsilon\rangle) = 0, \tag{7}$$

from which it is easy to show that the critical point is $\lambda_c = 1 - \langle\epsilon\rangle$. For uniform random distribution of $\epsilon$, $\langle\epsilon\rangle = 1/2$ and hence, $\lambda_c = 1/2$.

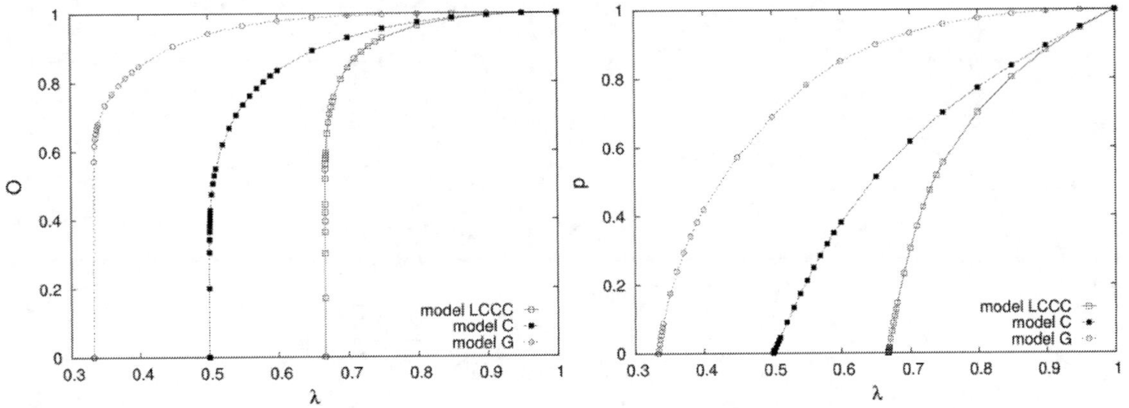


**Figure 1.** The phase diagrams for the 3 models. *Left*: Behavior of order parameter $O$. *Right*: Behavior of condensation fraction $p$.

*3.2. Global effect on LCCC model*

In the context of social opinion formation, global opinion often takes a vital role in influencing one's opinion. In that case, a person $i$, apart from being "influenced" stochastically by a person $j$, is also "influenced" stochastically by the average opinion of the entire society at that moment. Mathematically the dynamics can be represented by

$$\begin{aligned} o_i(t+1) &= \lambda[o_i(t) + \epsilon o_j(t)] + \epsilon' O(t) \\ o_j(t+1) &= \lambda[o_j(t) + \eta o_i(t)] + \eta' O(t), \end{aligned} \tag{8}$$

where $\epsilon$, $\epsilon'$, $\eta$ and $\eta'$ are random numbers, drawn from uniform distribution in $[0,1]$. In this case (model G hereafter), the *symmetry broken* phase $O \neq 0$ appears for $\lambda > 1/3$, and for $\lambda \leq 1/3$ the system is in a *symmetric* phase, with $O_i = 0\ \forall i$ and all individual agents have the opinion 0 (Fig. 1).

This transition point can again be explained by a mean-field approach. At the steady state, i.e. when $O$ reaches a steady value, Eqn. (8) can be written as,

$$o^* = \lambda(1 + \langle\epsilon\rangle)o^* + \langle\epsilon'\rangle o^* \tag{9}$$

from which it can be easily shown that $\lambda_c = 1/3$.

As in LCCC, the critical points of these two models have barely any effect due to change in the underlying geometry. In what follows we study the multi-agent dynamics of these models and their non-equilibrium relaxation.

## 4. Results

As mentioned above for the mean-field version of LCCC, the order parameter exponent is, $0.10 \pm 0.01$. The order parameter exponent for the model C is found to be $0.17 \pm 0.01$ for $O$ and again if one fits the behavior of $p$ in a similar form it comes out to be $0.98 \pm 0.02$, which is similar to the values found for LCCC. For the model G, we estimate the exponent $\beta = 0.081 \pm 0.001$ and for $p$ it is $0.85 \pm 0.01$.

We also calculate the relaxation behavior of the order parameters. An order parameter $X$ relaxes to equilibrium in time $t$ as

$$X(t) \propto \exp(-t/\tau)$$

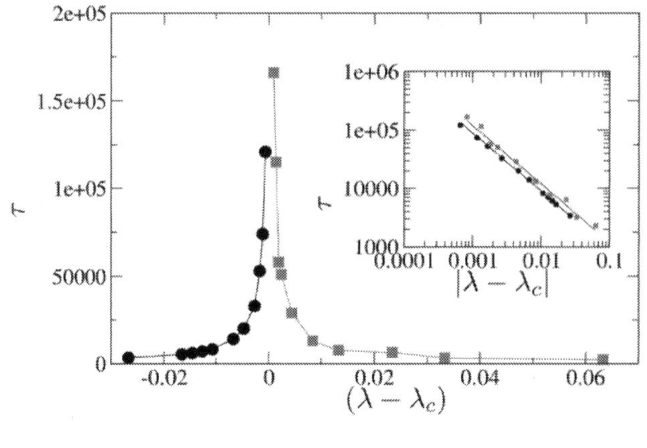


**Figure 2.** Behavior of relaxation time $\tau$ with $\lambda - \lambda_c$ from the order parameter $O$, for model LCCC (From Ref. [14]). The system size was $N = 500$.

for the symmetric phase (going to 0) and

$$X(t) \propto X_0[1 - \exp(-t/\tau)]$$

in the symmetry-broken phase, $X_0$ being the equilibrium value of $X$. Away from the critical point, we plot this relaxation behavior for different values of the parameter $\lambda$, to extract the value of $\tau$. We observe that the relaxation time diverges as $\tau \sim |\lambda - \lambda_c|^{-z}$ both below and above the critical point $\lambda_c$, with roughly the same exponent $z$. For LCCC (mean field), our estimate for $z$ is $0.97 \pm 0.01$ for $O$ while it is $1.10 \pm 0.01$ for $p$, which is consistent with another estimate ($1.16 \pm 0.03$ in Ref. [16]) but is different from that reported in Ref. [14]. For model C, our estimates for $z$ are $1.58 \pm 0.01$ for $O$ and $1.34 \pm 0.01$ for $p$. In case of model G, for $O$, $z = 1.2 \pm 0.1$ and for $p$, $z = 1.75 \pm 0.01$.

In the following subsection, we also report the study of a lattice version of the LCCC model. In this version, the agents are arranged on a $1d$ lattice, and a randomly chosen nearest neighbor pair update their opinions according to Eqn. (2). The critical behavior of this model is studied in detail.

## 5. Non-equilibrium relaxation

NER is a well established simulation strategy to investigate the phase transitions of systems in the thermodynamic limit (for a review see Ref. [17, 19]). In this method the system is allowed to relax from an initial non-equilibrium state to the equilibrium state. It turns out that with much smaller amount of systematic errors, the critical exponents can be estimated from this method.

All thermodynamic quantities show a relaxation in time, starting with initial values away from equilibrium. This temporal relaxation of the NER functions (above mentioned thermodynamic quantities) is used to determine the critical point accurately, as well as the critical exponents. The most important quantity in this study is of course the relaxation of the order parameter for the transition. We study the relaxation of both $O$ and $p$ at and near the critical point.

Here we first study the lattice version of the LCCC model in details, later we get back to the infinite range model as well. The dynamics is started from an initial condition with full order (i.e., all agents have same extreme opinion, $+1$ or $-1$). Away from criticality, the system is expected to relax exponentially to its equilibrium order parameter value. At the critical point, however, the order parameter will relax asymptotically following a power law of the form

$$O(t) \sim t^{-\delta}. \tag{10}$$

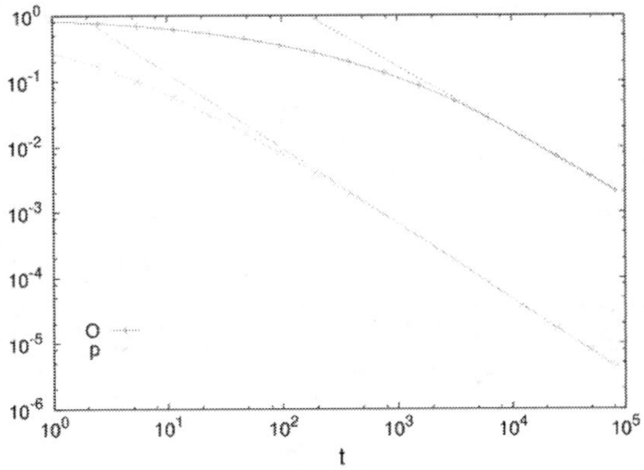


**Figure 3.** The relaxations of the two quantities $O$ (order parameter) and $p$ (condensation fraction) are shown at the critical point ($\lambda_c = 0.66679 \pm 0.00001$) for the 1$d$ version of LCCC model. The power-law decays fit to exponent $1.00 \pm 0.05$ for $O$ and $1.15 \pm 0.01$ for $p$. The system size was $N = 1200$.

Note that a similar power-law is also observed for the condensation fraction $p$. Fig. 3 shows the behavior of the two quantities at the critical point. From the slope of the log-log plot, the critical exponent $\delta$ can be found. From this figure we observe that the critical point for both the quantities are same, and also the critical exponent $\delta$ is very close for both these quantities ($\delta = 1.15 \pm 0.01$ for relaxation of $p$ and $1.00 \pm 0.05$ for $O$).

The critical point of the transition can be obtained with high accuracy by plotting the decay of the order parameter at and on either side of the critical point. Fig. 4 shows the variation of the order parameter $O$ for different $\lambda$ values near criticality. The accurately estimated critical point turns out to be $\lambda_c = 0.66679 \pm 0.00001$ for $N = 1200$, which is very close to the value (2/3) quoted in [14, 16]. A similar estimate was made with $p$ (Fig. 4), the estimate of the critical point turns out to be exactly the same.

From usual finite size scaling theory [17], the order parameter is expected to follow a scaling relation of the form

$$X(t) \approx t^{-\delta} \mathcal{F}(t^{1/\nu_{||}} \Delta), \tag{11}$$

where $\Delta = \lambda - \lambda_c$ and $\mathcal{F}$ is a universal scaling function of a form such that for large argument, the time dependence drops out ($\mathcal{F}(x) \sim x^{\delta \nu_{||}}$). From the data collapse of the off-critical relaxation (Insets of Fig. 4), one can obtain the correlation time exponent to be $\nu_{||} \approx 1.5 \pm 0.1$ for $p$ and $\nu_{||} \approx 1.2 \pm 0.1$ for $O$.

We carry out the above analysis for the fully connected version of the LCCC model. We only quote the main results of this version. The critical point is estimated to be $\lambda_c \approx 0.66659 \pm 0.00002$ for $N = 1200$, slightly different from the lattice version. The critical exponent $\delta$ for $p$ turns out to be $1.2 \pm 0.1$, again close to value with the lattice measurement. The off-critical scaling for $p$ yields the correlation time exponent $\nu_{||} \approx 1.1 \pm 0.1$, slightly different from the lattice counterpart.

We repeat the analysis for the fully connected version of model C. For $N = 512$, the critical point is found to be $0.500097 \pm 0.000001$, the same for the decay of $O$ and $p$, close to the mean-field estimate (1/2). We also get estimates of decay exponent $\delta$ as $0.500 \pm 0.005$ for $O$ and $0.521 \pm 0.005$ for $p$. Using the above scaling relation for the data collapse (Eqn.(11), we estimate values of exponent $\nu_{||}$ as $0.10 \pm 0.01$ for $O$ and $2.00 \pm 0.02$ for $p$ (Fig. 5).

The same analysis is repeated for the fully connected version of model G. For $N = 512$, the critical point is found to be $0.33338 \pm 0.00001$, the same for the decay of $O$ and $p$, close to the mean-field estimate (1/3). The estimates of the decay exponent $\delta$ are $0.585 \pm 0.001$ both

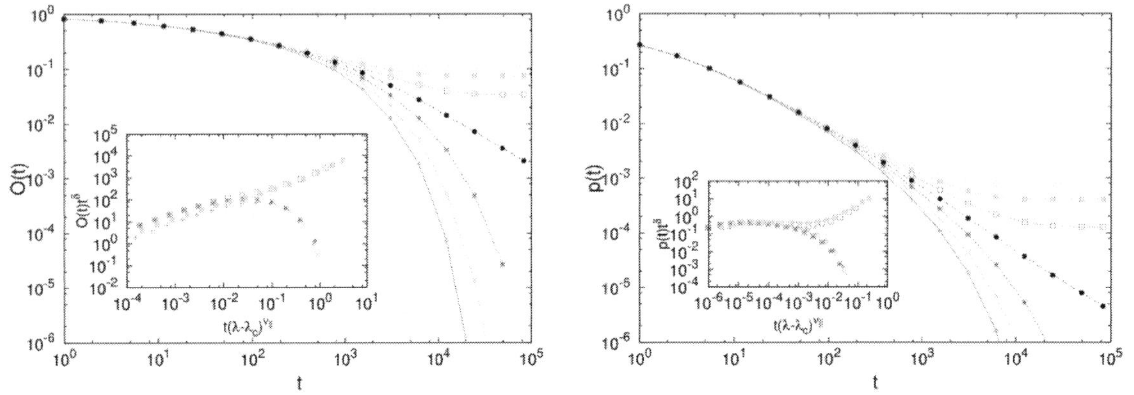


**Figure 4.** *Left*: Relaxation of the order parameter $O$ near $\lambda_c$ for the $1d$ version of LCCC model. The parameter $\lambda$ differs by 0.00001 for the successive sets. The power law is obtained (central curve) for $\lambda_c = 0.66679$. The system size for the simulation is $N = 1200$. Inset: Plot of $t(\lambda-\lambda_c)^{\nu_{||}}$ against $O(t)t^{\delta}$ using the previously obtained value of $\delta$. The data collapse is obtained for $\nu_{||} \approx 1.2\pm0.1$. *Right*: Same for the condensation fraction $p$. Inset: Plot of $t(\lambda-\lambda_c)^{\nu_{||}}$ against $p(t)t^{\delta}$ using the previously obtained value of $\delta$. The data collapse is obtained for $\nu_{||} \approx 1.5\pm0.1$.

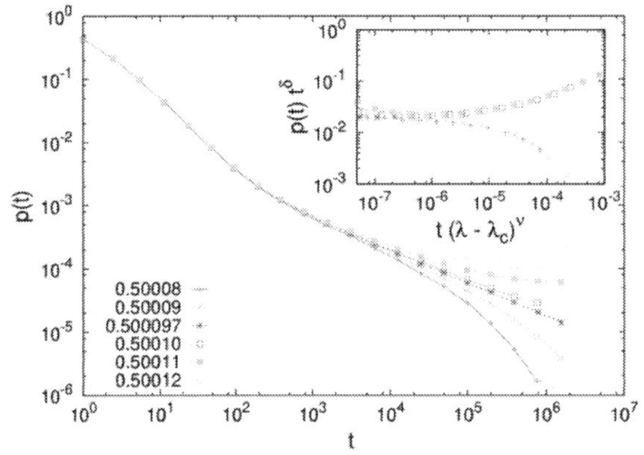


**Figure 5.** Relaxation of the condensation fraction $p$ near $\lambda_c$ for model C. The system size for the simulation was $N = 512$. Inset: Plot of $t(\lambda-\lambda_c)^{\nu_{||}}$ against $p(t)t^{\delta}$ using the previously obtained value of $\delta$. The data collapse is obtained for $\nu_{||} \approx 2.0\pm0.1$.

for $O$ and $p$. From the above scaling relation for the data collapse (Eqn.(11), the estimates of exponent $\nu_{||}$ are $1.6\pm0.1$ for $O$ and $\nu_{||} = 2.0\pm0.1$ for $p$. It is to be noted that these exponents violate the scaling laws.

In the quest of finding if there is a growing length scale in the system as the critical point is approached, we adopted the following strategy: we begin with all agents having opinion 0.0 except a particular agent was fixed at the negative extremity ($-1$) and was kept fixed forever in all subsequent updates. Now as the system is allowed to relax, it is expected that the agent with the rigid opinion will have a neighborhood of influence. This is a measure of the *correlation length* $\xi_c = N_-/N$, where $N_-$ is the number of agents with opinion $o_i < -\epsilon$, where $\epsilon$ is a very small number (in our simulation we have taken $\epsilon = 0.005$). This is expected to grow as the critical point is approached from the sub-critical regime, which is indeed seen from the Monte

Journal of Physics: Conference Series **297** (2011) 012004 doi:10.1088/1742-6596/297/1/012004

**Table 1.** Table comparing the different quantities for the 3 models

| Model | $\lambda_c$ (Mean field) | Measured quantity | $\beta$ | $z$ | $\delta$ | $\nu_{\|\|}$ |
|---|---|---|---|---|---|---|
| LCCC | 2/3 | $O$ | 0.10(1) | 0.97(1) | 1.00(5) | 1.2(1) |
| | | $p$ | 0.95(2) | 1.1(1) | 1.2(1) | 1.1(1) |
| C | 1/2 | $O$ | 0.17(1) | 1.58(1) | 0.500(5) | 0.10(1) |
| | | $p$ | 0.98(2) | 1.34(1) | 0.521(5) | 2.00(2) |
| G | 1/3 | $O$ | 0.081(1) | 1.2(1) | 0.585(1) | 1.6(1) |
| | | $p$ | 0.85(1) | 1.75(1) | 0.585(1) | 2.0(1) |

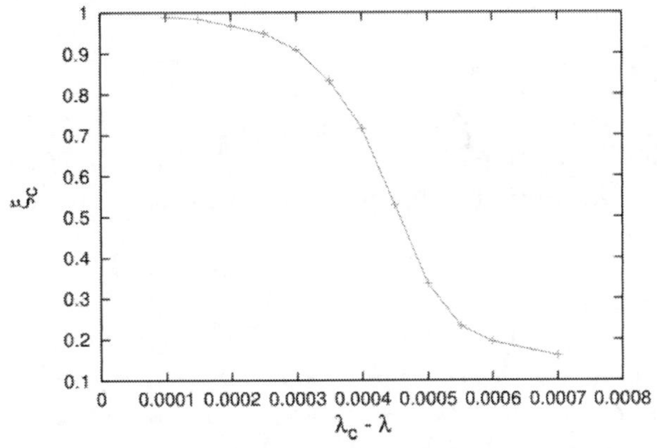


**Figure 6.** The correlation as a function of $(\lambda_c - \lambda)$ in 1$d$ $LCCC$ model for $N = 1000$.

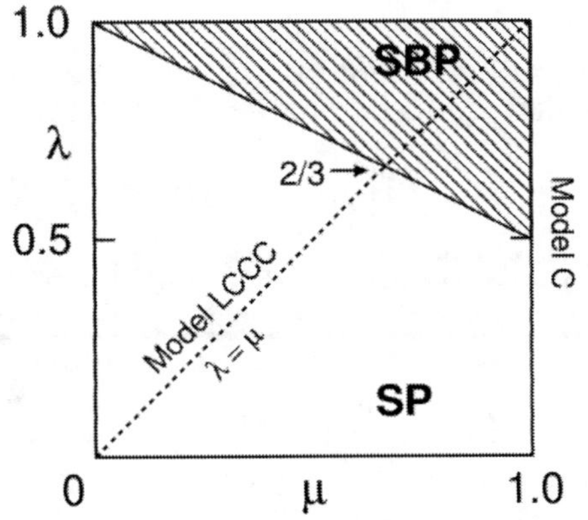


**Figure 7.** The schematic phase diagram of the generalised model. Distinct symmetric phase (SP) and symmetry broken phase (SBP) exist. The $\lambda = \mu$ line corresponds to the Model LCCC while $\mu = 1$ line corresponds to Model C.

Carlo simulations and will remain one in the super-critical regime. In Fig.6 we have shown the growth of length scale for the 1$d$ $LCCC$ model.

## 6. A generalised model

A generalised version of the LCCC model was introduced by Sen in Ref. [16] where the influencing parameter of the $i$th agent $\lambda_i$ is in general different from the conviction parameter

$\mu_i$. Here the binary opinion exchange is mathematically represented by

$$\begin{aligned} o_i(t+1) &= \lambda_i o_i(t) + \epsilon \mu_j o_j(t) \\ o_j(t+1) &= \lambda_j o_j(t) + \epsilon' \mu_i o_i(t) \end{aligned} \tag{12}$$

where the variables are as defined in previous sections. In the study $\lambda$ and $\mu$ were assumed to be homogeneous. The special case of $\lambda = \mu$ is the LCCC model. The limiting case of $\mu = 1$ corresponds to the model C (Fig. 7). In the generalised case the transition from symmetric to symmetry broken phase is determined by both $\lambda$ and $\mu$, the mean field phase boundary given by $\lambda = 1 - \mu/2$.

### *6.1. Non-universal behavior*

Along the phase boundary mentioned above, the critical behavior is reported to be strongly non-universal. Here also the relaxation time diverges close to the transition points along the phase boundary as $\tau \sim |\lambda - \lambda_c|^{-z}$ for corresponding values of $\mu_c$. But in this case $z$ varies with $\mu_c$ rather systematically, indicating a non-universal behavior. For $\mu_c = 0.4$, $z = 1.04 \pm 0.01$; $\mu_c = 2/3$, $z = 1.10 \pm 0.03$; $\mu_c = 0.9$, $z = 1.21 \pm 0.01$. This non-universal behavior is also present in the order parameter exponent $\beta$. For the order parameter $O$, $\beta = 0.079 \pm 0.001$ for $\mu_c = 0.4$ and $\beta = 0.155 \pm 0.001$ for $\mu_c = 0.9$.

The condensation fraction $p$ also shows similar behaviour as $O$. The relaxation time diverges close to the transition points along the phase boundary as $\tau_p \sim |\lambda - \lambda_c|^{-z_p}$ for corresponding values of $\mu_c$. The value of $z_p$ now only weakly varies with $\mu_c$ but is very close to $z$, indicating the existence of only one time scale. Also $p \sim (\lambda - \lambda_c)^{\beta_p}$ with $\beta_p \approx 0.91$ for $\mu_c = 0.4$, $\approx 0.95$ for $\mu_c = 2/3$ and $\approx 1.0$ for $\mu_c = 0.9$. Here also, although the non-universality is present, it is very weak.

## 7. A generalized map

Recently, an interesting map has been proposed by Chakrabarti in Ref. [20], limiting cases of which correspond to known kinetic models. It has the following form:

$$x(t+1) = \min\{(\alpha_1 + \epsilon_t \alpha_2)x(t) + \xi_t \alpha_3^n, \theta\} \tag{13}$$

where $\alpha_1$, $\alpha_2$ and $\alpha_3$ are linear functions of a single parameter $\alpha$, $(0 \leq \alpha_i \leq 1$ for $i = 1, 2, 3)$, with $-\infty \leq n \leq 1$ and $\theta$ taking value either a positive, finite value (take $\theta = 1$, for convenience) or $\infty$ (sufficiently large value) and $\epsilon_t, \xi_t$ are uniform in $[0, 1]$ and are independent.

(a) When $\alpha_1 = \alpha$ and $\alpha_2 = \alpha_3 = 1-\alpha$, $n = 1$, $\theta = \infty$, $x(t+1) = (\alpha + \epsilon(1-\alpha))x(t) + \xi(1-\alpha)$ for which the steady state distributions of $x$ are close to Gamma distributions for large *alpha*. For $\xi = \epsilon$, this reduces to $x(t+1) = \alpha x(t) + \epsilon(1-\alpha)(x(t)+1)$, closely representing the CC model.

(b) When $\alpha_1 = \alpha$ and $\alpha_2 = \alpha_3 = 1 - \alpha$, $-\infty \leq 0$, $\theta = \infty$, then $x(t+1) = \alpha + \epsilon(1-\alpha)x(t) + \xi(1-\alpha)^n$. It is observed that $x$ is distributed as a power law as $P(x) = x^{-\frac{n-2}{n-1}}$.

(c) When $\alpha_1 = \alpha_2 = \alpha$, $\alpha_3 = 0$, $n = 1$ and $\theta = 1$, then Eqn. (13) reduces to $x(t+1) = \min \alpha(1+\epsilon)x(t), 1$, which is nothing but the map version of the LCCC model as given in Eqn. (5).

## 8. A model with preference

Agents are endowed with opinion $o_i$ to begin with, with $o_i$ distributed random uniformly in $[-1, +1]$. The society is also parametrized by a conviction $\lambda$ (as in LCCC).

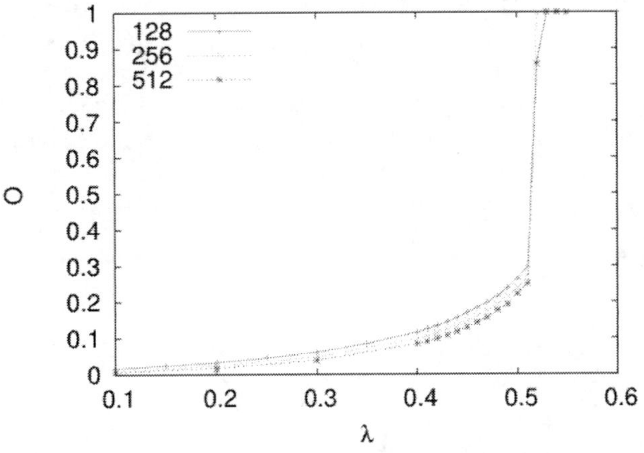


**Figure 8.** Behavior or order parameter $O$ for different system sizes.

The interactions are binary. A pair of agents $i$ and $j$ meet and finds out who has the stronger opinion (compare values of $o_i$ and $o_j$). If $o_i > o_j$, then

$$o_j = \lambda(o_j + \epsilon o_i), \tag{14}$$

and if $o_i < o_j$

$$o_i = \lambda(o_i + \epsilon o_j), \tag{15}$$

$\epsilon$ being the usual random number drawn from an uniform distribution in $[0, 1]$. Note that there is no dynamics if they have the same opinion (they agree).

One computes the usual order parameter $O = |\sum_i o_i|/N$ and also the fraction $p$ of agents at $\pm 1$. $O$ seems to undergo a phase transition at some $\lambda_c \simeq 0.52$ (Fig. 8) while $p$ is a step function at the same $\lambda_c$. These signatures tempt us to conclude that the phase transition here is discontinuous in nature.

## 9. Summary and Discussions

We have reviewed here the studies on kinetic exchange model for opinion formation [13, 14], its two variants (proposed here) and a generalized model [16]. In particular, we report the results from the study using non-equilibrium relaxation simulation technique to study the relaxation behavior of the order parameters for similar models.

Here the lattice version of the LCCC model was proposed. The critical points and the critical exponents $\delta$ and $\nu_{||}$ for $1d$ version were found using two quantities, i.e., the average opinion $O$ (order parameter) and the condensation fraction $p$. The critical points for both the quantities are exactly the same, however, the values of the exponents differ slightly. The mean field version gave same critical point and slightly different exponent values. Then a simpler version of the LCCC model was studied. And finally we also study the effect of the feedback of the global opinion formation in this model. A mean field calculation is made for both these variants, giving the critical points to be $1/2$ and $1/3$ respectively. These are again numerically verified within good accuracy. Analysing the NER functions of the order parameters, the dynamical critical exponent, the order parameter exponent were estimated.

Next, we discuss the generalized model with 'influence' and 'conviction' [16], and comment on the limiting cases which correspond to models LCCC and C. We also discuss the recently proposed map model, the special cases of which resemble some known kinetic exchange models in literature. The phase transitions observed in all the above mentioned cases has the unique feature that the disordered phase consists of individual opinions with values very near to 0. While in other phase transitions in the disordered phase the corresponding individual order parameters have finite positive or negative values but collectively resulting to a value of nearly 0. We also propose a modified model in the same spirit as the models discussed earlier, the only difference being that the agents are only influential over the ones with weaker opinion. However, agents who agree, do not influence each other. The model shows clear characteristics of a first order transition.

## Acknowledgments
The authors thank P. Sen and M. Marsili for discussions. B.K.C. acknowledges collaborations with A. S. Chakrabarti, A. Chakraborti, M. Lallouache.

## References
[1] *Econophysics and Sociophysics: Trends and Perspectives,* edited by Chakrabarti B K, Chakraborti A and Chatterjee A (2006) (Wiley-VCH, Berlin).
[2] Stauffer D 2005, in *Encyclopedia of Complexity and Systems Science Comp. Sc. Engg.* **5**, 71 ed Meyers R A (Springer, New York)
[3] Castellano C, Fortunato S and Loreto V (2009) *Rev. Mod. Phys.* **81** 591.
[4] Galam S, Gefen Y and Saphir Y (1982) *J. Math. Sociology* **9** 1.
[5] Galam S (2008) *Int. J. Mod. Phys. C* **19** 409
[6] Holley R A and Liggett T M (1975) *Ann. Probab.* **3** 643.
[7] Sznajd-Weron K and Sznajd J (2000) *Int. J. Mod. Phys. C* **11** 1157.
[8] Hegselmann R and Krause U (2002) *J. Artif. Soc. Soc. Simul.* **5** 2.
[9] Deffuant G, Neau D, Amblard D and Weisbuch G (2000) *Adv. Complex Syst.* **3** 87.
[10] Fortunato S (2005) *Int. J. Mod. Phys. C* **16** 17.
[11] Biswas S J, Chatterjee A and Sen P (2011), arXiv:1102.0902v2
[12] Biswas S, Sen P and Ray P, in this proceedings.
[13] Lallouache M, Chakraborti A and Chakrabarti B K (2010) *Sci. Cult.* **76** 485.
[14] Lallouache M, Chakrabarti A S, Chakraborti A and Chakrabarti B K (2010) *Phys. Rev. E* **82** 056112.
[15] Chakraborti A and Chakrabarti B K (2000) *Eur. Phys. J. B* **17**, 167; Chatterjee A, Chakrabarti B K and Manna S S (2004) *Physica A* **335**, 155; Chatterjee A and Chakrabarti B K (2007) *Eur. Phys. J. B* **60**, 135.
[16] Sen P (2011) *Phys. Rev. E* **83** 016108.
[17] Hinrichsen H (2000) *Adv. Phys.* **49**, 815.
[18] Patriarca M, Chakraborti A and Kaski K (2004) *Phys. Rev. E* **70**, 016104.
[19] Ozeki Y and Ito N (2007) *J. Phys. A: Math Theor.* **40**, R149.
[20] Chakrabarti A S, arXiv:1101.3617v1.

Journal of Physics: Conference Series **297** (2011) 012005 doi:10.1088/1742-6596/297/1/012005

# Order parameter fluctuations at a critical point - an exact result about percolation -

**Robert Botet**

Laboratoire de Physique des Solides Bât.510, CNRS UMR8502 - Université Paris-Sud, Centre d'Orsay, F-91405 Orsay, France

E-mail: botet@lps.u-psud.fr

**Abstract.** The order parameter of the system in the critical state, is expected to undergo large non-Gaussian fluctuations. However, almost nothing is known about the mathematical forms of the possible probability distributions of the order parameter. A remarkable exception is the site-percolation on the Bethe lattice, for which the complete order-parameter distribution has been recently derived at the critical point. Surprisingly, it appears to be the Kolmogorov-Smirnov distribution, well known in very different areas of mathematical statistics. In the present paper, we explain first how this special distribution could appear naturally in the context of the critical systems, under the assumption (still virtually unstudied) of the exponential distribution of the number of domains of a given size. In a second part, we present for the first time the complete derivation of the order-parameter distribution for the critical percolation model on the Bethe lattice, thus completing a recent publication [1] announcing this result.

## 1. Introduction

Phase transitions take a central place in statistical physics, because it is of prime importance in applications, and also because of the serious mathematical challenge. Such critical behavior happens when collective coherence spans suddenly the entire system for a definite value of the control parameter. The resulting correlations are infinitely-ranged and make the theoretical problem of the utmost difficulty. Only a very few exact results are known [2].

The order parameter is the key physical parameter to characterize the symmetry of the system state in any phase. This is a global macroscopic quantity, and its value may fluctuate because of the large number of different configurations that the system can access using energy fluctuations. However, the *local* order parameter, say: $\eta$, must be defined as soon as we have to deal with spatial fluctuations.

Then, the global order parameter is simply the sum of its local values, and one defines more generally the moments of the local order parameter, $\langle \eta^k \rangle$, where the notation $\langle \cdot \rangle$ stems for the average value over the statistical fluctuations of the order parameter, for a given value of the control parameter. The usual task consists in analyzing the singular behavior of some moments, when the value of the control parameter crosses the critical point. Instead, one should analyze the order-parameter distribution (OPD), $f(\eta)$, since all the moments are related to this function, namely:

$$\langle \eta^k \rangle = \int \eta^k f(\eta) d\eta$$

but this is much more complicated to deal with a function than with its moments, and it is a reason why this analysis is usually not carried out. A remarkable exception is the seminal analysis done by Landau [3], when he considered the density of free energy, $\mathcal{F}(\eta)/N$, of a system of $N$ components. The thermodynamic quantity $\mathcal{F}(\eta)/N$ is simply related to the OPD:

$$f(\eta) = \frac{1}{Z(\beta)} \, \mathrm{e}^{-\beta \mathcal{F}(\eta)/N} \ , \tag{1}$$

where $Z(\beta) = \int \exp(-\beta \mathcal{F}(\eta)/N) d\eta$ is the partition function of the system at the temperature $T = 1/k_B\beta$. This equation links clearly the OPD to the thermodynamics of the system.

Another exceptional model where this approach has been done is the Potts model, for which the density of free energy is known [4].

The present paper is essentially divided into two parts. In the first part, we discuss a relation between the fluctuations of the order parameter and the distribution of domains. This way of investigation is not well known though it could provide a different point of view of the problem, and could be tackled with computer simulations. Particular families of possible critical OPD appear naturally in this framework. The second part is about a recent result about the exact OPD at the critical point of the percolation problem on the Bethe lattice. Amazingly, this distribution is known in various other scientific areas not connected to criticality. This result shows in particular that the critical OPD is not necessary a very strange and complicated distribution.

## 2. Critical OPD for a system divided in domains.–

In a number of cases, the critical system is divided in areas of about the same value of the local order parameter. These areas are called 'domains'. Detailed knowledge of the distribution of the number of domains of a given size, allows in principle, to determine the OPD of the system. This way of study has not been investigated in details up to now, and we describe hereafter the corresponding framework.

### *2.1. Relation between the OPD and the distributions of the number of domains of given size.*

Let us consider a macroscopic system of size $N$, embedded in the $d$-dimensional space, which experiences a critical behavior for a definite value of the control parameter. We suppose that the system can, at any given time, be split into well-defined homogeneous domains, each of them characterized by a value of the local order parameter.

It is convenient to consider a generic example to illustrate the idea: the global magnetization per spin, $M_N/N$, of the $N$-spins Ising system is the sum of the magnetizations of the 'up'-spins domains minus the sum of the magnetizations of the 'down'-spins domains. If the spin 'up' corresponds to the value $\eta = +1$ of the local order parameter, and the spin 'down' to the value $-1$, one has:

$$m_N \equiv M_N/N = \frac{2}{N} \sum_{s=1}^{N} s n_s^{(+)} - 1 \ ,$$

where $n_s^{(+)}$ is the number of domains of 'up'-spins, of size $s$. That way, the variable $m_N$ is a number between -1 and 1. Clearly, the distribution of the magnetization is then directly related to the distribution, $n_s^{(+)}$, of the number of the domains of size $s$.

Let us be more precise. Let $\hat{f}_{M_N}(H)$ denote the Laplace transform of the probability distribution of $M_N$, i.e.:

$$\hat{f}_{m_N}(\lambda) = \int_0^\infty f_{m_N}(m) \mathrm{e}^{-\lambda m} dm \ .$$

This function corresponds, in the context of the statistical physics, to the ratio between the partition function $Z(\beta, H)$, in the field: $H = -\lambda/\beta$, and the partition function $Z(\beta) \equiv Z(\beta, 0)$ in the null field.

The variables $\{n_s^{(+)}\}_{s=1,\cdots,N}$ are weakly correlated, because the only constraint acting on these quantities is the conservation of the mass: $\sum_s s n_s^{(+)} + \sum_s s n_s^{(-)} = N$. Then, the function $\hat{f}_{M_N}(\lambda)$ is the product of the Laplace transforms of the $s n_s^{(+)}$-distributions:

$$\hat{f}_{m_N}(\lambda) = \mathrm{e}^{-\lambda} \prod_{s=1}^{N} \int_0^\infty f_{n_s^{(+)}}(n) \mathrm{e}^{-2\lambda s n/N} dn \ , \tag{2}$$

with $f_{n_s^{(+)}}$ the probability density of the random variable $n_s^{(+)}$. This relation allows to compute the OPD if we know the distributions of the numbers of the domains of a given size $s$.

### 2.2. *The case of the exponentially-distributed domain-number distribution*

At the critical point, the *average* distribution of the $n_s^{(+)}$ is known to be an inverse power-law [5]: $\langle n_s^{(+)} \rangle \simeq \alpha N s^{-\tau}$ with the exponent $\tau \geq 2$, and $\alpha$ a positive constant which can be found using the condition: $\lim_{N\to\infty} \langle m_N \rangle = 0$ at the critical point.

If we knew the probability distribution of the random variable $n_s^{(+)}$, one could deduce from (2) the OPD of the Ising model. Unfortunately, the probability distribution of the $n_s^{(+)}$ is generally unknown at the critical temperature [6], and the average values are only available [7]. However, we can make simple guess about the unknown distribution of the number of domains, to see which kind of OPD one could generally expect. The Gamma-distributions or the Poisson distribution can be studied this way.

Here, we consider the exponential distribution [8] as the simplest distribution for the number of domains of size $s$, that is:

$$f_{n_s^{(+)}}(n) = \frac{1}{\langle n_s^{(+)} \rangle} \mathrm{e}^{-n/\langle n_s^{(+)} \rangle} \ , \tag{3}$$

as the density of probability to find $n$ spins-'up' domains of size $s$. The limit distribution for $M_N$ can then be derived exactly for the integer values of $\tau$, as explained below.

Considering the exponential-distributions (3) and using the Fisher result $s\langle n_s^{(+)} \rangle / N \simeq \alpha s^{1-\tau}$, one obtains:

$$\hat{f}_{m_N}(\lambda) \simeq \mathrm{e}^{-\lambda} \left[ \prod_{s=1}^{N} (1 + 2\alpha\lambda s^{1-\tau}) \right]^{-1} \ . \tag{4}$$

Journal of Physics: Conference Series **297** (2011) 012005 doi:10.1088/1742-6596/297/1/012005

When the value of $\tau$ is an integer number, the asymptotic limit when $N \to \infty$ of the product above, is written more simply as the finite product:

$$\lim_{N\to\infty} \hat{f}_{m_N}(\lambda) = \mathrm{e}^{-\lambda} \prod_{j=1}^{\tau-1} \Gamma\left(1 - (2\alpha\lambda)^{1/(\tau-1)}\ \mathrm{e}^{i\pi(2j+1)/(\tau-1)}\right) \tag{5}$$

with $\Gamma$ the Gamma function in the complex plane. The Equ.(5) defines a one-parameter family of critical OPD, parametrized with the actual value of $\tau$.

Two particular cases are important to discuss: $\tau = 2$ and $\tau = 3$.

- The case $\tau = 2$.
  Using the relation:

  $$\mathrm{e}^{\gamma x}\Gamma(1+x) = \lim_{N\to\infty} \prod_{s=1}^{N} \frac{\mathrm{e}^{x/s}}{1+x/s} ,$$

  with $\gamma$ the Euler constant, the moment generating function of $m_N$ writes:

  $$\lim_{N\to\infty} \hat{f}_{m_N}(\lambda) \propto \Gamma(1 + 2\alpha\lambda) ,$$

  after proper rescaling. This generating function corresponds to the Gumbel distribution [9], that is (written in its simplest rescaled form):

  $$f_{Gumbel}(y) = \mathrm{e}^{y-\mathrm{e}^{y}} .$$

  This result might be related to a recent series of articles by Bramwell et al [10] in which it is argued that Gumbel-like distributions could be typical of the OPD of the critical 2d XY-model [11]. The analysis was both from numerical simulations and approximated theoretical arguments. Bertin [12] gave a different argument for the 2d XY-model at the Kosterlitz-Thouless transition [10], using an argument close to the one presented here. Actually, the result could be explained with the Fisher's result [5]: $\tau = 2 + 1/\delta$, with $\delta$ the critical exponent, known to be $\delta = 15$ in this model. Because the value of $\delta$ is that large, the critical OPD should not differ too much from the Gumbel distribution, but it could only be an approximation. Indeed, it has been proved recently that the Gumbel distribution could *not* be the correct OPD for the 2d XY-model [13], though it is indeed a good approximation for all values of the magnetization, but the extreme ones.
- The case $\tau = 3$.
  When $\tau$ is a general odd integer number, one can express the finite product (5) in a yet simpler form as:

  $$\lim_{N\to\infty} \hat{f}_{m_N}(\lambda) = \prod_{j=1}^{\tau-1} \left[\frac{\pi(\alpha\lambda)^{1/(\tau-1)}}{\sin\left(\pi(\alpha\lambda)^{1/(\tau-1)}\mathrm{e}^{i\pi(2j+1)/z}\right)}\right]^{1/2} .$$

  For the particular case $\tau = 3$, one finds the limit Laplace transform:

  $$\lim_{N\to\infty} \hat{f}_{m_N}(\lambda) = \mathrm{e}^{-\lambda} \frac{\sqrt{2\alpha\pi^2\lambda}}{\sinh\sqrt{2\alpha\pi^2\lambda}} ,$$

  which corresponds to the Kolmogorov-Smirnov distribution [14] (written here in a reduced form):

  $$f_{KS}(y) = \frac{d}{dy}\left[\sum_{n=-\infty}^{\infty} (-1)^n \mathrm{e}^{-n^2 y}\right] .$$

This result has been shown previously in various contexts. Probably Watson was the first one to mention it [15], in connection with the distribution of the excursions in the Brownian bridge problem [16]. One can also read Doob's paper [17].
In the physics of the critical systems, this distribution has been discovered in the OPD of the percolation-cluster size on the Bethe lattice, at the critical threshold [1]. A detailed derivation of this result is outlined in the next section.

## 3. Critical OPD for the site-percolation on the Bethe lattice.–

We shall derive now an exact critical OPD, which belongs to the family (4). The critical system is a mean-field percolation model. Note that, despite a number of available results [18], the distribution of the numbers of domains of given sizes is not known in this case, and we cannot conclude about a definite relation of this (possible) coincidence to the results developed in the preceding Section.

### *3.1. Scaling relations for the OPD*

A fundamental relation resulting from self-similarity of the system at the critical point, is the so-called *first scaling law* [19]. For the normalized OPD function $\mathcal{P}^{(N)}(\eta)$ of the finite system of size $N$, the first-scaling law writes:

$$\langle\eta\rangle_N \mathcal{P}^{(N)}(\eta) \sim \mathcal{F}_c\left(\frac{\eta}{\langle\eta\rangle_N}\right) .$$

with the scaling function $\mathcal{F}_c$ independent of the size of the lattice.
In terms of the Laplace transform of the OPD:

$$\hat{\mathcal{P}}^{(N)}(h) \equiv \int \mathcal{P}^{(N)}(\eta)\, \mathrm{e}^{-h\eta/\langle\eta\rangle_N} d\eta \ , \tag{6}$$

written here with the reduced variable $z_1 \equiv \eta/\langle\eta\rangle_N$, the first-scaling law is equivalent to the assertion that $\hat{\mathcal{P}}^{(N)}(h)$ tends to the Laplace transform of the scaling function, namely:

$$\hat{\mathcal{P}}^{(N)}(h) \to \int_0^\infty \mathcal{F}_c(z_1)\mathrm{e}^{-hz_1} dz_1 \ ,$$

when $N \to \infty$.
Note that the definition (6) leads to the small-$h$ expansion of the Laplace transform of the OPD under the form:

$$\hat{\mathcal{P}}^{(N)}(h) = 1 - h + \mathcal{O}(h^2) \ . \tag{7}$$

### *3.2. The Bethe lattice*

A finite Bethe lattice $\mathcal{B}_n$ of branching parameter $z$, is built from an origin site, $\mathcal{O}$. At the first generation ($n = 1$), $z$ sites are added, each linked to $\mathcal{O}$ with one bond. The $z$ sites are peripheral, while $\mathcal{O}$ becomes a bulk site. At the second generation ($n = 2$), $z - 1$ sites are added to every last-generation site, and so on. As $z(z-1)^{n-1}$ sites are added at the generation $n$, the total number of sites in $\mathcal{B}_n$ is:

$$N_{\mathcal{B}_n} = 1 + z\frac{(z-1)^n - 1}{z-2} \ ,$$

and this corresponds to $N_{\mathcal{B}_n} - 1$ bonds.

In this work, we consider cases where the branching parameter $z$ has a finite value larger than 3. Figure 1 shows the second-generation Bethe lattice $\mathcal{B}_2$, with the branching parameter $z = 3$.

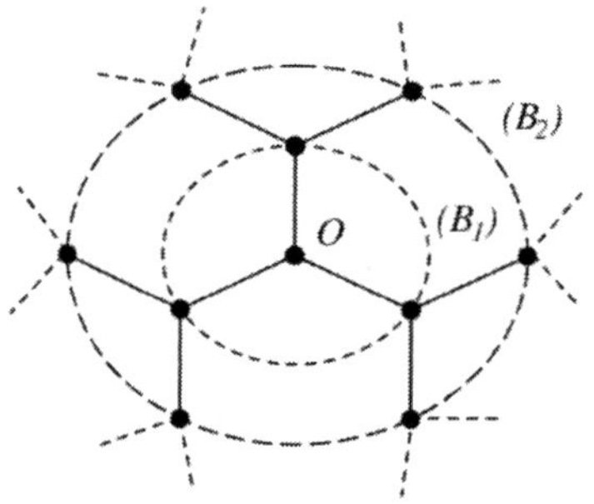


**Figure 1.** Sketch of the finite Bethe lattice of the second generation. It contains 10 sites, of which 6 are peripheral.

### *3.3. The site-percolation problem on the Bethe lattice*

We suppose that every site of the Bethe lattice has the probability $p$ to be occupied [20]. We denote $q \equiv 1 - p$, the probability that a site be vacant. A cluster is defined as an ensemble of occupied sites connected together by neighboring. The site-percolation problem on the infinite Bethe lattice is a second-order critical phenomenon [21]: transition occurs suddenly when a giant cluster (the percolation cluster) spans the lattice. This dramatic event occurs for the definite value $p = p_c(z) = 1/(z-1)$ of the occupation probability.

In this context, the order parameter for the transition, is the probability for a given site to belong to the percolation cluster. Clearly, it is 0 when the percolation cluster does not exist, and has a positive value when such a cluster appears in the system. In other words, the order parameter is the volume fraction of the percolation cluster.

We plan to analyze here the fluctuations of the order parameter. To complete this task, we have first to define properly the percolating cluster on the finite Bethe lattice. Then, we shall look precisely how the result for the finite lattice can be extended to the infinite size.

In the finite Bethe lattice, the percolation cluster is the cluster which links the lattice origin to *at least* one surface site. Within this definition, the percolation cluster is necessarily unique on the Bethe lattice since, by definition, it must include the origin, $\mathcal{O}$, of the lattice.

### *3.4. Probability distribution of the percolating-cluster size*

Journal of Physics: Conference Series **297** (2011) 012005 doi:10.1088/1742-6596/297/1/012005

The starting point of the derivation is the recurrence relation concerning the number, $g_{s,l}^{(n)}$, of possible different clusters of $\mathcal{B}_n$, containing $s$ sites, $\mathcal{O}$ being one of these sites, and $l$ sites of the cluster being peripheral.

Since any $l$ peripheral site of generation $n-1$ gives birth to $(z-1)l$ sites at generation $n$ (for any $n \geq 2$), one has the relations:

$$\begin{aligned} g_{s,l}^{(k)} &= \sum_{l'} \binom{(z-1)\,l'}{l} g_{s-l,l'}^{(k-1)} \qquad \text{for } k \geq 2 \ , \\ g_{s,l}^{(1)} &= \binom{z}{s-1} \delta_{s,l-1} \ , \end{aligned} \tag{8}$$

with $\delta_{i,j}$ the Kronecker symbol. In Equ.(8), the sum runs over all values of $l'$ such that: $l \leq (z-1)l' \leq (z-1)(s-l-1)$.

The rest of the derivation consists in transforming step by step these recurrence equations in the exact OPD for the percolation-cluster size.

- The generating function for the statistical weights $g_{s,l}^{(k)}$.–

  On the finite Bethe lattice $\mathcal{B}_n$ and for a given value of the occupation probability $p$, the probability that the cluster passing through $\mathcal{O}$ be of size $s$ with $l$ peripheral sites, is:

$$\mathcal{P}_{s,l}^{(n)} = g_{s,l}^{(n)}\, p^{\,s-1}\, q^{\,(z-2)s+2-(z-1)\,l} \ ,$$

  with the normalization $\sum_{s,l} \mathcal{P}_{s,l}^{(n)} \equiv 1$. It is convenient to consider the following generating function:

$$\Phi^{(n)}(x,y) \equiv \sum_{s,l} \mathcal{P}_{s,l}^{(n)}\, x^{\,s-1}\, y^{\,l} = q^{\,z} \sum_{s,l} g_{s,l}^{(n)}\, \left(p\, q^{z-2} x\right)^{s-1} \left(q^{1-z} y\right)^{l} \ . \tag{9}$$

  In particular, the value of $\Phi^{(1)}$ is explicitly:

$$\Phi^{(1)}(x,y) = (q + p\,x y)^{z} \ , \tag{10}$$

  and the normalization condition: $\Phi^{(n)}(1,1) = 1$, holds for any value of $p$ and of $n$.

- Definition of the probability distribution of the percolating-cluster size.–

  The percolating cluster, when it exists, is the cluster passing through $\mathcal{O}$, and connecting $\mathcal{O}$ to the periphery of $\mathcal{B}_n$. The probability of occurrence of a percolating cluster of size $s$ is then the probability that the cluster has size $s$ and any number $l$ of peripheral sites, minus the probability that the cluster has size $s$ but does not reach the periphery ($l = 0$), namely:

$$\mathcal{P}_{perco}^{(n)}(s) \equiv \sum_{l \neq 0} \mathcal{P}_{s,l}^{(n)} = \sum_{l} \mathcal{P}_{s,l}^{(n)} - \mathcal{P}_{s,0}^{(n)} \ .$$

- The Laplace transform of the distribution $\mathcal{P}_{perco}^{(n)}(s)$.–

Journal of Physics: Conference Series **297** (2011) 012005 doi:10.1088/1742-6596/297/1/012005

From the previous relation, and using (9), we can write down the discrete Laplace transform of the probability distribution $\mathcal{P}^{(n)}_{perco}(s)$ as:

$$\hat{\mathcal{P}}^{(n)}_{perco}(\lambda) \equiv \sum_s \mathcal{P}^{(n)}_{perco}(s)\,\mathrm{e}^{-\lambda s/\langle s_{perco}\rangle_n} / \sum_s \mathcal{P}^{(n)}_{perco}(s) \ , \tag{11}$$

$$= \mathrm{e}^{-\lambda/\langle s_{perco}\rangle_n} \frac{\Phi^{(n)}(\exp(-\lambda/\langle s_{perco}\rangle_n),1) - \Phi^{(n)}(\exp(-\lambda/\langle s_{perco}\rangle_n),0)}{\Phi^{(n)}(1,1) - \Phi^{(n)}(1,0)} \ , \tag{12}$$

with the average size of the percolating cluster: $\langle s_{perco}\rangle_n \equiv \sum_s s\mathcal{P}^{(n)}_{perco}(s)/\sum_s \mathcal{P}^{(n)}_{perco}(s)$. The running variable $\lambda$ is a positive real number in order to insure the uniform convergence of the series in (11).

- The average percolating-cluster size.–

  The definition (11) was written such that the expansion of $\hat{\mathcal{P}}^{(n)}_{perco}(\lambda)$ in powers of $\lambda$ begins as: $\hat{\mathcal{P}}^{(n)}_{perco}(\lambda) = 1 - \lambda + \mathcal{O}(\lambda^2)$, in accordance to (7). This constraint gives the value of the average size $\langle s_{perco}\rangle_n$ of the percolation cluster through the relation:

$$\langle s_{perco}\rangle_n = 1 + \left(\frac{\partial\Phi^{(n)}}{\partial x}(1,1) - \frac{\partial\Phi^{(n)}}{\partial x}(1,0)\right) / \left(\Phi^{(n)}(1,1) - \Phi^{(n)}(1,0)\right) \ . \tag{13}$$

So far, we obtained the Laplace transform of the probability distribution of the size of the percolation cluster for the site-percolation problem on the Bethe lattice at a given value (not necessary the critical value) of the occupation probability $p$. This Laplace transform is given by (12) and (13), in function of the generating functions $\Phi^{(n)}(x,y)$ defined in (9).

### *3.5. Asymptotic behavior of the generating function $\Phi^{(n)}(x_n, y)$*

In this Section, we shall use the recurrence equations (8) to study the asymptotic behavior of the $\Phi^{(n)}(x,y)$ at the critical point $p = p_c(z)$, when the generation number of the lattice, $n \to \infty$.

- Recurrence relation for the generating functions $\Phi^{(n)}$ .–

  Because of the relations (8), one has the recurrence equations:

$$\Phi^{(n)}(x,y) = \Phi^{(n-1)}(x,(q + p\,xy)^{z-1}) \ . \tag{14}$$

  Noting from (10) that the second argument in the right-hand term of (14) is $(\Phi^{(1)}(x,y))^{(z-1)/z}$, one deduces the general relation: $\Phi^{(n)}(x,y) = \Phi^{(n-j)}(x,(\Phi^{(j)}(x,y))^{(z-1)/z})$ for any value $j = 1, \cdots, n-1$. In particular, for $j = n-1$, the recurrence relation writes:

$$\Phi^{(n)}(x,y) = \left(q + p\,x\,(\Phi^{(n-1)}(x,y))^{(z-1)/z}\right)^z \ , \tag{15}$$

  with Equ.(10) as the initial condition.

- Asymptotic behavior of the generating functions $\Phi^{(n)}$ for $p = p_c(z)$.–

  From now, we consider the system at the critical value of the occupation probability $p = p_c(z) = 1/(z-1)$, and we wish to derive the asymptotic (*i.e.* $n \to \infty$) behavior of $\hat{\mathcal{P}}^{(n)}_{perco}(\lambda)$.

We study here the finite Bethe lattices of increasing size $k$, until a maximum size $k = n$, while all the other parameters are constant. Therefore, the $\Phi^{(n)}(x, y)$ are derived recursively from Equ.(15), written as:

$$\begin{aligned} \Phi^{(k)}(x_n, y) &= \left(q_c + p_c\, x_n\, (\Phi^{(k-1)}(x_n, y))^{(z-1)/z}\right)^z , \\ \Phi^{(0)}(x_n, y) &= y^{z/(z-1)} , \end{aligned} \tag{16}$$

where the initial condition was expressed conveniently for the index $k = 0$. In Equ.(16), the index $k$ runs from 1 to $n$, while the parameter $x_n = \mathrm{e}^{-\lambda/\langle s_{perco}\rangle_n}$ is a constant (*i.e.* independent of $k$).

In order to simplify Equ.(16), we introduce the auxiliary variable $\varphi_n(k)$ defined by:

$$\Phi^{(k)}(x_n, y) \equiv K_n^z \mathrm{e}^{-\varphi_n(k)/n} ,$$

where $K_n$ is the coefficient:

$$K_n \equiv \left(\frac{1}{(z-2)\, x_n} \frac{q_c}{p_c}\right)^{1/(z-1)} .$$

Equ.(16) writes then:

$$\varphi_n(k) = \frac{z}{2(z-2)} \frac{\kappa_n^2}{n} - zn \ln\left(\frac{z-2}{z-1} + \frac{1}{z-1}\, \mathrm{e}^{-\frac{z-1}{zn}\varphi_n(k-1)}\right) , \tag{17}$$

$$\varphi_n(0) = \begin{cases} zn \ln K_n & \text{if } y = 1 , \\ \infty & \text{if } y = 0 , \end{cases} \tag{18}$$

with

$$\kappa_n^2 = 2\, \frac{z-2}{z-1}\, n^2 \frac{\lambda}{\langle s_{perco}\rangle_n} . \tag{19}$$

In the relation (19), $x_n$ was replaced by its value: $\mathrm{e}^{-\lambda/\langle s_{perco}\rangle_n}$. The advantage of these auxiliary variables is that the parameter $x_n$ appears only in the constant term in Equ.(17) and the initial condition (18).

- General solution for the auxiliary sequences $\varphi_n(k)$.–

We suppose that one can develop the right-hand term of Equ.(17) in successive powers of $1/n$. This assumption - which will be checked later on the solution - needs the condition:

$$\lim_{n\to\infty} \varphi_n(n)/n = 0 . \tag{20}$$

When (20) is fulfilled, Equ.(17) writes, at the second order in $1/n$, as the difference equation:

$$\varphi_n(k) - \varphi_n(k-1) = \frac{z-2}{2zn}\left[\frac{z^2\kappa_n^2}{(z-2)^2} - \varphi_n^2(k-1)\right] . \tag{21}$$

Therefore, (21) is solved after considering the differential analog:

$$f'(u) = \frac{z-2}{2zn}\left[\left(\frac{z\kappa_n}{z-2}\right)^2 - f^2(u)\right] ,$$

with the initial conditions: $f(0) = zn \ln K_n$ for $y = 1$, and $f(0) = \infty$ for $y = 0$. This leads to the form:

$$\varphi_n(k) = \frac{z\kappa_n}{z-2}\,\frac{\rho\exp\left(\kappa_n k/n\right) - 1}{\rho\exp\left(\kappa_n k/n\right) + 1} \ ,$$

with $\rho$ a coefficient given by the initial conditions. Then, after using the identity:

$$\frac{\rho e^x - 1}{\rho e^x + 1} = \tanh(\frac{x}{2} + \frac{1}{2}\ln\rho) \ ,$$

the sequence $\varphi_n(k)$ can be expressed in terms of hyperbolic functions:

$$\varphi_n(k) = \begin{cases} \frac{z}{z-2}\,\kappa_n \tanh\left(\frac{\kappa_n}{2}\frac{k}{n} - a_0\right) & \text{if } y = 1 \ , \\ \frac{z}{z-2}\,\kappa_n \coth\left(\frac{\kappa_n}{2}\frac{k}{n}\right) & \text{if } y = 0 \ , \end{cases}$$

with $a_0$ the coefficient defined by:

$$a_0 = \frac{1}{2}\ln\frac{1 - \kappa_n/2n}{1 + \kappa_n/2n} \ . \qquad (22)$$

Note that $a_0$ tends to 0 as $\sim \sqrt{\lambda/\langle s_{perco}\rangle_n}$. Moreover, the assumption (20) can be rewritten in terms of $\kappa_n$ as $\lim_{n\to\infty}\kappa_n/n = 0$, which is well realized here.

At this stage, we obtained the asymptotic behavior of the generating function $\Phi^{(n)}(x_n, y)$ under the form:

$$\Phi^{(n)}(x_n, y) \equiv K_n^z e^{-\varphi_n(n)/n} \ ,$$

with the function $\varphi_n(n)$ given by:

$$\varphi_n(n) = \begin{cases} \frac{z}{z-2}\,\kappa_n \tanh\left(\frac{\kappa_n}{2} - a_0\right) & \text{if } y = 1 \ , \\ \frac{z}{z-2}\,\kappa_n \coth\left(\frac{\kappa_n}{2}\right) & \text{if } y = 0 \ , \end{cases} \qquad (23)$$

with $\kappa_n$ as in (19) and $a_0$ as in (22).

### *3.6. Exact expression for the limit distribution $\hat{\mathcal{P}}^{(\infty)}_{perco}(\lambda)$*

The explicit solutions (23) can be used in (12) to write the discrete Laplace transform of the probability distribution of $\mathcal{P}^{(n)}_{perco}(s)$ under the asymptotic form:

$$\hat{\mathcal{P}}^{(n)}_{perco}(\lambda) \ \simeq C\kappa_n\left(\coth(\tfrac{\kappa_n}{2}) - \tanh(\tfrac{\kappa_n}{2} - a_0)\right) \ . \qquad (24)$$

Then, the condition (7): $\hat{\mathcal{P}}^{(n)}_{perco}(\lambda) = 1 - \lambda + \mathcal{O}(\lambda^2)$, defines the two missing parameters: $C$ and $\langle s_{perco}\rangle_n$, namely:

$$C = \frac{1}{2} \ , \qquad (25)$$

and:

$$\langle s_{perco}\rangle_n \simeq \frac{z-2}{z-1}\frac{n^2}{3} \ . \qquad (26)$$

The equations (24), (25) and (26) characterize completely the limit scaled distribution $\hat{\mathcal{P}}^{(n)}_{perco}$ at the critical case point $p = p_c(z)$.

Considering the asymptotic dependences $\kappa_n \to \sqrt{6\lambda}$ and $a_0 \to 0$, when $n \to \infty$, one deduces that the Laplace transform of the probability distribution of the percolation-cluster size tends to the non-trivial distribution:

$$\hat{\mathcal{P}}^{(n)}_{perco}(\lambda) \to \hat{\mathcal{P}}^{(\infty)}_{perco}(\lambda) = \frac{\sqrt{6\lambda}}{\sinh\sqrt{6\lambda}} , \tag{27}$$

which is a universal function in the sense that it is independent on the value of branching parameter $z$.

### *3.7. Limit distribution*

As explained in details in [22], the function (27) is the Laplace transform of the Kolmogorov-Smirnov distribution [14]:

$$\mathcal{F}(z_1) = \frac{d}{dz_1}\left[\sum_{n=-\infty}^{\infty}(-1)^n \mathrm{e}^{-n^2\pi^2 z_1/6}\right] , \tag{28}$$

$$= \frac{d}{dz_1}\left[\sqrt{\frac{6}{\pi z}}\sum_{n=-\infty}^{\infty}\mathrm{e}^{-3(2n+1)^2/(2z_1)}\right] . \tag{29}$$

Both expressions are equivalent because of the Poisson summation formula [23].

The expression (29) gives the singular small-$z_1$ tail:

$$\mathcal{F}(z_1) \sim \frac{\mathrm{e}^{-3/(2z_1)}}{z_1^{5/2}} \qquad \text{for } z_1 \to 0 , \tag{30}$$

while (28) gives the exponential large-$z_1$ tail as:

$$\mathcal{F}(z_1) \sim \mathrm{e}^{-\pi^2 z_1/6} \qquad \text{for } z_1 \to \infty . \tag{31}$$

The Figure2 shows results of numerical simulations of the site-percolation on the $z = 3$-Bethe lattice at the critical point. We can see the convergence of the scaled OPD to the limit distribution.

In particular, all the moments of the limit distribution exist since both tails (30) and (31) go to zero faster than any power of $z_1$. The moments of the Kolmogorov-Smirnov distribution can be obtained readily in a closed form from the development of (27) as a series of Bernoulli numbers. Translated into moments of the order parameter, they write:

$$\frac{\langle\eta^k\rangle}{\langle\eta\rangle^k} = 2\frac{k!}{(2k)!}6^k\left|(2^{2k-1}-1)B_{2k}\right| ,$$

for any non-negative value of the integer exponent $k$.

## 4. Conclusion

The order parameter of the non-critical $N$-body system, is the sum of uncorrelated local variables, thus the Central Limit Theorem must apply and lead essentially to Gaussian fluctuations. On the other hand, the problem is not so simple at the critical point because of the long-ranged correlations which embrace the entire system. No general theory is available in the critical case, while the order-parameter fluctuations are expected to be non-Gaussian.

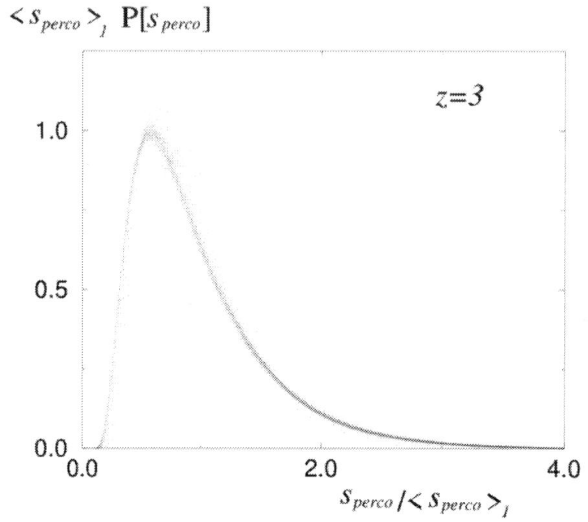


**Figure 2.** Numerical simulations of the OPD for the site-percolation model on the $z = 3$-Bethe lattice of generations $n = 2$ to 24. All the data are shown as dots.

We have seen in this work that the probability distribution of the order parameter is tightly linked to the statistics of the domain sizes. However, lack of knowledge about this statistics forbids presently any definite conclusion. Then, any exact result is of prime importance to understand which kind of fluctuations the order parameter can experience at the criticality.

The percolation on the Bethe lattice provides such unique information. For the first time, the complete derivation was presented in details in the present paper. It requires only standard tools of asymptotic analysis, and it can easily be extended to system states out of the criticality. As a matter of fact, the derivation (not presented here) for given values of the occupation probability different from $p_c(z)$ leads to the results expected from the Central Limit Theorem, as it should be. For this special model at the critical point, the OPD appears to be a known distribution (the Kolmogorov-Smirnov distribution). It does not depend on the value of the coordination number $z$ which defines the lattice. In this sense, this critical OPD is expected to be universal and characteristics of the mean-field percolation on loopless lattices. This is a first result concerning the possible universality of the critical OPD. Such a question, for the other systems at the critical point, is still pending.

## 5. Acknowledgments

This work was largely motivated by a fruitful collaboration with Marek Płoszajczak from GANIL (Caen).

[1] Botet R and Płoszajczak M 2005 *Phys. Rev. Lett.* **95** 185702
[2] e.g. Wu F Y 2009 *Exactly solved models: a journey in statistical mechanics : selected papers with Commentaries (1963-2008)* (Singapore: World Scientific)
[3] Landau L and Lifshitz E 1980 *Statistical Physics*, $3^{rd}$ edition, "Course of Theoretical Physics", vol.5, (Oxford: Butterworth-Heinemann)
[4] Wu F Y 1982 *Rev. Mod. Phys.* **54** 235
[5] Fisher M E 1967 *Physics* **3** 255; Fisher M E 1967 *Rep. Prog. Phys.* **30** 615

Journal of Physics: Conference Series **297** (2011) 012005 doi:10.1088/1742-6596/297/1/012005


[6] It is not the case for the systems far from the critical region. For example, in the case of the $d$-dimensional Ising system: $f_{n_s^{(+)}}(n) \sim \exp(-n^{(d-1)/d})$, see: Delyon F 1979 *J. Stat. Phys.* **21** 727

[7] Lukierska-Walasek K and Topolski K 2010 *Rev. Adv. Mater. Sci.* **23** 141; id. 2010 *Comp. Methods in Sciences and Technology* **16** 173

[8] Balakrishnan M and Basu A P 1995 *The Exponential distribution: theory, methods and applications*, (Amsterdam: Gordon and Breach)

[9] Johnson A and Kotz S 1970 *Distributions in statistics: continuous univariate distributions-2*, (New York: John Wiley and Sons)

[10] Bramwell S T *et al.* 2000 *Phys. Rev. Lett.* **84** 3744

[11] Kosterlitz J M and Thouless D J 1973 *Journal of Physics C: Solid State Physics* **6** 1181

[12] Bertin E 2005 *Phys. Rev. Lett.* **95** 170601

[13] Paredes R V and Botet R 2006 *Phys. Rev.* **E 74** 060102

[14] Kolmogorov A N 1933 *Giorn. Ist. Ital. Attuari* **4** 1

[15] Watson G S 1961 *Biometrika* **48** 109

[16] Vervaat W 1979 *Ann. Probab.* **7** 143

[17] Doob J 1949 *Ann. Math. Stat.* **20** 393

[18] Stauffer D and Aharony A 1994 *Introduction to percolation theory*, Revised Second Edition (London: Taylor and Francis)

[19] Botet R, Płoszajczak M and Latora V 1997 *Phys. Rev. Lett.* **78** 4593; Botet R and Płoszajczak M 2002 *Universal Fluctuations*, (Singapore: World Scientific)

[20] Efros A L 1987 *Physics and Geometry of Disorder: Percolation Theory* (Moscow: MIR).

[21] Braga G A, Sanchis R and Schieber T A 2005 *SIAM Review* **47** 349

[22] Biane Ph, Pitman J and Yor M 2001 *Bull. Amer. Math. Soc.* **38** 435

[23] Bellman R 1961 *A brief introduction to theta functions* (New York: Holt, Rinehart and Winston)

Journal of Physics: Conference Series **297** (2011) 012006 doi:10.1088/1742-6596/297/1/012006

# Finite sampling effects on generalized fluctuation-dissipation relations for steady states

**Juan Ruben Gomez-Solano, Artyom Petrosyan and Sergio Ciliberto**
Laboratoire de Physique, Ecole Normale Supérieure de Lyon, CNRS UMR 5672, 46, Allée d'Italie, 69364 Lyon CEDEX 07, France

E-mail: sergio.ciliberto@ens-lyon.fr

**Abstract.** We study the effects of the finite number of experimental data on the computation of a generalized fluctuation-dissipation relation around a nonequilibrium steady state of a Brownian particle in a toroidal optical trap. We show that the finite sampling has two different effects, which can give rise to a poor estimate of the linear response function. The first concerns the accessibility of the generalized fluctuation-dissipation relation due to the finite number of actual perturbations imposed to the control parameter. The second concerns the propagation of the error made at the initial sampling of the external perturbation of the system. This can be highly enhanced by introducing an estimator which corrects the error of the initial sampled condition. When these two effects are taken into account in the data analysis, the generalized fluctuation-dissipation relation is verified experimentally.

## 1. Introduction

Current theoretical developments in nonequilibrium statistical mechanics have led to significant progress in the study of systems around states far from thermal equilibrium. Systems in nonequilibrium steady states (NESS) are the simplest examples because the dynamics of their degrees of freedom $x$ under fixed control parameters $\lambda$ can be statistically described by time-independent probability densities $\rho_0(x, \lambda)$. NESS naturally occur in mesoscopic systems such as colloidal particles dragged by optical tweezeres, Brownian ratches and molecular motors because of the presence of nonconservative or time-dependent forces [1]. At these lengthscales fluctuations are important so it is essential to establish a quantitative link between the statistical properties of the NESS fluctuations and the response of the system to external perturbations. Around thermal equilibrium this link is provided by the fluctuation-dissipation theorem [2].

The generalization of the fluctuation-dissipation theorem around NESS for systems with Markovian dynamics has been achieved in recent years from different theoretical approaches [3, 4, 5, 6, 7, 8, 9, 10, 11, 12]. The different generalized formulations of the fluctuation-dissipation theorem link correlation functions of the fluctuations of the observable of interest $O(x)$ in the unperturbed NESS with the linear response function of $O(x)$ due to a small external time-dependent perturbation around the NESS. The observables involved in such relations are not unique but they are equivalent in the sense that they lead to the same values of the linear response function. These theoretical relations may be be useful in experiments and simulations to know the linear response of the system around NESS. Indeed the response can be obtaines from measurements entirely done at the unperturbed NESS of the system of interest without any need to perform the actual perturbation. Nevertheless, the theoretical equivalence of the different observables involved in those relations does not translate into equivalent experimental accessibility: *e.g.* strongly fluctuating observables such as instantaneous velocities may lead to large statistical errors in the measurements [13]. Besides, NESS quantities themselves such as local mean velocities, joint stationary densities and the stochastic entropy are not in general as easily measurable as dynamical observables directly related to the degrees of freedom [14]. Hence, before implementing the different fluctuation-response formulae in real situations it is important to test its experimental validity under very well controlled conditions and to assess the influence of finite data analysis. The experimental test of some fluctuation-dissipation relations has been recently done in Refs. [13, 14, 15, 16] for colloidal particles in toroidal optical traps.

In the present paper we discuss the effects of the finite number of experimental data on the determination of the linear response function around a NESS for a micron-sized system with Markovian dynamics: a Brownian particle in a toroidal optical trap. For this purpose we perform the respective data analysis on the measurements reported in [14, 16]. In Sect. 2 we briefly describe a generalized fluctuation-dissipation relation that has been derived for Markovian dynamics around a NESS exploiting the properties of the stationary density $\rho_0(x, \lambda)$. In Sect. 3 we recall the main features of a previous experiment that we use in the present paper for the data analysis. In Sect. 4 we discuss the two different kinds of finite-sampling effects that can appear in the computation of the differents terms involved in the fluctuation-dissipation relation. We show that the generalized fluctuation-dissipation relation is verified experimentally when performing a careful data analysis, which takes into account these effects. Finally we present the conclusion.

## 2. Hatano-Sasa relation and fluctuation-dissipation around NESS

The Hatano-Sasa relation provides a general identity for the transitions between either equilibrium or nonequilibrium steady states of Markovian systems [17]. In the following we will focus on a Langevin system with Markovian dynamics described by a steady state probability density $\rho_0(x, \lambda)$. When the system is subjected to a time-dependent variation of the control parameter $\lambda(t)$ between an initial time $t_i$ and a final time $t_f$, the Hatano-Sasa identity reads

$$\left\langle \exp\left(-\int_{t_i}^{t_f} dt\dot{\lambda}_\alpha(t)\frac{\partial\phi(x_t, \lambda(t))}{\partial\lambda_\alpha}\right)\right\rangle = 1, \tag{1}$$

where $\phi(x, \lambda) = -\ln\rho_0(x, \lambda)$ and the average $\langle \ldots \rangle$ is performed over an infinite number of realizations of a prescribed time-dependent protocol $\lambda(t)$. From Eq. (1), Prost *et al.* has directly derived a generalized fluctuation-dissipation relation that holds in the linear response regime around a NESS [7]

$$R_{\alpha\gamma}(t-s) = \frac{d}{dt}\left\langle \frac{\partial\phi(x_t, \lambda_{SS})}{\partial\lambda_\alpha}\frac{\partial\phi(x_s, \lambda_{SS})}{\partial\lambda_\gamma}\right\rangle_0. \tag{2}$$

In Eq. (2) $R_{\alpha\gamma}(t-s) = \delta\langle O(x_t)\rangle_h/\delta h_s|_{h=0}$ is the linear response function of the observable $O(x) = \partial\phi(x, \lambda_{SS})/\partial\lambda_\alpha$ due to a small external time-dependent perturbation $h_s = \lambda(s) - \lambda_{SS}$ around $\lambda_{SS} = \lambda(t_i)$ fixing an initial NESS at time $t_i$. The averages $\langle\ldots\rangle_h$ and $\langle\ldots\rangle_0$ are performed over the perturbed and unperturbed processes, respectively.

In experiments the formal average involved in Eq. (1) is not perfectly computed because of the finite number of independent realizations of $\lambda(t)$. Hence Eq. (1) allows one to estimate the experimental precision of (2) computed from a given number of experimental data provided that one can measure the observable $\partial\phi(x, \lambda_{SS})/\partial\lambda_\alpha$. In the next section we tackle this problem for the experimental trajectories of a colloidal particle in a toroidal optical trap.

## 3. Brownian particle in a toroidal optical trap

The Brownian motion of a colloidal particle in a toroidal optical trap has become an experimental model to study the generalization of the fluctuation-dissipation theorem around

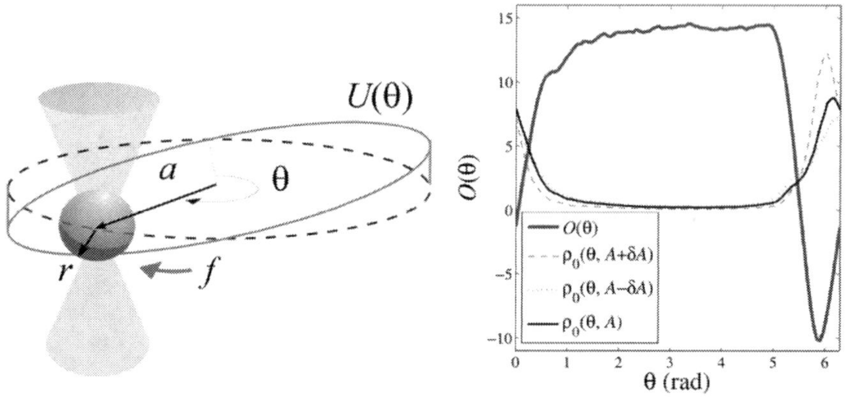


**Figure 1.** (a) Sketch of a Brownian particle in a toroidal optical trap subjected to a nonconservative force $f$ and undergoing a periodic potential $U(\theta)$.(b) Experimental profile of the observable $O(\theta)$ defined in Eq. (4) computed using the NESS densities $\rho_0(\theta, A+\delta A)$ and $\rho_0(\theta, A-\delta A)$ around $\rho_0(\theta, A)$ at fixed $F$.

a NESS [13, 14, 15, 16]. This is because it is a system with a single translational degree of freedom where one can easily tune its relevant control parameters. Our experiment has been already described in detail in [14, 16] so here we only explain it briefly. The Brownian motion of a spherical silica particle (radius $r = 1\,\mu$m) immersed in water is confined on a thin torus of major radius $a = 4.12\,\mu$m by a tightly focused laser beam rotating at 200 Hz. The water reservoir acts as a thermal bath at fixed temperature ($T = 20 \pm 0.5°$C) providing thermal fluctuations to the particle. The viscous drag coefficient at this temperature is $\gamma = 1.89 \times 10^{-8}$ kg s$^{-1}$. The rotation frequency of the laser is so high that it is not able to trap continuously the particle in the focus because the viscous drag force of the surrounding water quickly exceeds the optical trapping force. Consequently, at each rotation the beam only kicks the particle a small distance along the circle of radius $a$ exerting a nonconservative force $f = 66$ fN on it in the direction of the rotation. Thus, the particle motion is effectively confined on a circle: the angular position $\theta$ of its barycenter is the only relevant degree of freedom. In addition, a static light intensity profile (amplitude about 5% of the total laser intensity [14, 16]) is created along the circle acting as a periodic potential $U(\theta) = U(\theta + 2\pi)$ of amplitude $68.8k_BT$. Figure 1(a) depicts this experimental configuration. We track the 2D particle position by video microscopy in order to measure the time evolution of $\theta$. Thus for the experimentally accessible length and time scales the dynamics of $\theta_t$ is modeled by the first-order Langevin equation [13, 14, 15, 16, 18]

$$\dot{\theta} = -\partial_\theta H(\theta) + F + \xi, \tag{3}$$

where $H(\theta) = U(\theta)/(\gamma a)$ with amplitude $A = \max\{H(\theta)\} = 68.8k_BT/(\gamma a^2)$, $F = f/(\gamma a)$ and $\xi$ is a white noise process of zero mean and covariance $\langle \xi_t \xi_s \rangle = 2[k_BT/(\gamma a^2)]\delta(t-s)$. Under these fixed conditions the dynamics of $\theta_t$ settles into a NESS whose probability density function $\rho_0(\theta, A)$ is plotted in Fig 1(b) (solid black line).

## 4. Generalized fluctuation-dissipation relation

In the following analysis we take $x = \theta$ as the single degree of freedom and $\lambda = A$ as the main control parameter of the system. In this case the observable of interest involved in the

Journal of Physics: Conference Series **297** (2011) 012006 doi:10.1088/1742-6596/297/1/012006

fluctuation-dissipation relation (2) is

$$O(\theta) = -\frac{\partial \ln \rho_0(\theta, A)}{\partial A}. \tag{4}$$

The experimental profile of $O(\theta)$, computed as a discrete three-point derivative of $-\ln \rho_0(\theta, A)$ at two different NESS around $A$, is shown in Fig 1(b).

We focus on the response of the system after applying a small Heaviside perturbation $\delta A$ to $A$ at time $t_i$: $A \to A + \delta A$. In the experiment this dynamical procedure is done by suddenly switching the laser power modulation as explained in [14, 16]. This procedure yields the integrated response function, defined as

$$\chi(t - t_i) = \int_{t_i}^{t} R(t - s)\, \mathrm{d}s = \frac{\langle O(\theta_t) \rangle_{\delta A}}{\delta A}, \tag{5}$$

where the average $\langle \ldots \rangle_{\delta A}$ must be performed over the perturbed process at time $t$. Then the integrated version of the generalized fluctuation-dissipation relation (2) is in this case

$$\chi(t - t_i) = \langle O(\theta_t) O(\theta_{t_i}) \rangle_0 - \langle O(\theta_t) O(\theta_t) \rangle_0. \tag{6}$$

We now study the effects of a finite number of realizations of the perturbation $\delta A$ and the finite number of trajectories used to compute the averages $\langle \ldots \rangle_{\delta A}$ and $\langle \ldots \rangle_0$ in Eqs. (5) and (6) .

### *4.1. Statistical error*

As discussed in Sect. 2, the Hatano-Sasa relation (1) can be used to estimate the error of the experimental computation of Eq. (6) when performing $N < \infty$ independent realizations of $\delta A$ around the NESS. In the case of the dynamical process defined by the Heaviside perturbation $A \to A + \delta A$ at time $t_i$, as done in the experiment, Eq. (1) reads

$$\langle \exp\left[-\delta A O(\theta_{t_i})\right] \rangle = 1. \tag{7}$$

Eq. (7) only depends on the initial values $O(\theta_{t_i})$ when the system is still in NESS. Therefore, for a finite number of trajectories $N$ we introduce an estimator of the error of Eq. (7)

$$\Delta(N) = \left| \frac{1}{N} \sum_{j=1}^{N} \exp\left[-\delta A O_j(\theta_{t_i})\right] - 1 \right|, \tag{8}$$

where $O_j(\theta_{t_i})$ is the $j$-th sampled NESS initial condition. Fig. 2(a) shows the behavior of the error $\Delta(N)$ computed for $\delta A = 0.05A$ (the value realized in the dynamical experiment) using $N$ experimental values of $O(\theta)$ drawn from the NESS distribution. For small $N \lesssim 100$ the error is non-negliglible, $\Delta(N) \geq 4\%$, and this must be taken into account in the final accuracy of the generalized fluctuation-dissipation relation when comparing the left with the right-hand side of Eq. (6) using the experimenal data. Then as $N$ increases $\Delta(N)$ quickly converges to 0: for $N \geq 500$ the precision of Eq. (7) found in the experiment is better than 1%.

### *4.2. Effect of the initial sampled condition*

According to the Heaviside procedure for $\delta A$, in Eqs. (5) and (6) the initial condition $\theta(t_i)$ for the perturbed process is sampled from the NESS density $\rho_0(\theta, A)$. Then the integreated linear response function must formally satisfy the initial condition

$$\chi(0) = \frac{\langle O(\theta_{t_i}) \rangle_0}{\delta A} = 0, \tag{9}$$

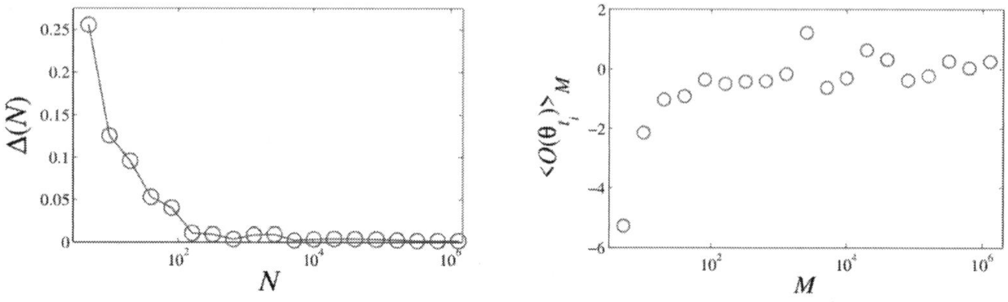


**Figure 2.** (a) Estimate of the error of the Hatano-Sasa relation (7) for a finite number $N$ of realizations of $\delta A$. (b) Estimate of the error of $\langle O(\theta_{t_i})\rangle_0$ using Eq. (10) for $M$ NESS data.

where the last equality is due to the normalization of $\rho_0(\theta_{t_i}, A)$. We are interested in the effect of a finite number $M < \infty$ of initial values $\theta(t_i)$ drawn from the inital NESS on the estimate of $\chi(t-t_i)$. It should be noted that in practice a small $M$ may significantly affect the computation of $\langle O(\theta_{t_i})\rangle_0$ in Eq. (9) because most of the positive values of $O(\theta)$ lie in the region where $\rho_0(\theta, A)$ is rarely sampled, as shown in Fig. 1(b). In Fig. 2(b) we plot some values of the finite average

$$\langle O(\theta_{t_i})\rangle_M = \frac{1}{M}\sum_{j=1}^{M} O_j(\theta_{t_i}), \tag{10}$$

where $O_j(\theta_{t_i})$ is the $j$-th initial condition at NESS. As expected, for small $M$, $\langle O(\theta_{t_i})\rangle_M < 0$ due to the fact that one samples mostly the negative values around the maximum of $\rho_0(\theta, A)$. The convergence to the theoretical value $\langle O(\theta_{t_i})\rangle_0 = 0$ is very slow: as $M$ increases $\langle O(\theta_{t_i})\rangle_M$ becomes very sensitive to $M$ and large positive values of $\langle O(\theta_{t_i})\rangle_M$ can be obtained. The general trend is around $\langle O(\theta_{t_i})\rangle_0 = 0$, though. Then even for large $M$ one must be careful with the computation of the integrated response function since a large initial error of $\chi(0)$ due to the use of the average $\langle\ldots\rangle_M$ may significantly propagate as $t$ increases.

In order to avoid the problem of the sensitivity to the initial condition, instead of using directly the average $\langle\ldots\rangle_M$ in Eq. (5), one can define an estimator $\chi_M(t-t_{t_i})$ satisfying the initial condition $\chi_M(0) = 0$ as required ideally by Eq. (9). In this way the propagation of the initial error given by $\langle O(\theta_{t_i})\rangle_M$ is suppressed at the beginning. An intuitive way to define $\chi_M$ can be outlined from the usual protocol to compute the integrated response function

$$\chi(t-t_i) = \frac{\langle O^{\delta A}(\theta_t)\rangle_{\delta A} - \langle O(\theta_{(t-t_i+t^*)})\rangle_0}{\delta A} \tag{11}$$

where the time $t^*$ is chosen such that $O(\theta_{t^*}) = O^{\delta A}(\theta_{t_i})$ and $O^{\delta A}(\theta_t)$ denotes the observable measured during the perturbed process. Notice that Eq.11 is justified by the fact that in the case of an infinite number of samples $\langle O(\theta_t)\rangle_0 = 0\ \forall t$, because of the time translational invariance of the NESS. In contrast when $M$ is finite, it is useful to take into account,in Eq.11, that $\langle\ldots\rangle_{\delta A}$ and $\langle\ldots\rangle_0$ are performed independently the first on the perturbed trajectory $O^{\delta A}(O(\theta_t)$ and the second on the unperturbed ones $O(\theta_{(t-t_i+t^*)})$, specifically

$$\chi_M(t-t_i) = \frac{1}{\delta A}\left[\frac{1}{M}\sum_{j=1}^{M} O_j^{\delta A}(\theta_t) - \frac{1}{L}\sum_{k=1}^{L} O_k(\theta_{(t-t_i+t^*)})\right] \tag{12}$$

where $L$ is the number of unperturbed trajectories such $O_k(\theta_{t^*}) = O_j^{\delta A}(\theta_{t_i})$. Therefore Eq.12 can be rewritten

$$\chi_M(t - t_i) = \frac{1}{\delta A}\frac{1}{M}\sum_{j=1}^{M}\left\{\frac{1}{L}\sum_{k=1}^{L}\delta O_{jk}(\theta_t)\right\} \tag{13}$$

where $\delta O_{jk}(\theta_t) \equiv O_j^{\delta A}(\theta_t) - O_k(\theta_{(t-t_i+t^*)})$ is the the instantaneous difference between a perturbed trajectory $O_j^{\delta A}(\theta_t)$ and an unperturbed one $O_k(\theta_t)$. An example of this procedure is depicted in Fig. 3(a), where $t_i$ has been set equal to zero. We see that for a given perturbed trajectory $O_j^{\delta A}(\theta_t)$ (thick dashed red line), obtained after that $\delta A$ has been applied, one should look for an unperturbed NESS trajectory $O_k(\theta(t))$ such that $O_k(\theta_{t^*}) = O_j^{\delta A}(\theta_{t_i})$ like the four unperturbed trajectories shown by the solid lines.

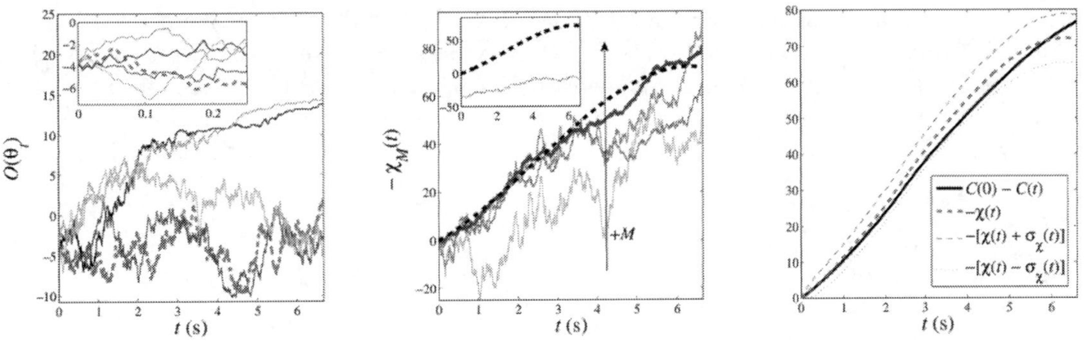


**Figure 3.** (a) Examples of perturbed (thick dashed red line) and unperturbed (solid lines) trajectories used to estimate $\chi(t)$ using Eq. (13). Inset: expanded view at short time. (b) Estimate of $-\chi(t)$ for $M = 50, 100, 250, 500$ and $L = 1$ (solid lines) and $M = 500$, $L = 200$ (dashed line). Inset: Comparison of $\chi_M(t)$ for $M = 500$, $L = 200$ with the poor estimate done by the uncorrected average of Eq. (10) for $M = 500$. (c) Comparison between the experimental $C(0) - C(t)$ and the best estimate of $-[\chi(t) \pm \sigma_\chi(t)]$.

In this way $\delta O_{jk}(\theta_{t_i}) = 0$ by construction and the estimator defined by Eq. (13) satisfies the condition $\chi_M(0) = 0$. For $M, L \to \infty$, $\chi_M(t - t_i)$ converges to $\chi$ defined by Eq. (5), because $\langle\langle O(\theta_t)\rangle\rangle_M \to 0$. In Fig. 3(b) we show $\chi_M(t)$, with $t_i$ redefined as $t_i = 0$, computed using Eq. (13) for different values of $M$ and for fixed $L = 1$ (solid lines) and $L = 200$ (dashed line). As $M$ increases for $L = 1$ the curves converge to a single profile which must correspond to that of $\chi(t)$ ideally given by Eq. (5). The additional conditional average done for $L = 200$ smoothes the slightly fluctuating profile for $M = 500$ (thick solid blue line) resulting in the thick dashed solid line. For comparison we also show in the inset of Fig. 3(b) the raw estimate of $\chi$ obtained using the average of Eq. (10) for the same $M = 500$ perturbed trajectories without correcting the effect of the initial sampling. In this case the propagation of the initial large error of $\langle O(\theta_0)\rangle_0$ gives rise to a very poor estimate of the integrated response function for $t > 0$.

### *4.3. Experimental test*

Finally we proceed to test the theoretical fluctuation-dissipation relation (6) for the experimental unperturbed NESS trajectories of the Brownian particle and those perturbed around NESS. For this purpose we compare the best estimate $\chi_M(t)$ of $\chi(t)$ done for $M = 500$ and $L_j = 200$ with the right-hand side of Eq. (6). The involved correlation function $C(t) = \langle O(\theta_t)O(\theta_0)\rangle_0$ on the right-hand side is computed using unperturbed NESS trajectories. In Fig. 3(c) we compare

$\chi_M(t)$ with $C(t) - C(0)$. Besides, one can estimate the statistical error of the experimental $\chi(t)$ at each $t \geq 0$ by computing the standard deviation of $\langle O^{\delta A}(\theta_t) - O(\theta_t)\rangle_M$ over the $L_j$ possible choices of the unperturbed $O_k(t)$. The standard deviation $\pm\sigma_\chi(t)$ obtained in this way is also shown in Fig. 3(c) showing that after following the careful procedure to estimate $\chi$ the relation $\chi(t) = C(t) - C(0)$ is verified by the finite experimental data. Note that without the finite correction $\langle\langle O(\theta_t)\rangle\rangle_M$ of the initial condition in Eq. (13) one would largely underestimate the direct measurement of the integrated response function leading to an apparent violation of Eq. (6). The results of the present paper are consistent with those of Refs. [14, 16] where two fluctuation-dissipation formulae equivalent to Eq. (6) but involving different observables from the one studied here are checked experimentally.

## 5. Conclusion

We have studied the influence of finite sampling in the computation of the linear response function of a Brownian particle in a toroidal optical trap around a NESS. We have shown that there are two different effects that may lead to a very poor estimate of the experimental linear response function when the data analysis is not performed carefully. This is an important point that must be assessed in general when applying in experiment and numerical simulations the different generalized fluctuation-dissipation formulae recently derived for NESS.

## References

[1] Reimann P 2002, *Phys. Rep.* **57** 361.
[2] Marini Bettolo Marconi U , Puglisi A , Rondoni L and Vulpiani A 2008,*Phys. Rep.* **461** 111.
[3] Harada T and Sasa S I 2005, *Phys. Rev. Lett.* **95**, 130602.
[4] Lippiello E , Corberi F , and M. Zannetti M 2005, *Phys. Rev. E* **71** 036104.
[5] Speck T and Seifert U 2006, *Europhys. Lett.* **74** 391.
[6] Chetrite R , Falkovich G, and K. Gawedzki K 2008, *J. Stat. Mech.* P08005.
[7] Prost J , Joanny J F , and Parrondo J M R 2009, *Phys. Rev. Lett.* **103** 090601.
[8] Chetrite R , and Gawedzki G 2009, *J. Stat. Phys.* **137** 890.
[9] Baiesi M , Maes C and Wynants B 2009 *Phys. Rev. Lett.* **103** 010602.
[10] Baiesi M , Maes C and Wynants B 2009, *J. Stat.Phys.* **137** 1094.
[11] Baiesi M ,Boksenbojm E , Maes C and Wynants B 2010, *J. Stat. Phys.* **139** 492.
[12] Seifert U and Speck T 2010, *Erophys. Lett.* **89** 10007.
[13] Mehl J , Blickle V , Seifert U , and Bechinger C 2010, *Phys. Rev. E* **82** 032401.
[14] Gomez-Solano J R , Petrosyan A , Ciliberto S , and Maes C 2011, *J. Stat. Mech.* P01008.
[15] Blickle V, Speck T , Lutz C , Seifert U , and Bechinger C 2007, *Phys. Rev. Lett.* **98**, 210601.
[16] Gomez-Solano J R , Petrosyan A , Ciliberto S, Chetrite R , Gawedzki G 2009, *Phys. Rev. Lett.* **103** 040601.
[17] Hatano T and Sasa S 2001, *Phys. Rev. Lett.* **86** 3463.
[18] Blickle V , Speck T , Seifert U , and Bechinger C , 2007, *Phys. Rev. E* **75** 060101(R).

Journal of Physics: Conference Series **297** (2011) 012007 doi:10.1088/1742-6596/297/1/012007

# Dynamics of a flexible polymer in planar mixed flow

**Dipjyoti Das**[1], **Sanjib Sabhapandit**[2] **and Dibyendu Das**[1]

[1] Department of Physics, Indian Institute of Technology Bombay, Powai, Mumbai - 400076, India

[2] Raman Research Institute, Bangalore 560080, India

E-mail: dipjyoti@phy.iitb.ac.in, sanjib@rri.res.in, dibyendu@phy.iitb.ac.in

**Abstract.** We present exact spatio-temporal correlation functions of a Rouse polymer chain submerged in a fluid having planar mixed flow, in the steady state. Using these correlators, determination of the time scale distribution functions associated with the first-passage tumbling events is difficult in general; it was done recently in *Phys. Rev. Lett.* **101**, 188301 (2008), for the special case of "simple shear" flow. We show here that the method used in latter paper fails for the general mixed flow problem. We also give many new estimates of the exponent $\theta$ associated with the exponential tail of the angular tumbling time distribution in the case of simple shear.

## 1. Introduction

Advancement in the techniques of fluorescence microscopy have led to direct observation and experimentation of single polymer chains (like a $\lambda$- DNA molecule) in fluid flows in the recent years [1, 2, 3, 4, 5, 6, 7, 8]. Several works on theory and simulations have also addressed this problem, mostly for the case of "simple shear" flow [9, 10, 11, 12, 13, 14, 15]. This problem has biological relevance [16] and in particular plays a crucial role in the blood clotting process [17]. But planar fluid flows need not be simple shear only, and more complicated flows which are either predominantly elongational or rotational have also been studied [9, 18, 19, 20].

In general, any planar flow of the form $\mathbf{v} = v_x\hat{x} + v_y\hat{y}$ can be considered as a linear superposition of a rotational component with vorticity $\omega = \frac{1}{2}(\frac{\partial v_y}{\partial x} - \frac{\partial v_x}{\partial y})$ and an elongational component with strain rate $\dot{\epsilon} = \frac{1}{2}(\frac{\partial v_y}{\partial x} + \frac{\partial v_x}{\partial y})$ [9]. The "mixed flow" refers to an intermediate flow between pure elongational flow ($\omega = 0$) which creates large chain deformation, and pure rotational flow ($\dot{\epsilon} = 0$) which is responsible for angular rotation. "Simple shear flow" is the limit where the magnitude of rotational component exactly equals that of elongatinal component i.e $|\omega| = |\dot{\epsilon}|$. In this type of flow a polymer tumbles as well as undergoes coil-stretch deformation. In Cartesian coordinates the force vector in a mixed flow can be represented as: $\mathbf{f}(t) \equiv (\dot{\gamma}y(t), \alpha\dot{\gamma}x(t), 0)$ where $\dot{\gamma}$ is a constant rate. The parameter $\alpha \in [-1, 1]$ controls the relative magnitude of elongational and rotational components. In the limit $\alpha \to 0$ this flow reduces to simple shear flow in the $X$ direction. On the other hand, the two extreme limits $\alpha \to +1$ and $\alpha \to -1$ correspond to pure elongational flow and pure rotational flow respectively (see Fig. 1).

A polymer in a flow field has been modeled theoretically as Rouse chain [21, 22] with FENE end-to-end constraint [19], as a FENE dumbbell [10, 11] and as a liner (but non-Rouse) chain with global contour length constraint[13]. Very recently we have shown [15] that analytically exact

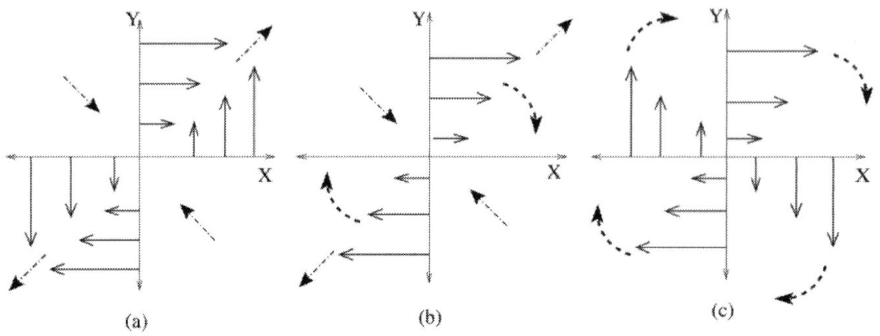


**Figure 1.** Schematic diagrams of different regimes of planar mixed flow. (a) The pure elongational flow ($\alpha = +1$). (b) Simple shear ($\alpha = 0$). (c) The pure rotational flow ($\alpha = -1$). Solid arrows represent the force vector field and dotted arrows represent the flow directions.

static distribution functions for the polar coordinates of the polymer's end-to-end vector $\mathbf{R}(t)$ are possible to be derived for a Rouse chain (without any nonlinear constraint) in simple shear ($\alpha = 0$). Furthermore for the latter flexible polymer, for the first time, we presented an analytical calculation for the angular tumbling time distribution function. This is very nontrivial as $\mathbf{R}(t)$ (although Gaussian) is a non-Markovian process, for which in general calculation of first-passage properties like tumbling are very difficult. It was shown that the probability density function (PDF) of angular tumbling time $\tau$, goes as $\exp(-\theta\tau/\tau_0)$ (where $\tau_0$ is the polymer's longest relaxation time in the absence of any external flow) and $\theta$ in general depends on the shear rate. In the limit of strong shear the exponent $\theta$ goes to a constant value and it was calculated (in Ref. [15]) by a method known as "independent interval approximation" (IIA) [23, 24, 25, 26]. The method of IIA was applicable in the limit of strong shear as the $X$-component of $\mathbf{R}(t)$ became a *smooth* process (defined later) in that limit.

In this paper we study the problem of a linear Rouse polymer in general mixed flow. We have derived the exact autocorrelation functions for the Cartesian components of $\mathbf{R}(t)$. We show that the $X$-component of $\mathbf{R}(t)$ is a *non-smooth* process and hence analytically IIA method is not applicable to find the exponent $\theta$ in the general mixed flow. In the special case of "simple shear", we have also investigated this exponent in more detail (than earlier [15]) and got few comparative estimates of constant $\theta$ using various available IIA schemes. For the general mixed flow we have numerical estimates of $\theta$ from observed exponential tail in the tumbling time distribution in our simulation.

## 2. General correlator for polymer in mixed flow

We study the linear Rouse model [21, 22] of a polymer chain of $N$ beads connected by harmonic springs (each has spring constant $k$) in a planar mixed flow. The position vector of $n$th bead ($n = 1, 2, ..., N$) at time $t$ is denoted by $\mathbf{r}_n(t)$. For internal beads ($n = 2, 3, ..., N-1$) the position vector evolves via the Langevin equation :

$$\frac{d\mathbf{r}_n(t)}{dt} = -k(2\mathbf{r}_n - \mathbf{r}_{n+1} - \mathbf{r}_{n-1}) + \mathbf{f}_n(t) + \boldsymbol{\eta}(n,t) \qquad (1)$$

where the vector $\mathbf{f}_n(t) \equiv (\dot{\gamma} y_n(t), \alpha\dot{\gamma} x_n(t), 0)^T$ denotes the planar mixed flow force field with the constant rate $\dot{\gamma}$. The vector $\boldsymbol{\eta}(n,t) \equiv (\eta_1(n,t), \eta_2(n,t), \eta_3(n,t))^T$ is the thermal white noise with zero mean and a correlator $\langle \eta_i(n,t)\eta_j(n',t')\rangle = \zeta\delta_{ij}\delta_{n,n'}\delta(t-t')$, where $i, j = 1, 2, 3$ and $n = 1, 2, ..., N$. The noise strength $\zeta$ is proportional to the temperature and in the Eq. (1) the force strengths are scaled by viscosity. It is convenient to introduce dimensionless so-called

Weissenberg number $\mathrm{Wi} = \dot{\gamma}\tau_0$, where $\tau_0 = N^2/k\pi^2$ is the longest relaxation time of the polymer in the absence of any external flow. Note that the two end-beads (for $n = 1$ and $n = N$) feel only one-sided interaction of the harmonic spring. The position vectors of these two beads can also be included into Eq. (1) if we add two hypothetical beads at $n = 0$ and $n = N + 1$ with the "free boundary conditions": $\mathbf{r}_0(t) = \mathbf{r}_1(t)$ and $\mathbf{r}_{N+1}(t) = \mathbf{r}_N(t)$.

In the large $N$ limit, the discrete index $n$ can be treated as a continuous variable $u$ and Eq. (1) then becomes

$$\frac{\partial \mathbf{r}(u,t)}{\partial t} = k\frac{\partial^2 \mathbf{r}(u,t)}{\partial u^2} + \mathbf{f}(u,t) + \boldsymbol{\eta}(u,t) \tag{2}$$

with the free boundary conditions : $\frac{\partial \mathbf{r}(u,t)}{\partial u}|_{u=0} = 0$ and $\frac{\partial \mathbf{r}(u,t)}{\partial u}|_{u=N} = 0$. In this continuum limit the noise correlator and the force field are given by $\langle \eta_i(u,t)\eta_j(u',t')\rangle = \zeta\delta_{ij}\delta(u-u')\delta(t-t')$ and $\mathbf{f}(u,t) \equiv (\dot{\gamma}y(u,t), \alpha\dot{\gamma}x(u,t), 0)^T$ respectively.

We solve Eq. (2) by Fourier cosine decomposition using the transformation :

$$\tilde{\mathbf{r}}_p(t) = \frac{1}{N}\int_0^N \cos\left(\frac{p\pi u}{N}\right)\mathbf{r}(u,t)\, du \qquad p = 0, 1, 2, ... \tag{3}$$

where each mode evolves independently. For the mode coordinates the problem reduces to the single particle problem (i.e a single bead connected to the origin by a harmonic spring in a planar mixed flow) with the modified spring constant $k_p = p^2/\tau_0$ and noise strength $\zeta' = \zeta/2N$:

$$\frac{\partial \tilde{\mathbf{r}}_p(t)}{\partial t} = -k_p\tilde{\mathbf{r}}_p(t) + \tilde{\mathbf{f}}_p(t) + \tilde{\boldsymbol{\eta}}_p(t) \tag{4}$$

where $\tilde{\mathbf{f}}_p(t) \equiv (\dot{\gamma}\tilde{y}_p(t), \alpha\dot{\gamma}\tilde{x}_p(t), 0)^T$ and $\tilde{\boldsymbol{\eta}}_p(t)$ are transformed force vector and white noise respectively. Now Eq. (4) which are coupled equations of the mode coordinates $\tilde{x}_p(t)$, $\tilde{y}_p(t)$ and $\tilde{z}_p(t)$ can be solved using the method of Laplace transform and we obtain the exact two-time correlators of the mode components.

Our main quantity of interest is the end-to-end vector defined as: $\mathbf{R}(t) \equiv \mathbf{r}(N,t) - \mathbf{r}(0,t)$. Using the inverse Fourier cosine transform we find that the correlators associated with the Cartesian components of $\mathbf{R}(t)$ are related to the correlators associated with the Cartesian mode components as below (for finite time increment $\tau \geq 0$) :

$$\langle R_i(t)R_j(t+\tau)\rangle = 16 \sum_{p=1,3,5,...}^{\infty} \langle \tilde{r}_{ip}(t)\tilde{r}_{jp}(t+\tau)\rangle \tag{5}$$

where $R_i$ and $\tilde{r}_{ip}$ are the Cartesian components of the end-to-end vector and the $p$th mode vector respectively. From the above equation various correlation functions associated with the Cartesian components of $\mathbf{R}(t)$ can be obtained explicitly.

### *2.1. Exact autocorrelation functions of the end-to-end vector*

Using Eq. (5) we have derived the exact autocorrelation functions of the Cartesian components of $\mathbf{R}(t)$ in the stationary state limit ($t \to \infty$) with a finite time increment $\tau \geq 0$, both in the $\alpha > 0$ and $\alpha < 0$ regimes separately.

Elongational regime ($\alpha > 0$): In this case the elongational effect dominates over the rotational effect. We note an important point that in the elongational regime the asymptotic correlators of $\mathbf{R}(t)$ exist only under the condition: $p^2 > \mathrm{Wi}\sqrt{\alpha}$ (for $p = 1, 3, 5, ...$). Thus the correlators diverge for the lower modes and hence one would expect huge instability in the polymer's configuration;

it will get infinitely stretched and will not attain a stable state. We give the results in the limit $t \to \infty$ below, and express them as functions of a scaled time-increment variable $T \equiv \tau/\tau_0$ :

$$
\begin{aligned}
\langle R_x(t)R_x(t+\tau_0 T)\rangle &= \frac{2\zeta\tau_0}{N}\sum_{p=1,3,5,\ldots}^{\infty} e^{-p^2T}\Big[\Big\{\cosh(\mathrm{Wi}\sqrt{\alpha}T)\left(\frac{p^2}{p^4-\mathrm{Wi}^2\alpha}+\frac{1}{p^2}\right) \\
&+\frac{\sinh(\mathrm{Wi}\sqrt{\alpha}T)\mathrm{Wi}\sqrt{\alpha}}{p^4-\mathrm{Wi}^2\alpha}\Big\}+\frac{1}{\alpha}\Big\{\cosh(\mathrm{Wi}\sqrt{\alpha}T)\left(\frac{p^2}{p^4-\mathrm{Wi}^2\alpha}-\frac{1}{p^2}\right) \\
&+\frac{\sinh(\mathrm{Wi}\sqrt{\alpha}T)\mathrm{Wi}\sqrt{\alpha}}{p^4-\mathrm{Wi}^2\alpha}\Big\}\Big]
\end{aligned} \tag{6}
$$

$$
\begin{aligned}
\langle R_y(t)R_y(t+\tau_0 T)\rangle &= \frac{2\zeta\tau_0}{N}\sum_{p=1,3,5,\ldots}^{\infty} e^{-p^2T}\Big[\Big\{\cosh(\mathrm{Wi}\sqrt{\alpha}T)\left(\frac{p^2}{p^4-\mathrm{Wi}^2\alpha}+\frac{1}{p^2}\right) \\
&+\frac{\sinh(\mathrm{Wi}\sqrt{\alpha}T)\mathrm{Wi}\sqrt{\alpha}}{p^4-\mathrm{Wi}^2\alpha}\Big\}+\alpha\Big\{\cosh(\mathrm{Wi}\sqrt{\alpha}T)\left(\frac{p^2}{p^4-\mathrm{Wi}^2\alpha}-\frac{1}{p^2}\right) \\
&+\frac{\sinh(\mathrm{Wi}\sqrt{\alpha}T)\mathrm{Wi}\sqrt{\alpha}}{p^4-\mathrm{Wi}^2\alpha}\Big\}\Big]
\end{aligned} \tag{7}
$$

$$
\begin{aligned}
\langle R_x(t)R_y(t+\tau_0 T)\rangle &= \frac{2\zeta\tau_0}{N}\sum_{p=1,3,5,\ldots}^{\infty} e^{-p^2T}\Big[\sqrt{\alpha}\Big\{\sinh(\mathrm{Wi}\sqrt{\alpha}T)\left(\frac{p^2}{p^4-\mathrm{Wi}^2\alpha}+\frac{1}{p^2}\right) \\
&+\frac{\cosh(\mathrm{Wi}\sqrt{\alpha}T)\mathrm{Wi}\sqrt{\alpha}}{p^4-\mathrm{Wi}^2\alpha}\Big\}+\frac{1}{\sqrt{\alpha}}\Big\{\sinh(\mathrm{Wi}\sqrt{\alpha}T)\left(\frac{p^2}{p^4-\mathrm{Wi}^2\alpha}-\frac{1}{p^2}\right) \\
&+\frac{\cosh(\mathrm{Wi}\sqrt{\alpha}T)\mathrm{Wi}\sqrt{\alpha}}{p^4-\mathrm{Wi}^2\alpha}\Big\}\Big]
\end{aligned} \tag{8}
$$

$$
\begin{aligned}
\langle R_y(t)R_x(t+\tau_0 T)\rangle &= \frac{2\zeta\tau_0}{N}\sum_{p=1,3,5,\ldots}^{\infty} e^{-p^2T}\Big[\sqrt{\alpha}\Big\{\sinh(\mathrm{Wi}\sqrt{\alpha}T)\left(\frac{p^2}{p^4-\mathrm{Wi}^2\alpha}-\frac{1}{p^2}\right) \\
&+\frac{\cosh(\mathrm{Wi}\sqrt{\alpha}T)\mathrm{Wi}\sqrt{\alpha}}{p^4-\mathrm{Wi}^2\alpha}\Big\}+\frac{1}{\sqrt{\alpha}}\Big\{\sinh(\mathrm{Wi}\sqrt{\alpha}T)\left(\frac{p^2}{p^4-\mathrm{Wi}^2\alpha}+\frac{1}{p^2}\right) \\
&+\frac{\cosh(\mathrm{Wi}\sqrt{\alpha}T)\mathrm{Wi}\sqrt{\alpha}}{p^4-\mathrm{Wi}^2\alpha}\Big\}\Big]
\end{aligned} \tag{9}
$$

$$
\langle R_z(t)R_z(t+\tau_0 T)\rangle = \frac{4\zeta\tau_0}{N}\sum_{p=1,3,5,\ldots}^{\infty}\frac{e^{-p^2T}}{p^2} \tag{10}
$$

$$
\langle R_z(t)R_x(t+\tau_0 T)\rangle = \langle R_z(t)R_y(t+\tau_0 T)\rangle = 0 \tag{11}
$$

Rotational regime ($\alpha < 0$): In this regime the amount of vorticity is more dominant than the elongational component. We find the correlators in the limit $t \to \infty$ to be the following :

$$\begin{aligned}\langle R_x(t)R_x(t+\tau_0 T)\rangle &= \frac{2\zeta\tau_0}{N}\sum_{p=1,3,5,\ldots}^{\infty} e^{-p^2T}\Big[\Big\{\cos(\mathrm{Wi}\sqrt{-\alpha}T)\left(\frac{p^2}{p^4-\mathrm{Wi}^2\alpha}+\frac{1}{p^2}\right)\\ &-\frac{\sin(\mathrm{Wi}\sqrt{-\alpha}T)\mathrm{Wi}\sqrt{-\alpha}}{p^4-\mathrm{Wi}^2\alpha}\Big\}+\frac{1}{\alpha}\Big\{\cos(\mathrm{Wi}\sqrt{-\alpha}T)\left(\frac{p^2}{p^4-\mathrm{Wi}^2\alpha}-\frac{1}{p^2}\right)\\ &-\frac{\sin(\mathrm{Wi}\sqrt{-\alpha}T)\mathrm{Wi}\sqrt{-\alpha}}{p^4-\mathrm{Wi}^2\alpha}\Big\}\Big] \qquad (12)\end{aligned}$$

$$\begin{aligned}\langle R_y(t)R_y(t+\tau_0 T)\rangle &= \frac{2\zeta\tau_0}{N}\sum_{p=1,3,5,\ldots}^{\infty} e^{-p^2T}\Big[\Big\{\cos(\mathrm{Wi}\sqrt{-\alpha}T)\left(\frac{p^2}{p^4-\mathrm{Wi}^2\alpha}+\frac{1}{p^2}\right)\\ &-\frac{\sin(\mathrm{Wi}\sqrt{-\alpha}T)\mathrm{Wi}\sqrt{-\alpha}}{p^4-\mathrm{Wi}^2\alpha}\Big\}+\alpha\Big\{\cos(\mathrm{Wi}\sqrt{-\alpha}T)\left(\frac{p^2}{p^4-\mathrm{Wi}^2\alpha}-\frac{1}{p^2}\right)\\ &-\frac{\sin(\mathrm{Wi}\sqrt{-\alpha}T)\mathrm{Wi}\sqrt{-\alpha}}{p^4-\mathrm{Wi}^2\alpha}\Big\}\Big] \qquad (13)\end{aligned}$$

$$\begin{aligned}\langle R_x(t)R_y(t+\tau_0 T)\rangle &= \frac{2\zeta\tau_0}{N}\sum_{p=1,3,5,\ldots}^{\infty} e^{-p^2T}\Big[-\sqrt{-\alpha}\Big\{\sin(\mathrm{Wi}\sqrt{-\alpha}T)\left(\frac{p^2}{p^4-\mathrm{Wi}^2\alpha}+\frac{1}{p^2}\right)\\ &+\frac{\cos(\mathrm{Wi}\sqrt{-\alpha}T)\mathrm{Wi}\sqrt{-\alpha}}{p^4-\mathrm{Wi}^2\alpha}\Big\}+\frac{1}{\sqrt{-\alpha}}\Big\{\sin(\mathrm{Wi}\sqrt{-\alpha}T)\left(\frac{p^2}{p^4-\mathrm{Wi}^2\alpha}-\frac{1}{p^2}\right)\\ &+\frac{\cos(\mathrm{Wi}\sqrt{-\alpha}T)\mathrm{Wi}\sqrt{-\alpha}}{p^4-\mathrm{Wi}^2\alpha}\Big\}\Big] \qquad (14)\end{aligned}$$

$$\begin{aligned}\langle R_y(t)R_x(t+\tau_0 T)\rangle &= \frac{2\zeta\tau_0}{N}\sum_{p=1,3,5,\ldots}^{\infty} e^{-p^2T}\Big[-\sqrt{-\alpha}\Big\{\sin(\mathrm{Wi}\sqrt{-\alpha}T)\left(\frac{p^2}{p^4-\mathrm{Wi}^2\alpha}-\frac{1}{p^2}\right)\\ &+\frac{\cos(\mathrm{Wi}\sqrt{-\alpha}T)\mathrm{Wi}\sqrt{-\alpha}}{p^4-\mathrm{Wi}^2\alpha}\Big\}+\frac{1}{\sqrt{-\alpha}}\Big\{\sin(\mathrm{Wi}\sqrt{-\alpha}T)\left(\frac{p^2}{p^4-\mathrm{Wi}^2\alpha}+\frac{1}{p^2}\right)\\ &+\frac{\cos(\mathrm{Wi}\sqrt{-\alpha}T)\mathrm{Wi}\sqrt{-\alpha}}{p^4-\mathrm{Wi}^2\alpha}\Big\}\Big] \qquad (15)\\ \langle R_z(t)R_z(t+\tau_0 T)\rangle &= \frac{4\zeta\tau_0}{N}\sum_{p=1,3,5,\ldots}^{\infty}\frac{e^{-p^2T}}{p^2} \qquad (16)\\ \langle R_z(t)R_x(t+\tau_0 T)\rangle &= \langle R_z(t)R_y(t+\tau_0 T)\rangle = 0 \qquad (17)\end{aligned}$$

Note that these correlators (Eqs. 12 - 17 ) are in fact the analytic continuation of the results in Eqs. 6 - 11 for $\alpha > 0$ to $\alpha < 0$. We also note that from the above list of the correlation functions we can obtain any static correlator by setting $T = 0$ in relevant equation. To investigate any dynamic properties (like tumbling) we need to keep $T \neq 0$.

### *2.2. Static distributions*

Since the Cartesian components of $\mathbf{R}(t) = (R_x(t), R_y(t), R_z(t))^T$ are all Gaussian variables, their joint PDF is also Gaussian and given by: $\mathcal{P}(R_x, R_y, R_z) = (2\pi)^{-3/2}|\mathbf{C}|^{-1/2}\exp(-\frac{1}{2}\mathbf{R}^T\mathbf{C}\mathbf{R})$, where $\mathbf{C}$ denotes the covariance matrix involving the static correlators. We also can obtain the joint PDF of the polar components of $\mathbf{R}(t)$ using the standard transformation: $\widetilde{\mathcal{P}}(R,\vartheta,\phi) = \mathcal{P}(R\cos\vartheta\cos\phi, R\cos\vartheta\sin\phi, R\sin\vartheta) \times R^2\cos\vartheta$. In particular the azimuthal angle ($\phi$) distribution $Q(\phi)$ provides some physical insight. Integrating $\widetilde{\mathcal{P}}(R,\vartheta,\phi)$ over $R$ we get the joint angular PDF $S(\vartheta,\phi)$ and again integrating it over $\vartheta$ we finally get the $\phi$-distribution, $Q(\phi)$.

In the elongational regime ($\alpha > 0$): The covariance matrix is:

$$\mathbf{C} = \begin{pmatrix} \langle R_x^2\rangle & \langle R_xR_y\rangle & \langle R_xR_z\rangle \\ \langle R_yR_x\rangle & \langle R_y^2\rangle & \langle R_yR_z\rangle \\ \langle R_zR_x\rangle & \langle R_zR_x\rangle & \langle R_z^2\rangle \end{pmatrix} = \begin{pmatrix} d & b & 0 \\ b & a & 0 \\ 0 & 0 & c \end{pmatrix} \quad (18)$$

and the determinant is $|\mathbf{C}| = af$, where $f = ad - b^2$; for the exact expressions of $d, ..., c$ see Appendix A. Finally the azimuthal angle distribution is :

$$Q(\phi) = \frac{|\mathbf{C}|}{2\pi c\sqrt{f}} \frac{1}{(a\cos^2\phi + d\sin^2\phi - b\sin 2\phi)} \quad (19)$$

The most probable value of azimuthal angle from the above equation is given by : $\phi_m = \frac{1}{2}\tan^{-1}\left(\frac{2b}{d-a}\right)$. Now, in the pure elongational limit $\alpha \to +1$ we get $d \approx a$ (see Appendix A) and therefore $\phi_m \to \pi/4$. So the polymer stays along the $\phi = \pi/4$ plane for most of the time in this case which is expected intuitively. We can calculate the full width at half maximum $\Delta\phi$ of $Q(\phi)$ near $\phi_m \sim \pi/4$ and it is given by : $\cos(2\Delta\phi) = 2 - (a/b)$. For $\mathrm{Wi} \to \infty$, we get $a/b \sim 1$ and hence $\Delta\phi \to 0$: so the $Q(\phi)$ is sharply localised which is quite expected since the pure elongational flow is vorticity free where the polymer hardly tumbles. But such a limit is not actually realisable as $\mathrm{Wi} \to \infty$ violates the mode stability condition $p^2 > \mathrm{Wi}\sqrt{\alpha}$, and so such a aligned polymer will experience an immediate catastrophic stretching.

In the rotational regime ($\alpha < 0$): In this case we write the covariance matrix as :

$$\bar{\mathbf{C}} = \begin{pmatrix} \bar{d} & \bar{b} & 0 \\ \bar{b} & \bar{a} & 0 \\ 0 & 0 & \bar{c} \end{pmatrix} \quad (20)$$

and again the explicit values of $\bar{a}, \bar{b}, \bar{c}$ and $\bar{d}$ are given in Appendix A. The azimuthal distribution is : $\bar{Q}(\phi) = \left(\frac{|\bar{\mathbf{C}}|}{2\pi\bar{c}\sqrt{\bar{f}}}\right)/(\bar{a}\cos^2\phi + \bar{d}\sin^2\phi - \bar{b}\sin 2\phi)$. In pure rotational limit $\alpha \to -1$, we get $\bar{d} \approx \bar{a}$ and $\bar{b} \to 0$, hence we have $\bar{Q}(\phi) = 1/2\pi$ =constant. Clearly, the pure vorticity dominated limit is quite expected to have a flat $\phi$ distribution. Moreover as $\alpha \to -1$, $\bar{d} \approx \bar{a} \approx \bar{c}$ and $\bar{b} \to 0$, leading to all eigenvalues of the matrix $\bar{\mathbf{C}}$ becoming identical. This means that the polymer approaches a spherical shape as $\alpha \to -1$ (pure rotational flow).

### *2.3. Dynamic properties*

Now we concentrate on the first-passage question of the "tumbling" process of a polymer in mixed flow. The tumbling event is either defined as the successive occurrence of coiled states (radial tumbling) [7, 10] or as successive crossing of a fixed plane, say $\phi = 0$, by the vector $\mathbf{R}(t)$ (angular tumbling) [5, 6, 10]. The definition of radial tumbling is subjective, since a tumbling event is marked by $|\mathbf{R}(t)|$ falling below a value $R_0$ that defines the coiled state. But the angular tumbling time $\tau$ is objective and is simply the interval between two successive zero crossings of the process $R_x(t)$. The PDF of $\tau$ asymptotically goes as $P(T) \sim \exp(-\theta T)$, where $T \equiv \tau/\tau_0$ is the scaled time-increment.

The mean density of zero crossings for a Gaussian stationary process is given by: $\rho = 1/\langle\tau\rangle = \sqrt{-A''(0)}/\pi$, where $A(\tau)$ is the normalised correlator of that process and $\tau$ denotes the intervals between zeros. The quantity $\rho$ is finite when $A'(0) = 0$ and $A''(0)$ is finite. Such processes are called "smooth" and for them one can apply IIA to find the nontrivial exponent of the asymptotic distribution of the intervals between zeros [23, 24].

In the present case we are only interested in angular tumbling of the polymer. In the elongational regime ($\alpha > 0$) angular tumbling is very rare event in the limit $\alpha \to +1$ due to sharp localisation of $\phi$-distribution for which $\phi$ hardly changes sign. To investigate the angular tumbling for $\alpha < 0$ we proceed with the autocorrelation function $C_{xx}(T) = \lim_{t\to\infty}\langle R_x(t)R_x(t+\tau_0 T)\rangle$ (see subsection 2.1) and consider the normalised correlator $A(T) = C_{xx}(T)/C_{xx}(0)$. To know the small $T$ behavior of $A(T)$ we first obtain the exact Laplace transform of $C_{xx}(T)$ :

$$\begin{aligned}\widetilde{C}_{xx}(s) \;=\; & \frac{2\zeta\tau_0}{N}\Big[\frac{1}{s}\frac{\pi}{8\sqrt{\beta}}\left(\tanh\frac{\pi\sqrt{\beta}}{2} + \tan\frac{\pi\sqrt{\beta}}{2}\right)\left(1+\frac{1}{\alpha}\right) + \frac{\pi^2}{8}\left(\frac{s}{s^2-\beta^2}\right)\left(1-\frac{1}{\alpha}\right) \\ & - \frac{\pi}{8\sqrt{s+\beta}}\left(\frac{2}{s+\beta} + \frac{\beta}{s(s+\beta)}(1+1/\alpha)\right)\tanh\left(\frac{\pi\sqrt{s+\beta}}{2}\right) \\ & - \frac{\pi}{8\sqrt{s-\beta}}\left(\frac{2}{s-\beta} - \frac{\beta}{s(s-\beta)}(1+1/\alpha)\right)\tanh\left(\frac{\pi\sqrt{s-\beta}}{2}\right)\Big] \qquad (21)\end{aligned}$$

where $\beta = \mathrm{Wi}\sqrt{\alpha}$. From the above equation, the expression of $\tilde{A}(s)$ in large $s$ limit (i.e $s \to \infty$) can be obtained easily and inverting that we derive the expression of A(T) for $T \to 0$ as below:

$$A(T) \;=\; 1 - \frac{2\zeta\tau_0}{NC_{xx}(0)}\sqrt{\pi}T^{1/2} - \frac{2\zeta\tau_0}{NC_{xx}(0)}\frac{\mathrm{Wi}^2(-\alpha)\pi^2}{16}\left(1-\frac{1}{\alpha}\right)T^2 + \cdots \qquad (22)$$

where $C_{xx}(0) = \langle R_x^2\rangle$ is the static correlator given in Appendix A. It is clear from the above equation that both $A'(0)$ and $A''(0)$ diverge. Thus $R_x(t)$ is a "non-smooth" process in the sense that the density of zero crossings is infinite and hence the method of IIA can not be applied.

Although IIA is inapplicable, we have simulated the Langevin dynamics of Eq. (1) and curves for $P(T)$ were obtained (see Fig. 2). The exponential tails fall more sharply ($\theta$ increases) with increasing negative $\alpha$. Furthermore the appearance of a peak (for $\alpha < 0$), signifying a characteristic cyclic tumbling, gets more prominent with increasing negative $\alpha$.

## 3. Special case of simple shear flow

The limiting case of "simple shear" ($\alpha = 0$) is the only solvable case for $P(T)$ using the method of IIA [15], that too for large shear rate $\mathrm{Wi} \to \infty$. In the limit $\alpha \to 0$, Eq. (22) yields

$$A(T) = 1 - \left[1 + \frac{\pi^4\mathrm{Wi}^2}{240}\right]^{-1}\frac{4}{\pi^{3/2}}T^{1/2} - \left[1 + \frac{\pi^4\mathrm{Wi}^2}{240}\right]^{-1}\frac{\mathrm{Wi}^2}{4}T^2 + \cdots \quad \text{as } T \to 0. \qquad (23)$$

Now we note that in the limit of strong shear ($\mathrm{Wi} \to \infty$) the singular contribution of the second term proportional to $\sqrt{T}$ in Eq. (23) is ignorable. In the limit of $T \to 0$, Eq. (23) has the limiting form $A(T) \approx 1 - (60/\pi^4)T^2 + \cdots$ as $T \to 0$. Hence, in the limit $\mathrm{Wi} \to \infty$ the process $R_x(t)$ becomes "smooth" [24] and IIA can be applied to find the constant $\theta$. Clearly we have $A'(0) = 0$ and $A''(0) = -120/\pi$. Thus we get finite density of zero crossings $\rho = \sqrt{120}/\pi^3$. If we assume $P(T)$ to be exponential not only asymptotically but also over the full range of $T$, the numerical value of $\rho$ gives a crude estimation of $\theta$ as below:

$$\theta \approx \rho \approx 0.353298\cdots$$

We note that the expression of $A(T)$ for any $T$ in the limit of $\mathrm{Wi} \to \infty$ is :

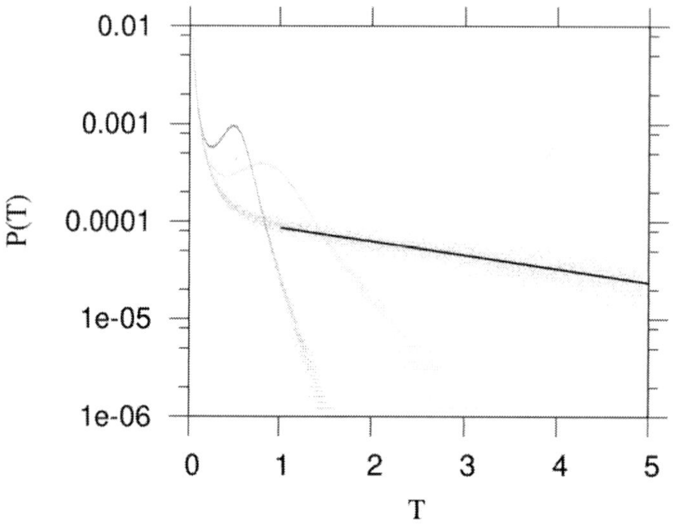


**Figure 2.** (color online) Linear-log plot of $P(T)$ versus $T$. Values of the parameter $\alpha$ are respectively: $\alpha = 0$ (the upper curve, with red dots), $\alpha = -0.1$ (the middle curve, with green dots) and $\alpha = -0.3$ (the lower curve, with blue dots). The fitted exponential function for the simple shear case ($\alpha = 0$) is : $0.00012 \times \exp(-0.324T)$ (where $\theta$ is used from method IV in section 3). All data are for $N = 10$ and $\dot{\gamma} = k = 1$.

$$A(T) = \frac{960}{\pi^6} \sum_{p=1,3,5,\ldots}^{\infty} e^{-p^2 T} \left( \frac{1}{p^6} + \frac{T}{p^4} \right) \tag{24}$$

Now we show below different estimates of $\theta$ from various IIA calculations.

**(I)** First we consider the "clipped" variable $\chi(t) = \mathrm{sgn}(R_x)$. Let $r(T) \equiv \frac{\langle \chi(t)\chi(t+\tau_0 T)\rangle}{\langle \chi(t)^2 \rangle}$ be the normalised asymptotic ($t \to \infty$) correlator of $\chi(t)$. One can express: $r(T) = \sum_{n=0}^{\infty} (-1)^n p_n(T)$, where $p_n(T)$ is the probability that the interval $T$ has $n$ zeros. The strategy is to write the joint PDF of $n$ successive intervals as the product of the distribution of single intervals by IIA [23, 26]. Then one ultimately finds a relation between the Laplace transforms of $P(T)$ and $r(T)$ (which are denoted by $\tilde{P}(s)$ and $\tilde{r}(s)$ respectively) as below:

$$\tilde{P}(s) = [1 - (\rho/2)s(1 - s\tilde{r}(s))]/[1 + (\rho/2)s(1 - s\tilde{r}(s))] \tag{25}$$

where $\rho$ is the density of zero crossings. The asymptotic PDF $P(T) \sim \exp(-\theta T)$ shows that $\tilde{P}(s)$ has a simple pole at $s = -\theta$ i.e the denominator in Eq. (25) has a simple zero at that point. Now, the correlator of the clipped process is given by: $r(T) = (2/\pi)\sin^{-1}(A(T))$ (this relation holds for any Gaussian process [27]), where $A(T)$ is given by Eq. (24) . Using this relation in Eq. (25) and equating the denominator to zero we get (with a transformed variable $x = e^{-T}$):

$$1 + (\rho/2)s \left[ 1 - \frac{2s}{\pi} \int_0^1 x^{s-1} \sin^{-1} \left( \frac{960}{\pi^6} \sum_{p=1,3,5,\ldots}^{\infty} x^{p^2} \left( \frac{1}{p^6} + \frac{\log x}{p^4} \right) \right) dx \right] = 0 \tag{26}$$

Solving the above equation numerically for $s = -\theta$ we obtain :

$$\theta \approx 0.356203 \cdots$$

**(II)** Two other IIA schemes are given in Ref. [23]. Let $U(T)dT$ be the conditional probability that a zero of $\chi(t)$ occurs between $(t+T)$ and $(t+T+dT)$, given that there is a zero crossing of $\chi(t)$ at time $t$. It has been shown by J. A. McFadden that $\tilde{P}(s)$ and $\tilde{U}(s)$, the Laplace transforms of $P(T)$ and $U(T)$ respectively, are related by : $\tilde{P}(s) = \tilde{U}(s)/[1+\tilde{U}(s)]$. The expression of $U(T)$ for a Gaussian process is given by [23] :

$$U = \frac{2}{\pi}\left(\frac{1}{H} + \tan^{-1}H\right)\frac{r''}{4\rho} \tag{27}$$

where

$$r'' = (2/\pi)M_{23}(1-A^2)^{-3/2} \quad ; \quad H = M_{23}\left(M_{22}^2 - M_{23}^2\right)^{-1/2}$$

$$M_{22} = -A''(0)\left(1-A^2\right) - A'^2 \quad ; \quad M_{23} = A''\left(1-A^2\right) + AA'^2$$

The dependence on $T$ is understood in the above equations. We note that the function $\tilde{U}(s)$ raises some computational difficulty as it does not exist at $s=0$, because $U(T)$ does not vanish as $T \to \infty$. Also one can observe that as $T \to \infty$, $U(T)dT \to \rho dT$. If a new variable is defined as [23]: $V(T) \equiv U(T) - \rho$, with Laplace transform $\tilde{V}(s) = \tilde{U}(s) - (\rho/s)$; then $\tilde{U}(s)$ can be substituted by $\tilde{V}(s)$ and the computation can be performed easily. Setting the denominator of $\tilde{P}(s)$ to zero we get the equation : $s + \rho + s\tilde{V}(s) = 0$. Solving this equation numerically for $s = -\theta$ we have :

$$\theta \approx 0.257136\cdots$$

**(III)** In the third scheme, instead of assuming that all successive zero-crossing intervals are independent, it was assumed in Ref. [23] that a given interval is independent of the sum of the $(2m+2)$ earlier intervals (for $m = 0, 1, 2, ...$). This is called "quasi independent interval approximation"(QIIA). By QIIA one gets : $\tilde{P}(s) = (\tilde{U}(s) + g(s))/[2 + \tilde{U}(s) - g(s)]$, where $g(s) = L\{r''(T)/4\rho\}$. Again solving numerically for the zero of the denominator of $\tilde{P}(s)$ (using the newly defined variable $\tilde{V}(s)$ instead of $\tilde{U}(s)$) at $s = -\theta$ we get :

$$\theta \approx 0.213558\cdots$$

**(IV)** In Ref. [15] another approximation $r(T) \approx A(T)$ was used that gives the value of $\theta$ very close to the numerics (see $\alpha = 0$ case in Fig. 2). This approximation yields :

$$\theta \approx 0.323558\cdots$$

We see that various estimates of $\theta$ differ slightly, but upto first decimal place (method III being an exception) the number is roughly $\approx 0.3$ .

## 4. Conclusion

We have exactly calculated the most general spatio-temporal correlation functions for a Rouse Polymer in planar mixed flow, in the steady state. Using the exact correlators of the mode coordinates, we explicitly write down the various correlators associated with the Cartesian components of the end-to-end vector $\mathbf{R}(t)$. The static correlators may be used to find the static probability distributions associated with the polar ($\vartheta$) and azimuthal ($\phi$) angles of $\mathbf{R}$. These static correlations give important information about the average shape of the polymer.

The more important concern in our paper has been the dynamics of angular tumbling of the vector $\mathbf{R}(t)$ in mixed flow, and the distribution of time scales associated with these first passage events. We find numerically that for rotationally dominated flows ($\alpha < 0$), the distribution has a peak and then an exponential tail for large times. We demonstrated that analytical estimates of the decay constant of the exponential tail can not be estimated using the method of IIA in $\alpha < 0$ cases, as the process $R_x(T)$ is not smooth for any limiting value of Wi.

The method of IIA was successfully applied by us [15] in the special case of "simple shear" ($\alpha = 0$), using a particular type of approximation (namely method IV in section 3). Here we have given many new analytical estimates of the number $\theta$, using many different types of approximations following the classic work of Ref. [23]. Although the numbers are slightly different from each other, up to the first decimal place $\theta \approx 0.3$.

### Acknowledgments
We are thankful to RRI for the travel support which allowed the collaborators to meet and discuss this project.

### Appendix A. Static correlators
In this appendix, we give the exact expressions of the static correlators, which we get by putting $T = 0$ in the equations (6) to (17).

The static correlators for $\alpha > 0$ :

Using the standard series sums [28]:

$$\sum_{m=1}^{\infty} \frac{1}{(2m-1)^2 - z^2} = \frac{\pi}{4z} \tan\frac{\pi z}{2} \quad ; \quad \sum_{m=1}^{\infty} \frac{1}{(2m-1)^2 + z^2} = \frac{\pi}{4z} \tanh\frac{\pi z}{2} \quad , \tag{A.1}$$

we simplify

$$d = \langle R_x^2 \rangle = \frac{2\zeta\tau_0}{N} \left[ (1 + 1/\alpha) \sum_{p=1,3,5,\ldots}^{\infty} \left( \frac{p^2}{p^4 - \mathrm{Wi}^2 \alpha} \right) + (1 - 1/\alpha) \frac{\pi^2}{8} \right] \tag{A.2}$$

to get :

$$d = \langle R_x^2 \rangle = \frac{2\zeta\tau_0}{N} \left[ (1 + 1/\alpha) \frac{\pi}{8x} \left( \tan\frac{\pi x}{2} + \tanh\frac{\pi x}{2} \right) + (1 - 1/\alpha) \frac{\pi^2}{8} \right] \tag{A.3}$$

Similar steps yield the following :

$$a = \langle R_y^2 \rangle = \frac{2\zeta\tau_0}{N} \left[ (1 + \alpha) \frac{\pi}{8x} \left( \tan\frac{\pi x}{2} + \tanh\frac{\pi x}{2} \right) + (1 - \alpha) \frac{\pi^2}{8} \right] \tag{A.4}$$

$$b = \langle R_x R_y \rangle = \frac{2\zeta\tau_0}{N} \left[ \frac{(1 + \alpha)}{\sqrt{\alpha}} \frac{\pi}{8x} \left( \tan\frac{\pi x}{2} - \tanh\frac{\pi x}{2} \right) \right] \tag{A.5}$$

$$c = \langle R_z^2 \rangle = \frac{\zeta\tau_0\pi^2}{2N} \tag{A.6}$$

where $x = \sqrt{\mathrm{Wi}}\alpha^{1/4}$.

The static correlators for $\alpha < 0$ :

$$\bar{d} = \langle R_x^2 \rangle = \frac{2\zeta\tau_0}{N} \left[ (1 + 1/\alpha) \frac{\pi}{8y} \frac{(\tan\frac{\pi y}{2}\mathrm{sech}^2\frac{\pi y}{2} + \tanh\frac{\pi y}{2}\sec^2\frac{\pi y}{2})}{(1 + \tan^2\frac{\pi y}{2}\tanh^2\frac{\pi y}{2})} + (1 - 1/\alpha) \frac{\pi^2}{8} \right] \tag{A.7}$$

$$\bar{a} = \langle R_y^2 \rangle = \frac{2\zeta\tau_0}{N} \left[ (1 + \alpha) \frac{\pi}{8y} \frac{(\tan\frac{\pi y}{2}\mathrm{sech}^2\frac{\pi y}{2} + \tanh\frac{\pi y}{2}\sec^2\frac{\pi y}{2})}{(1 + \tan^2\frac{\pi y}{2}\tanh^2\frac{\pi y}{2})} + (1 - \alpha) \frac{\pi^2}{8} \right] \tag{A.8}$$

$$\bar{b} = \langle R_x R_y \rangle = \frac{2\zeta\tau_0}{N}\left[\frac{(1+\alpha)}{\sqrt{|\alpha|}}\frac{\pi}{8y}\frac{\left(\tan\frac{\pi y}{2}\mathrm{sech}^2\frac{\pi y}{2} - \tanh\frac{\pi y}{2}\sec^2\frac{\pi y}{2}\right)}{(1+\tan^2\frac{\pi y}{2}\tanh^2\frac{\pi y}{2})}\right] \tag{A.9}$$

$$\bar{c} = \langle R_z^2 \rangle = \frac{\zeta\tau_0\pi^2}{2N} \tag{A.10}$$

where $y = \sqrt{\mathrm{Wi}/2}|\alpha|^{1/4}$. We checked that in the limit $\alpha \to 0+$ as well as $\alpha \to 0-$ the above static correlators smoothly go to the exact expressions of the static correlators in "simple shear" that are published in Ref. [15].

## References

[1] Smith D E and Chu S 1998 *Science* **281** 1335
[2] Smith D E, Babcock H P and Chu S 1999 *Science* **283** 1724
[3] Leduc P, Haber C, Bao G and Wirtz D 1999 *Nature* **399** 564
[4] Doyle P S, Ladoux B and Viovy J L 2000 *Phys. Rev. Lett.* **84** 4769
[5] Schroeder C M, Teixeira R E, Shaqfeh E S G and Chu S 2005 *Phys. Rev. Lett.* **95** 018301
[6] Teixeira R E, Babcock H P, Shaqfeh E S G and Chu S 2005 *Macromolecules* **38** 581-92
[7] Gerashchenko S and Steinberg V 2006 *Phys. Rev. Lett.* **96** 038304
[8] Gerashchenko S and Steinberg V 2008 *Phys. Rev.* E **78** 040801
[9] De Gennes P G 1974 *J. Chem. Phys.* **60** 5030
[10] Chertkov M, Kolokolov I, Lebedev V and Turitsyn K 2005 *J. Fluid Mech* **531** 251
[11] Celani A, Puliafito A and Turitsyn K 2005 *Europhys. Lett.* **70** 464
[12] Puliafito A and Turitsyn K 2005 *Physica* **211D** 9
[13] Winkler R G 2006 *Phys. Rev. Lett.* **97** 128301
[14] Buscalioni-Delgado R 2006 *Phys. Rev. Lett.* **96** 088303
[15] Das D and Sabhapandit S 2008 *Phys. Rev. Lett.* **101** 188301
[16] Chu S 2003 *Phil. Trans. R. Soc.* A **361** 689
[17] Schneider S W, Nuschele S, Wixforth A, Gorzelanny C, Kartz A A, Netz R R and Schneider M F 2007 *PNAS* **104** 7899
[18] Hur J S, Shaqfeh E S G, Babcock H P and Chu S 2002 *Phys. Rev.* E **66** 011915
[19] Dua A and Cherayil B J 2003 *J. Chem. Phys.* **119** 5696
[20] Woo N J and Shaqfeh E S G 2003 *J. Chem. Phys.* **119** 2908
[21] Rouse P E 1953 *J. Chem. Phys.* **21** 1272
[22] Doi M and Edwards S F 1986 *The Theory of Polymer Dynamics* (New York : Oxford)
[23] McFadden J A 1958 IRE *Trans. Info. Theory* **4** 14
[24] Majumder S N 1999 *Curr. Sci.* **77** 370
[25] Majumder S N and Sire C 1996 *Phys. Rev. Lett.* **77** 1420
[26] Majumder S N, Sire C, Bray A J and Cornell S J 1996 *Phys. Rev. Lett.* **77** 2867
[27] Rice S O 1944 *Bellsyst. Tech. J.* **23** 282
[28] Gradsteyn I S and Ryzhik I M 2000 *Tables of Integrals, Series and Products* (New York : Academic Press)

Journal of Physics: Conference Series **297** (2011) 012008 doi:10.1088/1742-6596/297/1/012008

# Non-equilibrium dynamics near a quantum multicritical point

**Ayoti Patra, Victor Mukherjee and Amit Dutta**

E-mail: ayoti@iitk.ac.in, victor@iitk.ac.in, dutta@iitk.ac.in
Department of Physics, Indian Institute of Technology Kanpur, Kanpur - 208016, India

**Abstract.** We study the non-equilibrium dynamics of a quantum system close to a quantum multi-critical point (MCP) using the example of a one-dimensional spin-1/2 transverse XY spin chain. We summarize earlier results of defect generenation and fidelity susceptibility for quenching through MCP and close to the MCP, respectively. For a quenching scheme which enables the system to hit the MCP along different paths, we emphasize the role of path on exponents associated with quasicritical points which appear in the scaling relations. Finally, we explicitly derive the scaling of concurrence and negativity for two spin entanglement generated following a slow quenching across the MCP and enlist the results for different quenching schemes. We explicity show the dependence of the scaling on the quenching path and dicuss the limiting situations.

## 1. Introduction

Recent years have seen a remarkable upsurge in the studies on the non-equilibrium dynamics of quantum phase transitions (QPTs) in quantum many body systems[1, 2, 3]. A zero temperature QPT, driven by quantum fluctuations, is associated with a fundamental change in the symmetry of the ground state of a quantum system. The possibility of experimental realization of QPT in ultracold atoms trapped in optical lattices, namely Mott insulator to superfluid transition, has also paved the way for rigorous theoretical investigations [4, 5]. A quantum critical point (QCP)[1, 2] is associated with a diverging length scale as well as a diverging time scale or the relaxation time of the quantum system arising due to a vanishing energy gap. In the vicinity of a QCP at $\lambda = 0$, the spatial correlation length $\xi$ diverges as as $\xi \sim |\lambda|^{-\nu}$, while the relaxation time $\xi_\tau$ (or the inverse of the energy gap) grows as $\xi_\tau \sim \xi^z$, where $\nu$ and $z$ are the quantum critical exponents characterizing the universality class of the QPT.

Divergences in length and time scales introduce non-trivial effects in the dynamics of a quantum system near a QCP. In particular, if a parameter of the Hamiltonian is tuned slowly so as to drive the system through a QCP, the dynamics fails to be adiabatic for any finite rate of variation of the parameter. This non-adiabaticity in the response of the system eventually leads to generation of defects in the final state. A quantitative analysis of defect production in a system following a quench can be carried out using the Kibble-Zurek (KZ) theory [6, 7] extended to quantum spin chains [8, 9, 10, 11, 12, 13, 14, 15, 16, 17, 18, 19]. The KZ analysis assumes that when a parameter of the Hamiltonian is varied in time $t$ at a uniform rate as $(t/\tau)$ through the QCP, where $\tau$ is the inverse rate of quenching, the system evolves adiabatically for the entire span of time except for the impulse region close to the QCP where the relaxation time of the system exceeds the rate of quenching; this leads to the generation of defects in the final state of a $d$-dimensional quantum Hamiltonian which scales with $\tau$ as $1/\tau^{\nu d/(\nu z+1)}$ [8, 9]. The scaling has also been extended to non-linear variation of parameters [16, 20].

There has also been recent developments in theoretical investigation of QPTs from the viewpoint of quantum information theory. Quantities like concurrence [21, 22], entanglement entropy[23, 24], fidelity susceptibility[25, 26, 27, 28, 29, 30, 31, 32, 33] and geometric phases[34, 35, 36, 37] have been shown to capture the non-analyticities associated with a QCP.

Fidelity is the measure of overlap of two neighbouring ground states of a quantum Hamiltonian in the parameter space. Fidelity susceptibility is the rate of change of the ground state under an infinitesimal variation of the parameters of the Hamiltonian. For two ground states $\psi_0(\lambda)$ and $\psi_0(\lambda+\delta\lambda)$ closely spaced in the parameter space ($\delta\lambda \to 0$), fidelity ($F$) is defined as[25]

$$F = |\langle\psi_0(\lambda)|\psi_0(\lambda+\delta\lambda)\rangle| \approx 1 - \frac{\delta\lambda^2}{2}\chi_F(\lambda) + \cdots \tag{1}$$

where the fidelity susceptibility $\chi_{\mathrm{F}}$ is the first non-vanishing term in the expansion of fidelity. The scaling behavior of $\chi_{\mathrm{F}}$ is well established near a QCP[26, 28, 29] which has recently been generalised to the case of a quantum MCP[38]. The symmetrty of the ground state of a quantum many-body system uhndergoes a change at QCP[1] leading to a sharp drop of fidelity. At the same time, the fidelity susceptibility ($\chi_{\mathrm{F}}$) usually diverges in a power law fashion with the system size as $\chi_{\mathrm{F}} \sim L^{2/\nu}$ [26, 27, 28, 29, 30, 31, 32, 39, 40, 41, 42, 43] where $\nu$ is the correlation length exponent asssociated with the QCP. In recent years a series of works on the connection between fidelity susceptibility and quantum criticality has been reported [26, 27, 28, 29, 30, 31, 32, 39, 40, 41, 42, 43]. Generation of two spin entanglement following a slow quench across a QCP has also been reported recently [44]. Recently, the two spin entanglement properties of a quantum system during time evolution after a sudden quench through a critical point have been studied [45].

In the present work we discuss the dynamics of quantum spin chains, namely transverse $XY$ model near a MCP [46]. The presence of minimum energy quasicritical points near a MCP results in strikingly different behaviors of measurable parameters, for example defect density following fast and slow quenches, fidelity susceptibility, etc. We summarize recent results on the defect generation and fidelity susceptibility in sections 2 and 3. In section 4, we study the scaling of concurrence [21] and negativity [47] following a path-dependent slow quench across a quantum MCP generalizing related work on passage through a QCP [44].

## 2. The model

We consider a one dimensional spin 1/2 $XY$ model in presence of a transverse field, given by

$$H = -\sum_j \left(J_x\sigma_j^x\sigma_{j+1}^x + J_y\sigma_j^y\sigma_{j+1}^y + h\sigma_j^z\right), \tag{2}$$

where $\sigma$'s are the Pauli spin matrices, $J_x$ and $J_y$, and $h$ are interaction strengths and the transverse field, respectively. Using Jordan-Wigner transformation the above Hamiltonian can be reduced to a direct product of $2\times 2$ Hamiltonians in the Fourier space given by the form [48, 50, 49, 51, 19]

$$H_k = \begin{bmatrix} h+\cos k & i\gamma\sin k \\ -i\gamma\sin k & -(h+\cos k) \end{bmatrix} \tag{3}$$

with $J = J_x + J_y$ is set equal to unity and anistropy $\gamma = J_x - J_y$; the energy spectrum is

$$\Lambda_k = \sqrt{(h+\cos k)^2 + \gamma^2\sin^2 k}. \tag{4}$$

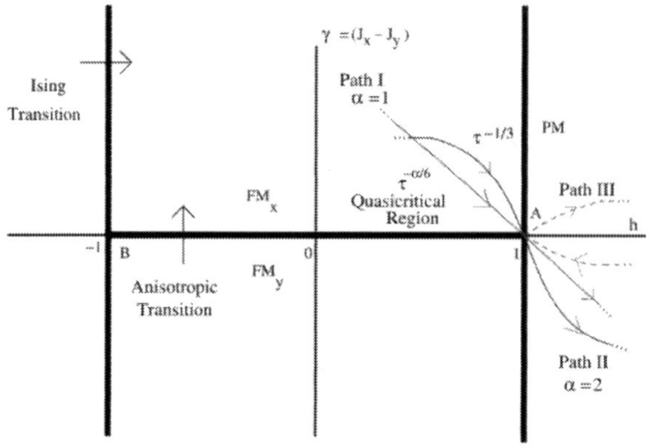


**Figure 1.** Phase diagram of $XY$ model in presence of transverse field. The horizontal bold line shows the anisotropic transition between ferromagnetic ordering in the $x$ direction and that in the $y$ direction, whereas, the vertical bold lines are lines of Ising transitions between paramagnetic and ferromagnetic phases. The different paths of quenching through the MCP A are shown.

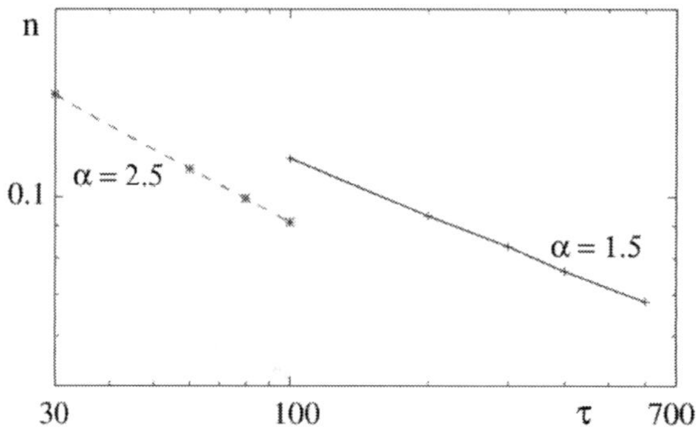


**Figure 2.** Plot of defect density as a function of $\tau$ for different values of $\alpha$. $n_{ex}$ varies $\tau^{-\alpha/6}$ for $1 < \alpha < 2$, and saturates to $n_{ex} \sim \tau^{-1/3}$ for $\alpha > \alpha_c = 2$.

Eq. 4 shows that the gap vanishes for the mode $k = 0(\pi)$ for $h = -1(+1)$ indicating ferromagnetic to paramagnetic QPT with critical exponents $\nu = z = 1$. Similarly, the horizontal bold line indicate quantum transitions between two ferromagnetically ordered phases (see Fig. 1) with the same set of critical exponents. The Ising and anistropic critical lines meet at multicritical points ($h = \pm 1, \gamma = 0$) where the critical exponents are $\nu = \nu_{mc} = 1/2$, $z = z_{mc} = 2$, respectively.

If the transverse field $h$ is varied as $t/\tau$ sweeping the system across the Ising critical points, a defect density scaling as $n \sim \tau^{-1/2}$ has been derived; this is in complete agreement with the KZ scaling noting that $\nu = z = 1$ and $d = 1$ [8, 9, 10, 11, 13]. Identical scaling relation has been found for quenching across both Ising and the anisotropic critical points by varying the interaction term [14]. On the other hand, a similar quench through the MCP produces a much slower decay defect density given by $n_{ex} \sim \tau^{-1/6}$ [14, 52] in sharp contrast with the KZ prediction $n_{ex} \sim \tau^{-1/4}$. A subsequent investigation on this apparent violation of KZ scaling revealed the existence of minimum energy gap points occuring away from the MCP [52] where the excitation scales with the momentum $\Delta_k \sim k^{z_{qc}}$ thereby defining an effective dynamical exponent $z_{qc}$ which turned out to be equal to three for the present model. For a system reducible to non-interacting two-level problems, a modified KZ scaling relation of the form $n_{ex} \sim \tau^{2z_{qc}}$ was proposed for quenching across a MCP [52].

Journal of Physics: Conference Series **297** (2011) 012008 doi:10.1088/1742-6596/297/1/012008

The above apparent violation of KZ mechanism led to a series of works on the non-equilibrium properties of the system near a MCP [38, 53, 54]. It was shown that the dynamical response of the system stems from the presence of quasicritical points where the energy gap is minimum [53, 54] as the energy gap is not necessarily minimum at the MCP. Deng et al, [53] used a non-linear two parameter quenching scheme $|\gamma(t)| = h(t) - 1 = |t/\tau|^\alpha sgn(t)$ so that the MCP ($h = 1, \gamma = 0$) is reached at a time $t = 0$. In the case $\alpha = 1$ (linear quenching), the defect density was exactly calculated using the Landau-Zener [56] tunneling formula and the relation $n_{ex} \sim \tau^{-1/6}$ was retrieved. It was argued that quasicritical points characterized by a sharp decrease in energy lie on the ferromagnetic side of the MCP and the system encounters them whenever the transverse field satisfies the relation $h = -k^2/2$ for modes close to the critical modes during its temporal evolution. The minimum excitation at a quasi critical point $\Delta_k \sim k^3$, as compared to $\Delta_k \sim k^2$ at the MCP. This defines the effective dynamical exponent $z_{qc} = 3$ as was reported in the reference [52]. For a generic $\alpha$ and a model with $\nu_{mc}z_{mc} \neq 1$ adiabatic perturbation theory was used to derive the scaling relation [53]

$$n_{ex} \sim \tau^{-\frac{d\alpha\nu_{qc}}{\alpha\nu_{qc}z_{qc}+1}} \tag{5}$$

where $\nu_{qc} = \nu_{mc}z_{mc}/z_{qc}$

Mukherjee and Dutta [54] introduced a quenching scheme

$$h(t) = 1 + |\gamma(t)|^\alpha sgn(t); \quad \alpha > 0 \;\text{ and }\; \gamma(t) = -t/\tau, \tag{6}$$

with time varying from $-\infty$ to $\infty$ so that the system can apprach the MCP through different paths characterized by the parameter $\alpha$. It was shown that whether the system encounters qusicrtical points depend crucially on the path and so do the effective critical exponents $z_{qc}$ and $\nu_{qc}$. For the present model, it was shown that the system hits quasicritical points only up to a limiting path given by $\alpha = 2$ while for $\alpha > 2$, the energy gap is minimum right at the MCP. Within the linearization approximation [55], the effective dynamical exponent was found to be $z_{qc} = 2/\alpha + 1$ leading to a scaling relation $n_{ex} \sim \tau^{-\alpha/6}$ so that for $\alpha = 1$ (linear path), we retrive the exponent 1/6. On the other hand, in the limit $\alpha \to 2$ the scaling saturates to $n_{ex} \sim \tau^{-1/3}$ indicating that for $\alpha > 2$, the quenching scheme essentially boils down to sweeping the system across the MCP along the gapless Ising critical line [15]. In fact, it should be reiterated that for $\alpha > 2$, system does not cross any quasi-critical point which shows that the scaling and the effective exponents are indeed path-dependent and $z_{qc} = z_{mc} = 2$. Using adaibatic perturbation theory, the generic scaling form was obtained as

$$n_{ex} \sim \tau^{-\frac{\alpha d}{2z_{qc}\alpha + z_{mc}(1-\alpha)}}, \tag{7}$$

for $\nu_{mc}z_{mc} = 1$. For a generic model with $\nu_{mc}z_{mc} \neq 1$ Eq. (7) gets modified to

$$n_{ex} \sim \tau^{-\frac{\alpha d\nu_{qc}}{[\alpha(z_{qc}\nu_{qc}+1)+\nu_{qc}z_{mc}(1-\alpha)]}}, \tag{8}$$

Eq. (7) includes an additional term when compared to the scaling form given in Eq. (5) which arises due to the linearization process.

The exoponents associated with the quasi critical points namely, $z_{qc}$ and $\nu_{qc}$, can be understood using a two parameter scaling of the energy gap near the MCP given by

$$\Delta_k = k^{z_{mc}} f\left(\frac{\lambda_1}{k^{1/\nu_1}}, \frac{\lambda_2}{k^{1/\nu_2}}\right) \tag{9}$$

where $k = |\vec{k}|$ and in the present case $\lambda_1 = 1-h$ and $\lambda_2 = \gamma$ with exponents $\nu_1 = 1/2$ and $\nu_2 = 1$, respectively. For the path $\lambda_1 = \lambda_2$, the quasi-critical points typically appear when $\lambda_1/k^{1/\nu_1} \simeq 1$ so that $\Delta_k \sim k^{z_{mc}+1/\nu_1+1/\nu_2}$ yielding the quasi-critical dynamical exponent given by

$$z_{qc} = z_{mc} + \frac{1}{\nu_1} + \frac{1}{\nu_2}, \tag{10}$$

so that $z_{qc} = 3$ for $z_{mc} = 2, \nu_1 = 1/2, \nu_2 = 1$ as argued previously. For a non-linear path $\lambda_1 = \lambda_2^{\alpha}$, the exponent $\nu_1$ gets renormalized to $\alpha\nu_1$ [16] leading to the relation

$$z_{qc} = z_{mc} + \frac{1}{\alpha\nu_1} + \frac{1}{\nu_2}, \tag{11}$$

which reduces to $z_{qc} = 2/\alpha + 1$ for the transverse $XY$ chain.

Finally, the MCP can also be approached along a path lying entirely in the paramagnetic phase using the quenching scheme $h(t) = 1 + |\gamma(t)|^{\alpha}$, $\gamma = t/\tau$ [53, 54] (path III in Fig. 1). The defect density for the transverse $XY$ chain is $n_{sp} \sim \tau^{-(4-\alpha)/(2(1+\alpha))}$ which can be generalised using adiabatic perturbation theory to the form

$$n_{sp} \sim \tau^{-\frac{(d+d_0(\alpha))\alpha\nu_{qc}}{\alpha\nu_{mc}z_{mc}+1}}. \tag{12}$$

Eq.(12) shows that for these special paths, there is a dimensional shift in KZ scaling relation arising due to shift in the centre of the impulse region so that the non-critical modes contribute to the defect generation [53]. For the present model, this dimensional shift $d_0(\alpha) = 2(2-\alpha)/\alpha$ does also depend on path and vanishes in the limit of $\alpha \to 2$.

## 3. Fidelity susceptibility near a quantum multicritical point

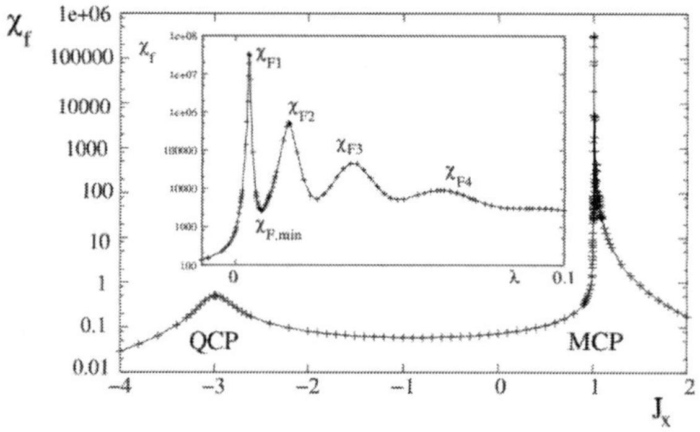


**Figure 3.** Plot of fidelity susceptibility as a function of $J_x$ for the critical and multicritical points. $\chi_f$ diverges as $L$ near the QCP, while an MCP results in oscillatory $\chi_f$, with the maximum values scaling as $L^5$. Inset Shows the oscillations in $\chi_f$ close to the MCP.

The scaling of the fidelity susceptibility and defect generated following a rapid quench near a quantum MCP has been studied recently[38] . At a QCP, $\chi_f(=\chi_F/L^d)$ varies as $L^{2/\nu-d}$ where $L$ is the system size, while away from the QCP $\chi_f$ scales as $\chi_f \sim \lambda^{\nu d-2}$, where $\lambda$ measures the deviation from the QCP. Studies on fidelity susceptibility near MCP in Hamiltonian (2) along a linear path parametrized by $\lambda_1 = \lambda_2 = \lambda$, shows a strikingly different behavior, with $\chi_f$ exhibiting an oscillatory variation with $\lambda$; it peaks whenever a quasi-critical point at

$\lambda = k^{1/\nu_1}/2 = k^2/2$ is encountered while no peak is observed right at the MCP. The maximum value of any peak scales as

$$\chi_f \sim L^{2/\nu_{qc}-d}, \tag{13}$$

leading to $\chi_f \sim L^5$ for transverse field $XY$ chain. For a general path $\lambda_1 = \lambda_2^\alpha$ with $\alpha < \alpha_c$, $\nu_{qc} = (2/\alpha + 1)^{-1}$, hence Eq. (13) reduces to

$$\chi_f \sim L^{4/\alpha - 1}. \tag{14}$$

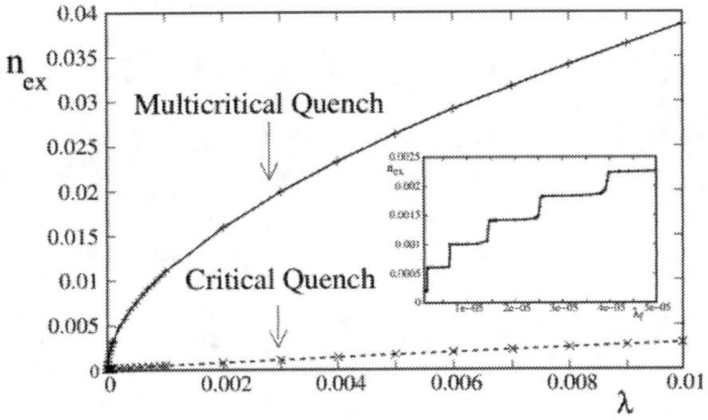


**Figure 4.** Plot of kink density following a sudden quench starting from the MCP. Sudden jumps in $n_{ex}$ are observed whenever the system encounters a quasi-critical point

The defect density generated following a fast quench of small amplitude $\lambda$ starting from a QCP has been argued to satisfy the scaling relation [32]

$$n_{ex} \sim \lambda^2 \chi_f \sim \lambda^{\nu d}. \tag{15}$$

A similar approach near MCP showed a violation of the the simple relation (Eq. 15) when system in quenched suddenly into the ferromagnetic region. Further, sudden increase of kink density is observed at quasicritical points. However, the coarse grained kink density still scales with $\lambda$ as $n_{ex} \sim \lambda^{\nu d}$.

## 4. Geneation of Concurrence and Negativity following a slow quench

In this section, we study generation of two-spin entanglement in the system (2) for a slow quench through the MCP and quantify it in terms of the measures concurrence [21] and negativity[47]. These quantities or their derivatives show peaks or discontinuous jumps at the QCP signalling the singularities associated with them. We generalise the results obtained by Sengupta and Sen [44] for passage through a QCP to the present case of a MCP. We employ different linear and non-linear quenching schemes parametrized by an exponent $\alpha$ (as discussed previously) to study the scaling behaviors of concurrence and negativity of the ground state in the final state of the system.

Sengupta and Sen [44] argued that for the passage through a QCP by a linear or non-linear variation of parameters, entanglement is generated only between spins separated by even number of lattice spacings. Their study also established the existence of a critical quench rate $\tau_c^{-1}$ such that for a fatser quench rate $\tau^{-1} > \tau_c^{-1}$, no such two-spin entanglement is generated. Our study reveals that the above remains valid even for quenching through a MCP; the scalings of concurrence and negativity are related to the scaling of defect density which as discussed previously depends on $\tau$ and also the path $(\alpha)$. In the limiting situations ($\tau \to 0$ or $\tau \to \infty$) system remains unentangled, and carries only the information about its initial ground state.

Concurrence and negativity for two spins at the sites $i$ and $i+n$ separated by $n$ lattice spacings can be derived from the eigenvalues of their reduced density matrix given by [57]

$$\rho^n = \begin{pmatrix} a_+^n & 0 & 0 & b_1^n \\ 0 & a_0^n & b_2^n & 0 \\ 0 & b_2^{n*} & a_0^n & 0 \\ b_1^{n*} & 0 & 0 & a_-^n \end{pmatrix}, \tag{16}$$

where the matrix elements are related to the two-spin correlation function as follows:

$$\begin{aligned} a_\pm^n &= \frac{1}{4}\langle(1 \pm \sigma_i^z)(1 \pm \sigma_{i+n}^z)\rangle, \\ a_0^n &= \frac{1}{4}\langle(1 \pm \sigma_i^z)(1 \mp \sigma_{i+n}^z)\rangle, \\ b_{1(2)}^n &= \langle\sigma_i^- \sigma_{i+n}^{-(+)}\rangle, \end{aligned} \tag{17}$$

with

$$\beta_n = \int_0^\pi \frac{dk}{\pi} p_k \cos(nk), \tag{18}$$

where $p_k$ is the probability of defect formation when the system is quenched through multi-critical point at an inverse quench rate $\tau$ along a path characterized by $\alpha$. From Ref. [13], utilizing the expression for $\beta_n$, we get

$$\begin{aligned} \langle\sigma_i^z\rangle &= 1 - 2\beta_0, \\ \langle\sigma_i^z \sigma_{i+n}^z\rangle &= \langle\sigma_i^z\rangle^2 - 4\beta_n^2, \end{aligned} \tag{19}$$

Noting that $p_k$ is invariant under $k \to \pi - k$ and hence $\beta_n = 0$ for odd $n$, it can be readily shown from Eq. (17) that $b_1^n = \langle\sigma_i^- \sigma_{i+n}^-\rangle = 0$ for all $n$ as the correlations between two fermionic creation or annihilation operators vanish. The transformations $\sigma_n^x \to (-1)^n \sigma_n^x$, $\sigma_n^y \to (-1)^n \sigma_n^y$ and $\sigma_n^z \to \sigma_n^z$ leave the Hamiltonian (2)(the $Z_2$ symmetry) and as a consequence $p_k$ is also invariant . As a consequence, $b_2^n = \langle\sigma_i^- \sigma_{i+n}^+\rangle$ vanishes for odd $n$ so that the two-spin entanglement will not be generated for spins separated by odd lattice spacing [13]. This can also be deduced from the fact that the spin chain forms even and odd sublattice structure as explained in Ref. [13]. Correlation functions (19) can be evaluated by expressing the correlators as polynomials of pair correlators of Majorana Fermions [13, 48]. Using Eqs. (17), (18) and (19), we obtain the matrix elements of (16) as

$$\begin{aligned} \langle\sigma_i^+ \sigma_{i+2}^-\rangle &= \beta_2\langle\sigma_i^z\rangle, \\ \langle\sigma_i^+ \sigma_{i+4}^-\rangle &= (\beta_4\langle\sigma_i^z\rangle + 2\beta_2^2)\langle\sigma_i^+ \sigma_{i+2}^-\rangle, \\ \langle\sigma_i^+ \sigma_{i+6}^-\rangle &= \left[\beta_6\langle\sigma_i^+ \sigma_{i+2}^-\rangle + 4\beta_2(\beta_4^2 - \beta_2^2 + \beta_4\langle\sigma_i^z\rangle)\right] \times \left[\langle\sigma_i^z\rangle\{\langle\sigma_i^+ \sigma_{i+2}^-\rangle - 4(\beta_2^2 + \beta_4^2)\} - 16\beta_2^2\beta_4^2\right], \end{aligned} \tag{20}$$

With $b_1^n = 0$, the matrix $\rho^n$ reduces to the form

$$\rho^n = \begin{pmatrix} A & 0 & 0 & 0 \\ 0 & B & C & 0 \\ 0 & C & G & 0 \\ 0 & 0 & 0 & D \end{pmatrix}.$$

Journal of Physics: Conference Series **297** (2011) 012008 doi:10.1088/1742-6596/297/1/012008

| S. no. | Quenching Scheme | Defect density ($n_d$) | $\mathcal{C}^n$ | $\mathcal{N}^n$ |
|---|---|---|---|---|
| 1. | $h(t) = h\|\frac{t}{\tau}\|^\alpha sgn(t)$ | $\sqrt{\frac{\alpha}{\tau}}$ | $\sqrt{\frac{\alpha}{\tau}}$ | $\frac{\alpha}{\tau}$ |
| 2. | $h(t) - 1 = \frac{t}{\tau}, \gamma(t) = \frac{t}{\tau}$ | $\frac{1}{\tau^{\frac{1}{6}}}$ | $\frac{1}{\tau^{\frac{1}{6}}}$ | $\frac{1}{\tau^{\frac{1}{3}}}$ |
| 3. | $h(t) - 1 = \|\frac{t}{\tau}\|^\alpha sgn(t), \gamma(t) = \|\frac{t}{\tau}\|^\alpha sgn(t)$ | $(\frac{\alpha}{\tau})^{\frac{1}{4+\frac{2}{\alpha}}}$ | $(\frac{\alpha}{\tau})^{\frac{1}{4+\frac{2}{\alpha}}}$ | $(\frac{\alpha}{\tau})^{\frac{1}{2+\frac{1}{\alpha}}}$ |
| 4. | $h(t) - 1 = \|\frac{t}{\tau}\|^\alpha sgn(t), \gamma(t) = \|\frac{t}{\tau}\|$ | $(\frac{\alpha}{\tau})^{\frac{\alpha}{6}}$ | $(\frac{\alpha}{\tau})^{\frac{\alpha}{6}}$ | $(\frac{\alpha}{\tau})^{\frac{\alpha}{3}}$ |
| 5. | $h = 1, \gamma = \frac{t}{\tau}$ | $(\frac{1}{\tau})^{\frac{1}{3}}$ | $(\frac{1}{\tau})^{\frac{1}{3}}$ | $(\frac{1}{\tau})^{\frac{2}{3}}$ |

**Table 1.** Comparative study of different quenching schemes

For a matrix of this form, substituting the elements from Eq. (17) the expressions for concurrence and negativity simply become [44, 21]

$$\begin{aligned} \mathcal{C}^n &= \max\{0, 2(|b_2^n| - \sqrt{a_+^n a_-^n})\}, \\ \mathcal{N}^n &= \max\{0, \frac{1}{2}\left[a_+^n + a_-^n - \sqrt{(a_+^n - a_-^n)^2 + 4|b_2^n|^2}\,\right]\}. \end{aligned} \tag{21}$$

The scalings of concurrence and negativity of spins at site $i$ and $i+n$ therefore depends on the scalings of $a_0^n$, $a_\pm^n$ and $b_2^n$ as shown below for $n = 2$;

$$\begin{aligned} a_\pm^2 &= \frac{1}{4}\langle(1 \pm \sigma_i^z)(1 \pm \sigma_{i+2}^z)\rangle, \\ &= \frac{1}{4}\left[1 \pm 2\langle\sigma_i^z\rangle + \langle\sigma_i^- \sigma_{i+2}^-\rangle\right], \\ &= \frac{1}{4}\left[1 \pm (2 - 4\beta_0) + (1 - 2\beta_0)^2 - 4\beta_2^2\right], \end{aligned} \tag{22}$$

which simplifies to

$$a_+^2 = (1 - \beta_0)^2 - \beta_2^2, \tag{23}$$
$$a_-^2 = \beta_0^2 - \beta_2^2. \tag{24}$$

On the other hand, $a_0$ is given by

$$\begin{aligned} a_0^2 &= \frac{1}{4}\langle(1 \pm \sigma_i^z)(1 \mp \sigma_{i+2}^z)\rangle, \\ &= \frac{1}{4}\left[1 - \langle\sigma_i^z \sigma_{i+2}^z\rangle\right], \\ &= \beta_0(1 - \beta_0) - \beta_2^2. \end{aligned} \tag{25}$$

while

$$b_2^2 = b_2^{2*} = \langle\sigma_i^- \sigma_{i+2}^+\rangle = \beta_2(1 - 2\beta_0) \tag{26}$$

From Eq. (18), it is clearly seen that $\beta_n$ scales in the same way for all $n$. Retaining the highest order terms of Eqs. (24),(25) and (26), it can be shown that concurrence $\mathcal{C}^n \sim \beta_n$ whereas $\mathcal{N}^n \sim \beta_n^2$. On the other hand, the defect density $n_{ex} \sim \beta_n$ as seen from Eq. (18) thus establishing a direct relation with the scalings of concurrence and negativity.

By employing various quenching schemes to drive the system through the MCP, we get scaling relations of concurrence and negativity as tabulated in Table (1) (also see Fig. (1)). Scheme (1) summarises results for a non-linear queching of the transverse field through the Ising critical point as reported in reference [44] while schemes (2),(3) [53] and (4) [54] denote quenching across the MCP where quasicritical points play a dominant role. Finally, in the scheme (5) [15] the system swept along a gapless Ising transition line across the MCP. In the scheme 4, which we have introduced in section 2., the quasicritical exponents and consequently the scaling of $\mathcal{C}_n$ and $\mathcal{N}^n$ depend on path up to the limting path $\alpha = 2$ beyond which scaling relations for scheme (4) converges to those for scheme (5) as shown in the case of defect density. From Eq. (21), we note the existence of a critical quench rate when $|b_2^n|$ equals $\sqrt{a_+^n a_-^n}$, above which $\mathcal{C}^n = \mathcal{N}^n = 0$ and no two-spin entanglement is generated. Interestingly, the critical rate does not depend on the path.

## 5. Conclusion

We have summarized recent results on the non-equilibrium dynamics of a quantum spin chain, mainly the defect density following a slow quench across a MCP and the scaling of fidelity susceptibility close to it. In the final section, we have calculated the scaling of the concurrence and negativity for a multicritical slow quantum quench. We show that scaling of both the afore mentioned quantities with the rate of quenching depends crucially on path for a path dependent quenching shceme discussed here. The scaling exponent saturates to a constant value beyond a limiting path when the system does no longer encounter any quasicritical point.

## Acknowledgments

AD and VM thank oraganizers of Statphys-VII, Kolkata, India for giving opportunity to present their work and Uma Divakaran, A. Polkovnikov and Diptiman Sen for collaboration in related works. The present study was partially supported by CSIR, New Delhi, India .

## References

[1] Sachdev S 1999 *Quantum Phase Transitions* (Cambridge University Press, Cambridge)
[2] Chakrabarti B K , Dutta A and Sen P 1996 *Quantum Ising Phases and transitions in transverse Ising Models*, m41 (Springer, Heidelberg)
[3] Continentino M A 2001 *Quantum Scaling in Many-Body Systems* (World Scientific)
[4] Greiner M, Mandel O, Hansch T W, and Bloch I 2002 *Nature (London)* **415** 39
[5] Jaksch D, Bruder C, Cirac J I, Gardiner C W, and Zoller P 1998 *Phys. Rev. Lett.* **81** 3108
[6] Kibble T W B 1976 *J. Phys. A* **9** 1387
[7] Zurek W H 1985 *Nature (London)* **317** 505 ; 1996 *Phys. Rep.* **276** 177
[8] Zurek W H , Dorner U and Zoller P 2005 *Phys. Rev. Lett.* **95** 105701
[9] Polkovnikov A 2005 *Phys. Rev. B* **72** 161201(R)
[10] Damski B 2005 *Phys. Rev. Lett.* **95** 035701
[11] Dziarmaga J 2005 *Phys. Rev. Lett.* **95** 245701
[12] Damski B and Zurek W H 2006 *Phys. Rev. A* **73** 063405
[13] Cherng R W and Levitov L S 2006 *Phys. Rev. A* **73** 043614
[14] Mukherjee V, Divakaran U, Dutta A and Sen D 2007 *Phys. Rev. B* **76** 174303
[15] Divakaran U, Sen D and Dutta A 2008 *Phys. Rev. B* **78** 144301
[16] Sen D, Sengupta K, and Mondal S 2008 *Phys. Rev. Lett.* **101** 016806
[17] Sengupta K, Sen D, and Mondal S 2008 *Phys. Rev. Lett.* **100** 077204
[18] Dziarmaga J, 2010 *Advances in Physics* **59** 1063
[19] Dutta A, Divakaran U, Sen D, Chakrabarti B K, Rosenbaum T F and Aeppli G 2010 *arXiv:1012.0653*
[20] Barankov R and Polkovnikov A 2008 *Phys. Rev. Lett.* **101** 076801

[21] Hill S and Wooters W K 1997 *Phys. Rev. Lett.* **78** 5022 ; Wooters W K 1998 *ibid* **80** 2245
[22] Osterloh A, Amico L, Falci G and Fazio R 2002 *Nature* **416** 608
[23] Vidal G, Latorre J I, Rico E and Kitaev A 2003 *Phys. Rev. Lett.* **90** 227902
[24] Kitaev A and Preskill J 2006 *Phys. Rev. Lett.* **96** 110404
[25] Zanardi P, Paunkovic N 2006 *Phys. Rev. E* **74** 031123
[26] Venuti L and Zanardi P 2007 *Phys. Rev. Lett.* **99** 095701
[27] Zanardi P, Giorda P and Cozzini M 2007 *Phys. Rev. Lett.* **99** 100603
[28] Gu S -J 2010 *Int. J. Mod. Phys. B* **24** 4371

[29] You W L, Li Y -W and Gu S -J 2007 *Phys. Rev. E* **76** 022101; Chen S, Wang L, Gu S -J and Wang Y 2007 *Phys. Rev. E* **76** 061108 ; Gu S -J, Kwok H -M, Ning W -Q and Lin H -Q 2008 *Phys. Rev. B* **77** 245109 ; Chen S, Wang L, Gu S -J and Wang Y 2008 *Phys. Rev. A* **77** 032111; Yang S, Gu S -J, Sun C -P and Lin H -Q 2008 *Phys. Rev. A* **78** 012304; Yu W -C, Kwok H -M, Cao J and Gu S -J 2009 *Phys. Rev. E* **80** 021108; Wang Z, Ma T, Gu S -J and Lin H -Q 2010 *Phys. Rev. A* **81** 062350
[30] Zhou H -Q and Barjaktarevic J P 2008 *J. Phys. A: Math. Theor.* **41** 412001; Zhou H -Q, Orus R and Vidal G 2008 *Phys. Rev. Lett.* **100** 080601 ; Li B, Li S -H and Zhou H -Q 2009 *Phys. Rev. E* **79** 060101(R); Zhao J -H and Zhou H -Q 2009 *Phys. Rev. B* **80** 014403; Zhou H -Q, Zhao J -H, Li B 2008 *J. Phys. A: Math. Theor.* **41** 492002
[31] Schwandt D, Alet F and Capponi S 2009 *Phys. Rev. Lett.* **103** 170501
[32] De Grandi C, Gritsev V and Polkovnikov A 2010 *Phys. Rev. B* **81** 012303
[33] Polkovnikov A, Sengupta K, Silva A, Vengalattore M 2010 *arXiv:1007.5331*
[34] Carollo A C M and Pachos J K 2005 *Phys. Rev. Lett.* **95** 157203
[35] Zhu S -L 2006 *Phys. Rev. Lett.* **96** 077206
[36] Hamma A 2006 *arXiv:quant-ph/0602091*
[37] Patra A, Mukherjee V and Dutta A 2011 *J. Stat. Mech.* P03026
[38] Mukherjee V, Polkovnikov A, and Dutta A 2011 *Phys. Rev. B* **83** 075118
[39] Albuquerque A F, Alet F, Sire C, and Capponi S 2010 *Phys. Rev. B* **81** 064418
[40] Gritsev V and Polkovnikov A 2009 *arXiv:0910.3692*
[41] Gu S -J and Lin H -Q 2009 *Euro. Phys. Lett.* **87** 10003
[42] Jacobson N T, Garnerone S, Haas S, and Zanardi P 2009 *Phys. Rev. B* **79** 184427; Rezakhani A T, Abasto D F, Lidar D A, Zanardi P 2010 *Phys. Rev. A 82* 012321; Garnerone S, Abasto D, Haas S, and Zanardi P 2009 *Phys. Rev. A* **79** 032302 Garnerone S, Abasto D, Haas S, and Zanardi P 2009 *Phys. Rev. Lett.* **102** 057205; Lu X M, Sun Z,Wang X, and Zanardi P 2008 *Phys. Rev. A* **78** 032309; Venuti L , Saleur H and Zanardi P 2009 *Phys. Rev. B* **79** 092405 ; Abasto D, Hamma A, and Zanardi P 2008 *Phys. Rev. A* **78** 010301(R) ; Cozzini M, Ionicioiu R and Zanardi P 2007 *Phys. Rev. B* **76** 104420; Cozzini M, Giorda P, and Zanardi P 2007 *Phys. Rev. B* **75** 014439 ; Venuti L, Cozzini M, Buonsante P, Massel F,Bray-Ali N, and Zanardi P 2008 *Phys. Rev. B* **78** 115410
[43] Rams M M and Damski B 2011 *Phys. Rev. Lett.* **106** 055701
[44] Sengupta K and Sen D 2009 *Phys. Rev. A* **80** 032304
[45] Sen(De) A, Sen U, Lewenstein M 2005 *Phys. Rev. A.* **72** 052319; Wichterich H and Bose S 2009 *ibid* **79** 060302(R)
[46] Damle K and Sachdev S 1996 *Phys. Rev. Lett.* **76** 4412
[47] Peres A 1996 *Phys. Rev. Lett.* **77** 1413
[48] Lieb E, Schultz T, and Mattis D 1961 *Ann. Phys. (NY)* **16** 407
[49] Barouch E, McCoy B M and Dresden M 1970 *Phys. Rev. A.* **2** 1075; Barouch E and McCoy B M 1971 *Phys. Rev. A* **3** 786
[50] Kogut J B 1979 *Rev. Mod. Phys.* **51** 659
[51] Bunder J E and McKenzie R H 1999 *Phys. Rev. B.* **60** 344
[52] Divakaran U, Mukherjee V, Dutta A and Sen D 2009 *J. Stat. Mech.* P02007
[53] Deng S, Ortiz G and Viola L 2009 *Phys. Rev. B* **80** 241109(R)
[54] Mukherjee V and Dutta A 2010 *Eur. Phys. Lett.* **92** 37004
[55] Mondal S, Sengupta K, and Sen D 2009 *Phys. Rev. B* **79** 045128
[56] Zener C 1932 *Proc. Roy. Soc. London,Ser A* **137** 696 ; Landau L D and Lifshitz E M 1965 *Quantum Mechanics: Non-relativistic Theory*, 2nd Ed., (Pergamon Press, Oxford)
[57] Syljuasen O F 2003 *Phys. Rev. A* **68** 060301(R)

Journal of Physics: Conference Series **297** (2011) 012009 doi:10.1088/1742-6596/297/1/012009

# Explosive percolation in graphs

**Santo Fortunato**

Complex Networks and System Lagrange Laboratory, ISI Foundation, Viale S. Severo 65, 10133, Torino, Italy

E-mail: fortunato@isi.it

**Filippo Radicchi**

Chemical and Biological Engineering, Northwestern University, Evanston, IL 60208, USA

E-mail: f.radicchi@gmail.com

**Abstract.** Percolation is perhaps the simplest example of a process exhibiting a phase transition and one of the most studied phenomena in statistical physics. The percolation transition is continuous if sites/bonds are occupied independently with the same probability. However, alternative rules for the occupation of sites/bonds might affect the order of the transition. A recent set of rules proposed by Achlioptas et al. [*Science* **323**, 1453 (2009)], characterized by competitive link addition, was claimed to lead to a discontinuous connectedness transition, named "explosive percolation". In this work we survey a numerical study of the explosive percolation transition on various types of graphs, from lattices to scale-free networks, and show the consistency of these results with recent analytical work showing that the transition is actually continuous.

## 1. Introduction

Phase transitions lie at the heart of the modern development of statistical physics [1, 2]. They are changes in the state of order of a system and can be classified based on their properties in the immediate proximity of the critical point. Phase transitions are *continuous* (or second-order), if the order parameter changes continuously across the two phases, with an infinite correlation length and consequent power law decay of correlations and divergence of higher moments of the order parameter at the critical point. Otherwise one speaks of *discontinuous* (first-order) phase transitions, which are typically characterized by a discontinuous jump of the order parameter at the critical point.

The purely geometric process known as *random percolation* [3] offers a paradigmatic example of a continuous phase transition. The starting point is a graph, e. g. a lattice. The sites or the links of the lattice are occupied independently with some probability $p$. Nearest-neighboring occupied sites/links form structures called *clusters*. For low values of the occupation probability $p$ just a few small clusters are formed, but if $p$ increases the number and size of the clusters will increase as well. When the occupation probability exceeds a critical value $p_c$, a macroscopic cluster, occupying a finite fraction of the total number of sites/links, emerges. This macroscopic structure is called *percolation cluster* and its relative size $P$, the *percolation strength*, is the order parameter of the transition: $P = 0$ indicates the phase with only microscopic clusters, whereas $P > 0$ indicates the phase with (at least) one macroscopic cluster. Random percolation has been studied on lattices, random graphs [4] and scale-free networks [5, 6, 7, 8, 9]. Analytical and numerical studies have proved that the percolation transition is continuous, without exceptions. This however holds for random percolation. It cannot be excluded *a priori* that alternative processes of occupation of sites/links might lead to different types of geometric transitions.

In a recent paper [10], Achlioptas et al. have introduced a special set of rules, in which links are occupied as a result of a competitive process between pairs of links (*Achlioptas processes*). The idea is to slow down the process of cluster growth, by inserting links leading to the merge of small clusters. This can be done in several ways. Achlioptas et al. focused on the so-called *product rule* (PR): given a pair of links, randomly selected among those which are not yet occupied, one occupies the link merging the two clusters with smaller product size (Fig. 1). One could consider variants of this rule, like taking the sum instead of the product, or just the minimum size of the clusters of each pair. Also, the competition can be extended to more than two links. In any case, the result of such processes is a slow growth of the cluster sizes, which causes a delay in the onset of the percolation transition. On the other hand, since the density of links and, consequently, of clusters at the onset is higher than for random percolation, it is natural to expect that the percolation cluster has a very rapid growth. This is indeed confirmed by numerical studies; in fact, the growth of the percolation cluster is so quick that the percolation strength $P$ appears to vary discontinuously at the onset. The sudden jump in the order parameter has motivated the name "explosive percolation". In the last two years, Achlioptas processes have been extensively studied and meanwhile a lot is known about the explosive percolation transition [11, 12, 13, 14, 15, 16, 17, 18, 19, 20, 21, 22].

The main issue concerned the order of the transition: is it continuous or discontinuous? Achlioptas et al. claimed that it is discontinuous, and this is what has mostly attracted the attention of scholars. As a matter of fact, it was soon shown that, despite the alleged jump of the order parameter at the threshold $p_c$, the explosive percolation transition has peculiar features of continuous phase transitions, like power law distributions of cluster sizes [15, 17, 19] and power law scaling of the mean cluster size at $p_c$ [12, 13, 15]. Indeed, in recent works by Nagler et al. [21] and da Costa et al. [22] it was proven that the transition is actually continuous, and that the jump of the order parameter is only apparent, and due to the very small critical exponent of the order parameter $\beta$.

In this paper we summarize the finite size scaling analysis of Ref. [15], and verify that the results are indeed consistent with the analytical findings of da Costa et al.. In Section 2 we introduce finite size scaling, in Section 3 we discuss the results for different types of graphs. A summary is reported in Section 4.

## 2. Finite size scaling

Finite size scaling [23] is a well-known technique adopted in numerical studies of phase transitions. For continuous phase transitions, the correlation length is infinite at the critical threshold $p_c$, so every variable $X$ is scale-independent in the proximity of that point and has a power law form,

$$X \sim |p - p_c|^{\omega}, \tag{1}$$

where $\omega$ is a critical exponent. If the system has a finite size $N$, the variable $X$ near the threshold has the following scaling form

$$X = N^{-\omega/\nu} F\left[(p - p_c)\ N^{1/\nu}\right]. \tag{2}$$

In Eq. 2, $\nu$ is a critical exponent and $F$ a universal function. Exactly at the critical point $(p = p_c)$ the variable displays the simple scaling $X \sim N^{-\omega/\nu}$, which can be used to extract the exponents' ratio $\omega/\nu$, by using systems of different sizes. Moreover, if $p_c$, $\nu$ and $\omega$ are known,

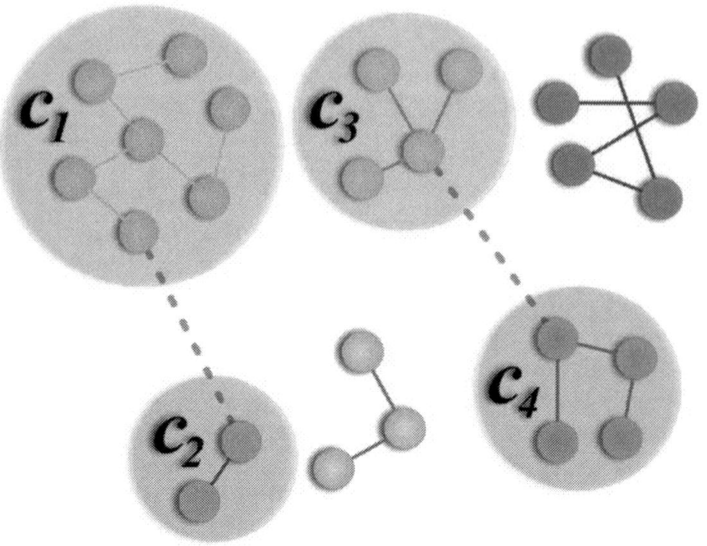


**Figure 1.** Scheme of an Achlioptas process with product rule. Two links (dashed lines) are selected at random among all possible pairs of non-adjacent nodes. The link joining the pair of clusters with the smaller product size is eventually added to the system. Here the winning link is that between clusters $c_1$ and $c_2$ (product size $7 \cdot 2 = 14 < 4 \cdot 4 = 16$).

the expression $XN^{\omega/\nu}$ as a function of $(p - p_c)\ N^{1/\nu}$ is just the universal function $F$, which is independent of $N$, so curves referring to different system sizes collapse.

We investigated the two main variables of percolation [3], i.e. the *percolation strength* $P$ and the *average cluster size* $S$. The percolation strength $P$, as we have said above, is the order parameter of the transition, and equals the relative size of the percolating cluster(s) with respect to the total system size $N$. While on lattices there are operative criteria to define a percolating cluster (e.g. if it runs from one edge of the lattice to the opposite one, say), on generic graphs this is not the case, so $P$ is defined as the relative size of the largest connected cluster. The scaling ansatz of the percolation strength is

$$P = N^{-\beta/\nu} F^{(1)} \left[(p - p_c)\ N^{1/\nu}\right]. \tag{3}$$

The average cluster size $S$ is defined as

$$S = \frac{\sum_s n_s s^2}{\sum_s n_s s}, \tag{4}$$

where $n_s$ is the number of clusters of size $s$ per node. The sums run over all possible values of $s$ except for the one of the largest cluster. The scaling ansatz of $S$ is

$$S = N^{\gamma/\nu} F^{(2)} \left[(p - p_c)\ N^{1/\nu}\right]. \tag{5}$$

The universal functions $F^{(1)}$ and $F^{(2)}$ of Eqs. (3) and (5) are not the same, but they are related.

In random percolation, the probability distribution $P(s)$ of cluster sizes (except the largest), decays at $p_c$ as the power law $P(s) \sim s^{-\tau}$ with the cluster size $s$. We have computed the cluster size distribution $P(s)$ at $p_c$ and measured the Fisher exponent $\tau$. For a given system, $P(s)$ is

Journal of Physics: Conference Series **297** (2011) 012009 doi:10.1088/1742-6596/297/1/012009

related to $n_s$ by the relation $P(s) = Nn_s/n_c$, where $n_c$ is the total number of "finite" clusters. We shall use the symbol $n_s$ to indicate $P(s)$ as well, but in the plots $n_s$ is normalized as $P(s)$, for consistency.

The percolation threshold $p_c$ is localized in two independent ways. The first method exploits the scaling of the pseudo-critical points $p_c(N)$

$$p_c = p_c(N) + bN^{-1/\nu} . \qquad (6)$$

By using several system sizes, one can perform a fit with the three parameters $b$, $\nu$ and $p_c$. The pseudocritical point for a system with finite size $N$ can be defined in several ways, for us it indicates the value of $p$ at which the average cluster size $S$ peaks.

An alternative procedure relies on Eq. (3). The percolation strength $P$ is plotted as a function of the system size $N$ for a given value of $p$. Since, for $p = p_c$ the scaling follows a power law, the correct value of the percolation threshold can be determined by finding the value of $p$ which yields the best power law fit.

## 3. Results

### *3.1. Implementing Achlioptas processes with product rule*

The starting point is just a graph with $N$ nodes and no links. Links are added one by one, according to the competitive rule previously described, i.e. by selecting each time a pair of links at random and picking the one yielding the smaller product for the sizes of the clusters it merges. For scale-free networks the situation is a bit more involved, and we describe it in Section 3.4. The procedure goes on until the desired density of links $p$ is reached. We defined $p$ as the number of links of the graph divided by the total number of links present in the graph when it has been "completed", i.e., after the addition of the last link. All graphs considered here are "sparse", i.e., the average degree $\langle k \rangle$, expressing the ratio between (twice) the number of links and the number of nodes $N$, does not depend on $N$. At time $t$ of the growth process there are exactly $t$ links in the system: their density $p$, according to our definition, is then $t/(N\langle k \rangle)$.

### *3.2. Lattices*

Achlioptas processes on the square lattice were first studied by Ziff [11], who found similar properties for the explosive transition as Achlioptas et al. had found for Erdös-Rényi random graphs. The criterion to assess the nature of the transition was the same one proposed by Achlioptas et al., namely the scaling with the system size $N$ of the transition window $\Delta p = p_2 - p_1$, where $p_2$ is the lowest value of $p$ for which $P > 0.5$ and $p_1$ the lowest value of $p$ for which $P > 1/\sqrt{N}$. In a recent paper Ziff has performed a finite size scaling analysis as well [18]. The results of our analysis are shown in Fig. 2. The order parameter $P$ yields a very small exponents' ratio $\beta/\nu$, compatible with zero [0.07(3)] (Fig. 2, left), which is what one would expect to find for a discontinuous transition. The average cluster size $S$, instead, has a non-trivial power law scaling at $p_c$, with exponent $\gamma/\nu = 1.7(1)$ (Fig. 2, center). This had first been observed by Cho *et al.* in scale-free networks [12]. Fig. 2 (right), showing the distribution of sizes $n_s$ for all clusters except the largest one, provides an explanation of the scaling of $S$. The distribution is a clear power law [exponent 1.9(1)], which is incompatible with a classic discontinuous transition, as it usually occurs for continuous transitions. Since $n_s$ is a power law, all variables derived from $n_s$, including the average cluster size $S$, have power law scaling.

In 3d-lattices, the general picture is consistent with that in two dimensions (Fig. 3). Again, the scaling at $p_c$ of the order parameter $P$ yields a very small exponent $\beta/\nu = 0.02(2)$, compatible with zero (Fig. 3, left). Still, $S$ scales with an exponent $\gamma/\nu = 2.1(1)$ (Fig. 3, center), again due to the power law shape of the distribution of cluster sizes (Fig. 3, right). The Fisher exponent $\tau = 1.99(4)$ is compatible with the value we measured in two dimensions [1.9(1)].

Journal of Physics: Conference Series **297** (2011) 012009 doi:10.1088/1742-6596/297/1/012009

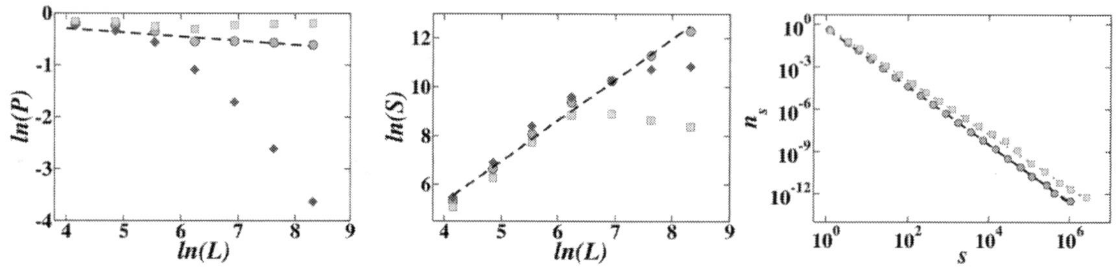

**Figure 2.** Explosive percolation on $2d$-lattices. (Left) Percolation strength $P$ as a function of the lattice side $L$ for three different values of the occupation probability: $p = 0.5256$ (violet diamonds), $p = 0.5266$ (orange circles) and $p = 0.5276$ (grey squares). The dashed line indicates the best fit obtained at the critical point $p = p_c = 0.5266(2)$, from which we get $\beta/\nu = 0.07(3)$. (Center) The average cluster size $S$ as a function of the lattice side $L$ for the same values of $p$ used for $P$. The dashed line has slope $\gamma/\nu = 1.7(1)$. (Right) Cluster size distributions measured at $p_c$ for PR (grey squares) and random percolation (orange circles). The exponents are $\tau = 1.9(1)$ for PR (red dotted line), while for random percolation $\tau = 2.05(1)$ (black dashed line). Simulations have been performed on lattices with side $L = 4096$.

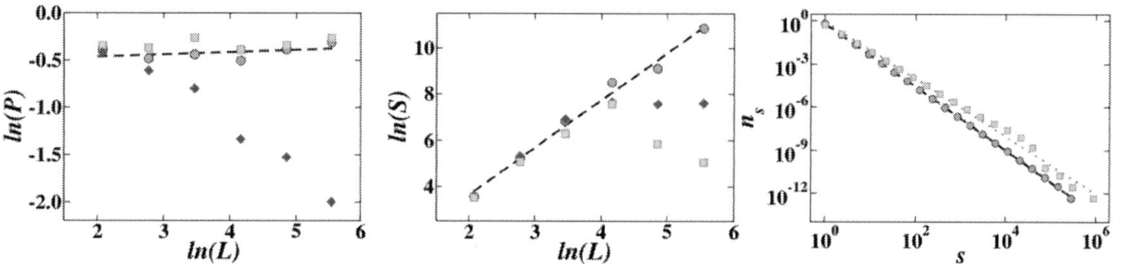

**Figure 3.** Explosive percolation on $3d$-lattices. (Left) Percolation strength $P$ as a function of the lattice side $L$ for three different values of the occupation probability: $p = 0.3866$ (violet diamonds), $p = 0.3876$ (orange circles) and $p = 0.3886$ (grey squares). The dashed line indicates the best fit obtained at the critical point $p = p_c = 0.3876(2)$, from which we get $\beta/\nu = 0.02(2)$. (Center) The average cluster size $S$ as a function of the lattice side $L$ for the same values of $p$ used for $P$. The dashed line has slope $\gamma/\nu = 2.1(1)$. (Right) Cluster size distributions at the critical point for PR (grey squares) and random percolation (orange circles). The exponents of the power law fits are $\tau = 1.99(4)$ for PR (red dotted line) and $\tau = 2.20(1)$ for random percolation (black dashed line). Simulations have been performed on lattices with side $L = 256$.

### *3.3. Erdös-Rényi graphs*

Fig. 4 presents the results of our analysis. For PR we see again a flat profile of the order parameter $P$ with $N$ ($\beta/\nu = 0.02(1)$), consistent with a discontinuous transition, along with a power law scaling of the average cluster size $S$, with exponent $\gamma/\nu = 0.48(4)$. The exponent $\tau = 2.08(5)$ (Fig. 4, right) is still compatible with the values found for PR on lattices, both in two and three dimensions (see Section 3.2 and Table 1).

### *3.4. Scale-free networks*

Many natural, social and man-made systems, if represented as graphs, display common features. The most striking is a broad distribution of the degree of the nodes, with a large majority of

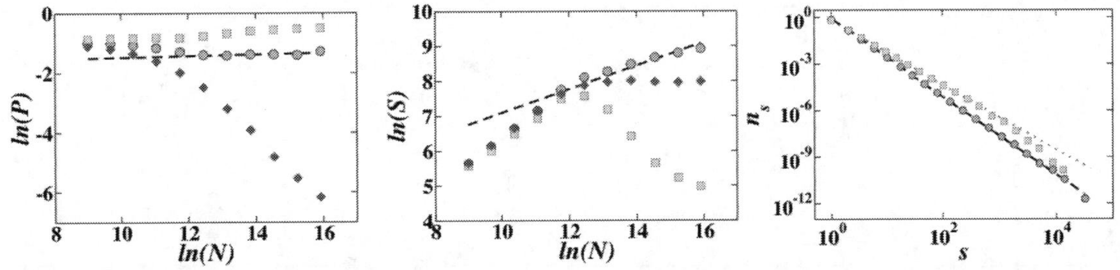

**Figure 4.** Explosive percolation on Erdös-Rényi random graphs. (Left) Percolation strength $P$ as a function of the network size $N$ for three different values of the occupation probability: $p = 0.8872$ (violet diamonds), $p = 0.8882$ (orange circles) and $p = 0.8892$ (grey squares). The dashed line indicates the best fit obtained at the critical point $p = p_c = 0.8882(2)$, from which we get $\beta/\nu = 0.02(1)$. (Center) The average cluster size $S$ as a function of the network size $N$ for the same values of $p$ used for $P$. The dashed line has slope $\gamma/\nu = 0.48(4)$. (Right) Cluster size distributions $n_s$ measured at the critical point for PR (grey squares) and random percolation (orange circles). The exponents are $\tau = 2.08(5)$ (red dotted line) for PR and $\tau = 2.51(2)$ (black dashed line) for random percolation. Simulations have been performed on graphs with $N = 8\,192\,000$.

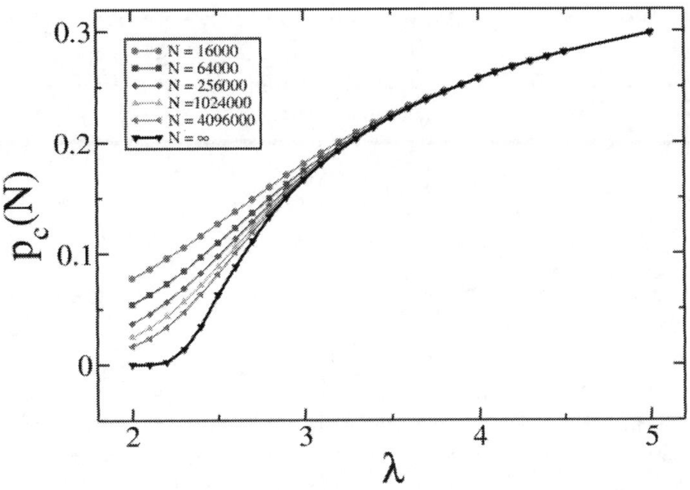


**Figure 5.** Explosive percolation on random SF networks. Percolation threshold $p_c(N)$ as a function of the degree exponent $\lambda$ for different network sizes $N$. The black line is the infinite size limit extrapolation of the critical threshold, performed by applying Eq. (6). The percolation threshold vanishes for $\lambda < \lambda_c \sim 2.3$, while it is non-zero for $\lambda > \lambda_c$.

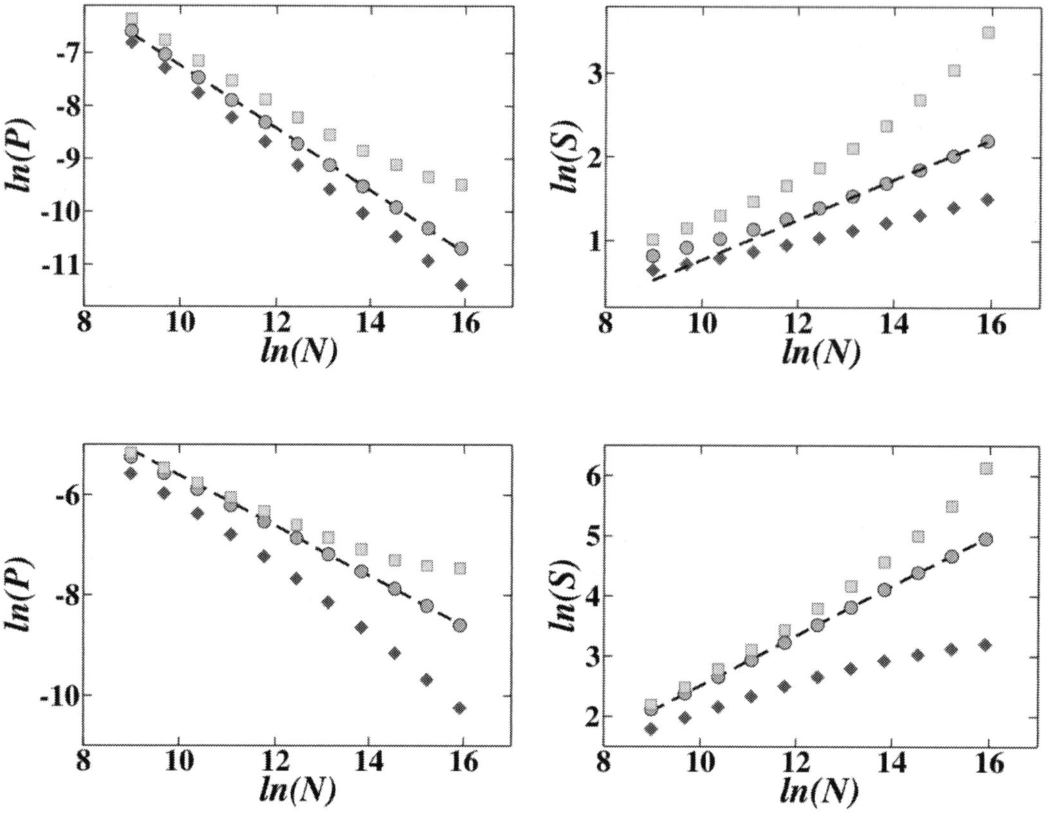


**Figure 6.** Explosive percolation on SF networks for degree exponents $\lambda = 2.5$ (top diagrams) and $\lambda = 2.8$ (bottom diagrams). (Left) Percolation strength $P$ as a function of the system size $N$ for three different values of the occupation probability. Top: $p = 0.0529$ (violet diamonds), $p = 0.0629$ (orange circles) and $p = 0.0729$ (grey squares). Bottom: $p = 0.1229$ (violet diamonds), $p = 0.1329$ (orange circles) and $p = 0.1349$ (grey squares). The dashed lines indicate the best fits obtained at the critical points $p = p_c = 0.0629(1)$ (top) and $p = p_c = 0.1329(1)$ (bottom), from which we get $\beta/\nu = 0.59(1)$ (top) and $\beta/\nu = 0.50(1)$ (bottom). (Right) The average cluster size $S$ as a function of the network size $N$ for the same values of $p$ used for $P$. The dashed lines have slopes $\gamma/\nu = 0.24(1)$ (top) and $\gamma/\nu = 0.42(1)$ (bottom). Simulations have been performed on graphs with $N = 8\,192\,000$.

nodes having low degree and a small subset of nodes having high degree. The tails of the degree distributions are often well fitted by power laws, which lack a characteristic scale, justifying the name "scale-free networks" given to such systems [24, 25, 26, 27, 28, 29, 30, 31]. We indicate the exponent of the power law degree distribution with $\lambda$. Nodes with large degree, called "hubs", have a fundamental role for the structure and dynamics of networks. For instance, they "keep" a large portion of the system in the same connected component, guaranteeing stability and compactness to the structure. In particular, in random scale-free (SF) networks with degree exponent $\lambda < 3$, there are many hubs and a very small fraction of links (vanishing in the limit of systems of infinite size) manages to keep a macroscopic fraction of nodes of the graph together in the same connected component. This amounts to saying that the percolation threshold is zero [5, 6, 7, 8, 9].

The Achlioptas process with product rule for SF networks is implemented as follows. We

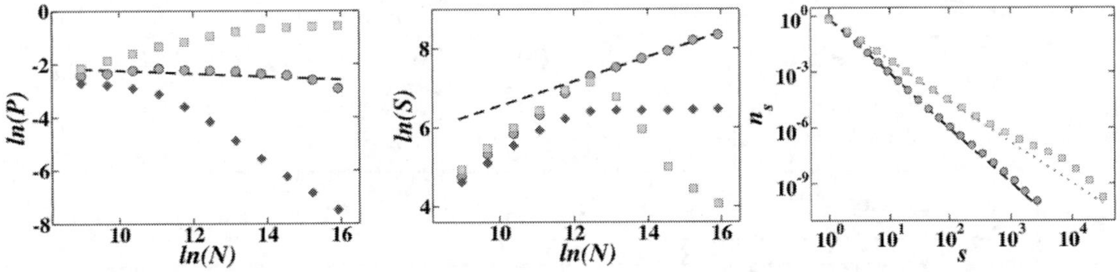


**Figure 7.** Explosive percolation on SF networks for degree exponent $\lambda = 3.5$. (Left) Percolation strength $P$ as a function of the system size $N$ for three different values of the occupation probability: $p = 0.2214$ (violet diamonds), $p = 0.2224$ (orange circles) and $p = 0.2234$ (grey squares). The dashed line indicates the best fit obtained at the critical point $p = p_c = 0.2224(2)$, from which we get $\beta/\nu = -0.06(3)$. (Center) The average cluster size $S$ as a function of the network size $N$ for the same values of $p$ used for $P$. The dashed line has slope $\gamma/\nu = 0.40(9)$. (Right) Cluster size distributions $n_s$ measured at the critical point for PR (grey squares) and random percolation (orange circles). The exponents are $\tau = 2.2(1)$ (red dotted line) for PR and $\tau = 2.94(1)$ (black dashed line) for random percolation. Simulations have been performed on graphs with $N = 8\,192\,000$.

start with a set of $N$ nodes and a given degree sequence $\{k_1, k_2, \ldots, k_N\}$. The degrees of the sequence are extracted from a power law distribution with exponent $\lambda$. The average degree $\langle k \rangle$ is set equal to 5. Every node $i$ holds $k_i$ stubs (half-links), where $k_i$ is the degree of $i$. By attaching such stubs in pairs one builds a network with the desired power law degree distribution with exponent $\lambda$. If stubs are randomly connected we would have the simple configuration model [32], which yields a random SF network. For the Achlioptas process at each iteration two pairs of stubs are selected and the PR determines which pair of stubs has to be eventually joined in a link (the PR applies as shown in Fig. 1).

The first remarkable result concerns the threshold. For random percolation it is zero for $\lambda < 3$ and non-zero for $\lambda > 3$ [5]. However, for explosive percolation, the threshold is non-zero already for $\lambda > \lambda_c \sim 2.3$ (Fig. 5) [12, 13].

In Fig. 6 we show the scalings at $p_c$ of $P$ and $S$ for $\lambda = 2.5$ and 2.8. Here we see an important difference with the cases of lattices and Erdös-Rényi random graphs, namely that the scaling of $P$ at $p_c$ is non-trivial, as $P$ decreases with $N$ as a power law in both cases. This is very different from what one would find for a discontinuous transition, in which $P$ would be approximately constant with $N$. For $S$ we also find power law scaling (Fig. 6, right panels), as we had previously seen for lattices and Erdös-Rényi random graphs. Overall, the transition looks like a standard continuous transition.

For $\lambda > 3$, however, the scenario changes. In Fig. 7 we show the results of our finite size scaling analysis for SF networks with exponent $\lambda = 3.5$. For random percolation on SF networks it is well known [33] that $\beta/\nu = 1/(\lambda - 1)$ and $\gamma/\nu = (\lambda - 3)/(\lambda - 1)$ for $3 \leq \lambda \leq 4$. For $\lambda > 4$, we are in the mean field limit and the exponents are independent of the degree exponent $\lambda$: $\beta/\nu = \gamma/\nu = 1/3$. These values coincide with the exponents for the percolation transition of Erdös-Rényi random networks. SF networks are very similar to Erdös-Rényi random networks in the limit $\lambda \to \infty$. The explosive percolation transition looks as on lattices and ER random networks. The scaling of $P$ at $p_c$ is trivial (Fig. 7, left), with $\beta/\nu = -0.06(3)$, which is essentially zero. The power law scaling of $S$ at $p_c$ is non-trivial (Fig. 7, center), with $\gamma/\nu = 0.40(9)$. The Fisher exponent $\tau = 2.2(1)$ (Fig. 7, right).

| System | Model | $p_c$ | $\beta/\nu$ | $\gamma/\nu$ | $\tau$ |
|---|---|---|---|---|---|
| $2d$-lattice | RP | 0.5 | 0.11(1) | 1.76(1) | 2.05(1) |
| | PR | 0.5266(2) | 0.07(3) | 1.7(1) | 1.9(1) |
| $3d$-lattice | RP | 0.2488(3) | 0.48(1) | 2.0(1) | 2.20(1) |
| | PR | 0.3876(2) | 0.02(2) | 2.1(1) | 1.99(4) |
| ER network | RP | 0.5 | 0.33(1) | 0.34(1) | 2.51(2) |
| | PR | 0.8882(2) | 0.02(1) | 0.48(4) | 2.08(5) |
| SF network $\lambda = 2.5$ | RP | 0 | – | – | – |
| | PR | 0.0629(1) | 0.59(1) | 0.24(1) | 2.15(2) |
| SF network $\lambda = 2.8$ | RP | 0 | – | – | – |
| | PR | 0.1329(1) | 0.50(1) | 0.42(1) | 2.13(6) |
| SF network $\lambda = 3.5$ | RP | 0.078(1) | 0.38(1) | 0.15(2) | 2.94(1) |
| | PR | 0.2224(2) | −0.06(3) | 0.40(9) | 2.2(1) |

**Table 1.** Percolation thresholds and critical exponents for random percolation (RP) and Achlioptas process with product rule (PR).

## 4. Summary

We have found two different classes of phase transitions for explosive percolation, that we shall call class A and B. Class A includes ER random networks (the original system studied by Achlioptas et al. [10]), lattices and SF networks with degree exponent $\lambda > 3$. Here we have observed an apparent saturation of the order parameter $P$ at $p_c$ with the size of the system $N$, due to an exponent $\beta$ that is very small, compatible with zero. Class B includes SF networks with $\lambda < 3$, where there seems to be no room for a discontinuous transition, as the results of the finite size scaling analysis are fully compatible with a standard continuous transition. Still, apart from the anomalous scaling behavior of the order parameter $P$ at $p_c$ observed in class A, we found that the other percolation variables display power law scaling in all cases, without exceptions, just as in second-order phase transitions. The most striking feature is the fact that the size distribution $n_s$ of the "finite" clusters at $p_c$ is a power law, not exponential or Gaussian as one expects from first-order phase transitions.

Da Costa et al. [22] have recently proved that an aggregation process in the same spirit as Achlioptas processes leads to a continuous phase transition, characterized by an exponent $\beta$ for the order parameter $P$ that is very small, $\beta \sim 0.0555$. They also demonstrated that the fact that the transition is continuous for the special process they considered enforces the same type of transition for Achlioptas processes too, with generally different but still small values for $\beta$. This is fully consistent with what we have found for the systems of class A. Da Costa et al. have derived important relations for the critical exponents. For instance, $\beta/\nu = \beta/(4\beta + 1)$ and the Fisher exponent $\tau = 2 + \beta/(3\beta + 1)$. For $\beta \sim 0.0555$ the exponents' ratio $\beta/\nu \sim 0.0455$. This value is compatible, within errors, with the values of $\beta/\nu$ computed for the systems of class A (see Table 1). So, the apparent saturation of the order parameter $P$ with the system size $N$ could be indeed due to the smallness of the exponent $\beta$. Moreover, the Fisher exponent $\tau$ for every transition we have investigated is very close to 2. From the relation $\tau = 2 + \beta/(3\beta + 1) < 2 + \beta$, and the fact that $\beta$ is small, we see that $\tau$ will always be very close to 2, just as we found. The difference between $\tau$ and 2 is hard to determine numerically. Likewise, the values of $\tau$ derived from different Achlioptas processes can be hardly distinguished from each other. We remark that this holds for lattices, Erdös-Rényi graphs and SF networks with degree exponent $\lambda > 3$, while the explosive percolation transition on SF networks with $\lambda < 3$ appears as something very different ($\beta$ is not small here), and deserves further investigations in the future.

We have also found that there is a non-zero explosive percolation threshold for SF networks for $\lambda > \lambda_c \sim 2.3$ (Fig. 5), while the threshold for random percolation is zero as long as $\lambda \leq 3$. Cho *et al.* [12] suggested that this is due to the non-random addition of links during the Achlioptas process, because of which the degree distribution of the system during the growth deviates from that imposed by construction, which will be eventually reached at the end of the process. This has been indeed verified numerically [12, 15].

## Acknowledgments

S. F. gratefully acknowledges ICTeCollective, grant number 238597 of the European Commission.

## References

[1] Stanley HE 1971, *Introduction to Phase Transitions and Critical Phenomena,* (New York: Oxford University Press)
[2] Binney JJ, Dowrick NJ, Fisher AJ and Newman MEJ 1992, *The Theory of Critical Phenomena: an Introduction to the Renormalization Group,* (New York: Oxford University Press)
[3] Stauffer D and Aharony A 1994, *Introduction to Percolation Theory,* (London: Taylor & Francis)
[4] Erdös P and Rényi A 1960 *Publ. Math. Inst. Hungar. Acad. Sci.* **5** 17
[5] Cohen R, Erez K, ben-Avraham D and Havlin S 2000 *Phys. Rev. Lett.* **85** 4626
[6] Newman MEJ, Strogatz SH and Watts DJ 2001 *Phys. Rev. E* **64** 026118
[7] Pastor-Satorras R, Vazquez A and Vespignani A 2000 *Phys. Rev. Lett.* **87** 258701
[8] Dorogovtsev SN, Goltsev AV and Mendes JFF 2008 *Rev. Mod. Phys.* **80** 1275
[9] Vazquez A and Moreno Y 2004 *Phys. Rev. E* **67** 015101(R)
[10] Achlioptas D, D'Souza RM and Spencer J 2009 *Science* **323** 1453
[11] Ziff RM 2009 *Phys. Rev. Lett.* **103** 045701
[12] Cho YS, Kim JS, Park J, Kahng B and Kim D 2009 *Phys. Rev. Lett.* **103** 135702
[13] Radicchi F and Fortunato S 2009 *Phys. Rev. Lett.* **103** 168701
[14] Friedman EJ and Landsberg AS 2009 *Phys. Rev. Lett.* **103** 255701
[15] Radicchi F and Fortunato S 2010 *Phys. Rev. E* **81** 036110
[16] Moreira AA, Oliveira EA, Reis SDS, Herrmann HJ and Andrade JS 2010 *Phys. Rev. E* **81** 040101(R)
[17] D'Souza RM and Mitzenmacher M 2010 *Phys. Rev. Lett.* **104** 195702
[18] Ziff RM 2010 *Phys. Rev. E* **82** 051105
[19] Cho YS, Kim S-W, Noh JD, Kahng B and Kim D 2010 *Phys. Rev. E* **82** 042102
[20] Araújo NAM and Herrmann HJ 2010 *Phys. Rev. Lett.* **105** 035701
[21] Nagler J, Levina A and Timme M 2010 *Nat. Phys* in press
[22] da Costa RA, Dorogovtsev SN, Goltsev AV and Mendes JFF 2010 *Phys. Rev. Lett.* **105** 255701
[23] Landau DP and Binder K 2000, *A Guide to Monte Carlo Simulations in Statistical Physics,* (Cambridge: Cambridge University Press)
[24] Albert R and Barabási AL 2002 *Rev. Mod. Phys.* **74** 47
[25] Dorogovtsev SN and Mendes JFF 2002 *Adv. Phys.* **51** 1079
[26] Newman MEJ 2003 *SIAM Rev.* **45** 167
[27] Pastor-Satorras R and Vespignani A 2004, *Evolution and Structure of the Internet: A Statistical Physics Approach,* (New York: Cambridge University Press)
[28] Boccaletti S, Latora V, Moreno Y, Chavez M and Hwang DU 2006 *Phys. Rep.* **424** 175
[29] Caldarelli G 2007, *Scale-free networks,* (Oxford: Oxford University Press)
[30] Barrat A, Barthélemy M and Vespignani A 2008, *Dynamical processes on complex networks,* (Cambridge: Cambridge University Press)
[31] Cohen R and Havlin S 2010, *Complex networks: Structure, Robustness and Function,* (Cambridge: Cambridge University Press)
[32] Molloy M and Reed BA 1995 *Random Struct. Algorithms* **6** 161
[33] Cohen R, ben-Avraham D and Havlin S 2002 *Phys. Rev. E* **66** 036113

Journal of Physics: Conference Series **297** (2011) 012010 doi:10.1088/1742-6596/297/1/012010

# Scale-free coordination number disorder and multifractal size disorder in weighted planar stochastic lattice

**M. K. Hassan[1], M. Z. Hassan[2] and N. I. Pavel[1]**

1 Department of Physics, University of Dhaka, Curzon Hall, Dhaka 1000, Bangladesh
2 Institute of Computer Science, Bangladesh Atomic Energy Commission, Dhaka 1000, Bangladesh

E-mail: khassan@univdhaka.edu

**Abstract.**
The square lattice is perhaps the simplest cellular structure. In this work, however, we investigate the various structural and topological properties of the kinetic and stochastic counterpart of the square lattice and termed them as kinetic square lattice (KSL) and weighted planar stochastic lattice (WPSL) respectively. We find that WPSL evolves following several non-trivial conservation laws, $\sum_i^N x_i^{n-1} y_i^{\frac{4}{n}-1} = \text{const.}\ \forall\ n$, where $x_i$ and $y_i$ are the length and width of the $i$th block. The KSL, on the other hand, evolves following only one conservation law, namely the total area, although one find three apparently different conserved integrals which effectively the total area. We show that one of the conserved quantity of the WPSL obtained either by setting $n = 1$ or $n = 4$ can be used to perform multifractal analysis. For instance, we show that if the $i$th block is populated with either $p_i \sim x_i^3$ or $p_i \sim y_i^3$ then the resulting distribution in the WPSL exhibits multifractality. Furthermore, we show that the dual of the WPSL, obtained by replacing each block with a node at its center and common border between blocks with an edge joining the two vertices, emerges as a scale-free network since its degree distribution exhibits power-law $P(k) \sim k^{-\gamma}$ with exponent $\gamma = 5.66$. It implies that the coordination number distribution of the WPSL is scale-free in character as we find that $P(k)$ also describes the fraction of blocks having $k$ neighbours.

## 1. Introduction

The formation of planar cellular structures have attracted considerable interest among scientists in general and physicists in particular because cellular structures are ubiquitous in nature. Especially, the class of non-equilibrium systems with apparent disordered patterns have come under close scrutiny. Examples of cellular structures include acicular texture in martensite growth, tessellated pavement on ocean shores, agricultural land division according to ownership, grain texture in polycrystals, cell texture in biology, soap froths and so on [1, 2, 3, 4]. Most existing theoretical models which can mimic such systems are typically formed by tessellation, tiling, or subdivision of a plane into domains of contiguous and nonoverlapping cells. For instance, Voronoi lattice is formed by partitioning a plane into mutually exclusive convex polygons, Apollonian packing on the other hand is formed by tiling a plane into contiguous and non-overlapping disks, etc. These models have found widespread applications in physics and biology [5, 6]. Significant progress has already been made in understanding various properties of both Voronoi lattice and Apollonian packing.

A fact worth mentioning is that most of the existing models for generating cellular structures have one thing in common and that is: the number of neighbours of a cell never exceeds the number of sides it has. Another important property of the existing cellular structures is that they have small mean coordination number where it is almost impossible to find cells that have significantly higher or fewer neighbours than the average number of neighbours. However, geologists and soil scientists observe an abundance of cellular patterns ranging from topographical ground formations and underwater stone arrangements where number of neighbours may exceeds the number of sides of the cell. Typically, these structures emerge through evolution and in the process each cell may acquire more neighbours than the number of sides the corresponding cell has. Such disordered structures can be of great interest in physics provided they have some global topological and geometrical properties since they can mimic kinetic disordered medium on which one can study percolation and random walk problems.

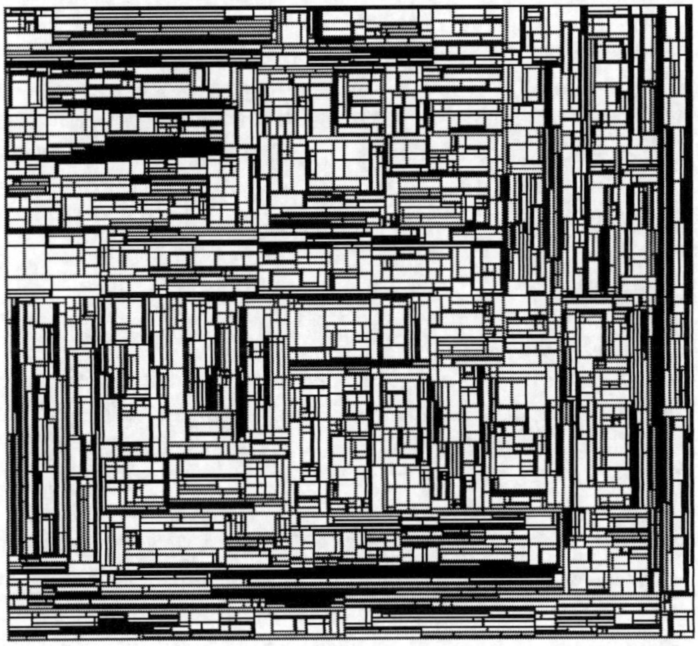

**Figure 1.** A snapshot of the weighted planar stochastic lattice containing 30001 blocks.

Perhaps, the square lattice is the simplest example of the cellular structure where every cell has the same size and the same coordination number. Its construction starts with an initiator, say a square of unit area, and a generator that divides it into four equal parts. In the next step and steps thereafter the generator is applied to all the available blocks which eventually generates a square lattice. In this talk, we intend to address the following questions. Firstly, what if the generator is applied to only one of the available block at each step by picking it preferentially with respect to the areas? Secondly, what if we use a modified generator that divides the initiator randomly into four blocks instead of four equal parts and apply it to only one of the available block at each step, which are again picked preferentially with respect to their respective areas [7]? Our primary focus, however, will be on the later case which results in the tiling of the initiator into increasingly smaller mutually exclusive rectangular blocks. The process is so simple that even a few steps by hand on a piece of paper can lead to the conclusion that the blocks in the resulting lattice may have more neighbours than the number of sides of a cell. We term the resulting structure as *weighted planar stochastic lattice* (WPSL) since the spatial randomness is incorporated by the modified generator and also the time is incorporated in it by the sequential application of the modified generator. The definition of the model may appear too simple but the results it offers are found to be far from simple. To illustrate the type of systems expected, we show a snapshot of the resulting weighted planar stochastic lattice taken during the evolution (see figure 1). We intend to investigate its topological and geometrical properties in an attempt to find some order in this seemingly disordered lattice.

The model in question can also describe the kinetics of fragmentation in two dimensions as it can be defined as follows [8, 9]. At each time step one may assume that a seed is being nucleated randomly on the initiator and hence the greater the area of the block, the higher is the probability that it contains the seed. Upon nucleation two orthogonal cracks parallel to the sides of the block are grown until intercepted by existing cracks which divides it randomly into four rectangular fragments. In reality though, fragments are produced in fracture of solid objects by propagation of interacting cracks resulting in the fragments of arbitrary size and shape which makes it a formidable mathematical problem. However, the present model can be considered as the minimum model which should be capable of capturing the essential generic aspects of the underlying mechanism. In addition, Ben-Naim and Krapivsky have noted that the WPSL can also describe kinetics of martensite formation. A clear testament to this can be found if one compares figure 1 of our work and figure 2 of the work of Rao *et al.* in the context of martensite formation [1, 10]. Yet another application, albeit a little exotic, is the random search tree problem in computer science [11].

A truly disordered lattice where the coordination number disorder and the block size disorder are introduced naturally is perhaps long over due. On the other hand, where there is a disorder physicists have the natural tendency to look for an order in it. To this end, we invoke the idea of multifractality to quantify the size disorder and the idea of scale-invariance to quantify the coordination number disorder. Firstly, we characterize the blocks of the lattice by the size of their respective length $x$ and width $y$. Furthermore, we identify each blocks by labelling them as $i = 1, 2, ..., N$ and the corresponding length and width by $(x_1, y_1)$, $(x_2, y_2)$, ..., $(x_N, y_N)$ respectively. We then show that the dynamics of the WPSL is governed by infinitely many conservation laws, namely the numerical value of $\sum_i^N x_i^{n-1} y_i^{4/n-1}$ remains the same regardless of the lattice size $N$ and the value of $n$ where $n = 2$ corresponds to the conservation of total area. Of the non-trivial conservation laws, we single out either $\sum_i^N y_i^3$ by setting $n = 1$ or $\sum_i^N x_i^3$ by setting $n = 4$ as we can use it to perform multifractal analysis. For instance, we show that if the blocks are populated according to the cubic power of their own length (or width) then the distribution of the population in the WPSL emerges as multifractal. On the other hand, if the blocks are characterized by the coordination number or the number of neighbours $k$ with whom the block have common sides then we show that the coordination number distribution function $\rho(k)$, the probability that a block picked at random have coordination number $k$, have power-law tail. A power-law distribution function is regarded as scale-free since it looks the same regardless of the scale we use to look at it.

The organization of this talk is as follows. In section 2, we give the exact algorithm of the model. In section 3, we give the geometric properties of the WPSL in an attempt to quantify the annealed size disorder and at the same time we proposed kinetic square lattice in an attempt to look for a possible origin of multifractality. In section 4, we discuss the various structural topological properties of the WPSL and its dual are discussed in order to quantify the annealed coordination number disorder. Finally, section 5 gives a short summary of our results.

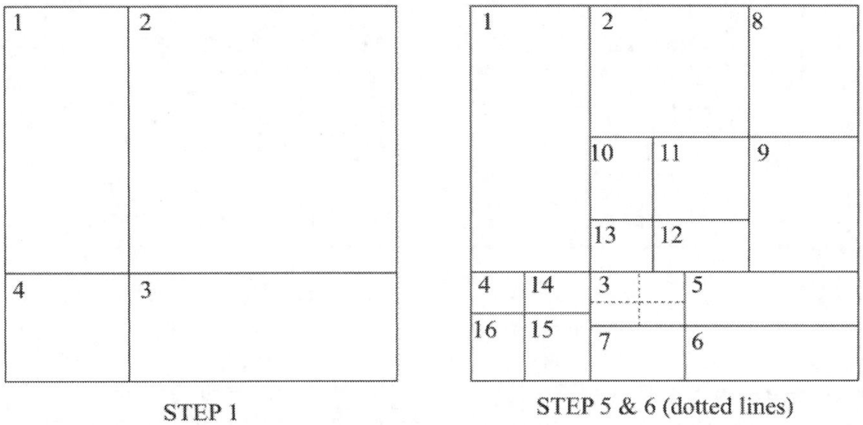


**Figure 2.** Schematic illustration of the first few steps of the algorithm.

## 2. Algorithm of the weighted planar stochastic lattice (WPSL)

Perhaps an exact algorithm can provide a better description of the model than the mere definition. In step one, the generator divides the initiator, say a square of unit area, randomly into four smaller blocks. The four newly created blocks are then labelled by their respective areas $a_1, a_2, a_3$ and $a_4$ in a clockwise fashion starting from the upper left block (see Fig. 2). In step two and thereafter only one block is picked at each step with probability equal to their respective area and then it is divided randomly into four blocks. In general, the $j$th step of the algorithm can be described as follows.

**(i)** Subdivide the interval $[0, 1]$ into $(3j - 2)$ subintervals of size $[0, a_1]$, $[a_1, a_1 + a_2]$, ..., $[\sum_{i=1}^{3j-3} a_i, 1]$ each of which represent the blocks labelled by their areas $a_1, a_2, ..., a_{(3j-2)}$ respectively.

**(ii)** Generate a random number $R$ from the interval $[0, 1]$ and find which of the $(3i - 2)$ subinterval contains this $R$. The corresponding block it represents, say the $p$th block of area $a_p$, is picked.

**(iii)** Calculate the length $x_p$ and the width $y_p$ of this block and keep note of the coordinate of the lower-left corner of the $p$th block, say it is $(x_{low}, y_{low})$.

**(iv)** Generate two random numbers $x_R$ and $y_R$ from $[0, x_p]$ and $[0, y_p]$ respectively and hence the point $(x_R + x_{low}, y_R + y_{low})$ mimicking a random nucleation of a seed in the block $p$.

**(v)** Draw two perpendicular lines through the point $(x_R + x_{low}, y_R + y_{low})$ parallel to the sides of the $p$th block mimicking orthogonal cracks parallel to the sides of the blocks which stops growing upon touching existing cracks and divide it into four smaller blocks. The label $a_p$ is now redundant and hence it can be reused.

**(vi)** Label the four newly created blocks according to their areas $a_p$, $a_{(3j-1)}$, $a_{3j}$ and $a_{(3j+1)}$ respectively in a clockwise fashion starting from the upper left corner.

**(vii)** Increase time by one unit and repeat the steps (i) - (vii) *ad infinitum.*

### 3. Geometric properties of WPSL

In general, the distribution function $C(x, y; t)$ describing the blocks of the lattice by their length $x$ and width $y$ evolves according to the following kinetic equation [8, 9]

$$\frac{\partial C(x,y;t)}{\partial t} = -C(x,y;t)\int_0^x\int_0^y dx_1dy_1F(x_1,x-x_1,y_1,y-y_1) + 4\int_x^\infty\int_y^\infty C(x_1,y_1;t)F(x,x_1-x,y,y_1-y)dx_1dy_1, \quad (1)$$

where kernel $F(x_1, x_2, y_1, y_2)$ determines the rules and the rate at which the block of sides $(x_1 + x_2)$ and $(y_1 + y_2)$ is divided into four smaller blocks whose sides are the arguments of the kernel. The first term on the right hand side of equation (1) represents the loss of blocks of sides $x$ and $y$ due to nucleation of seed of crack on one such block from which mutually perpendicular cracks are grown to divide it into four smaller blocks. Similarly, the second term on the right hand side represents the gain of blocks of sides $x$ and $y$ due to nucleation of seed of crack on a block of sides $x_1$ and $y_1$ ensuring that one of the four new blocks have sides $x$ and $y$. Let us now consider the case where the generator divides the initiator randomly into four smaller rectangles and apply it to only one of the available squares thereafter by picking preferentially with respect to their areas. It effectively describes the random sequential nucleation of seeds with uniform probability on the initiator. Within the rate equation approach this can be ensured if one chooses the following kernel

$$F(x_1, x_2, y_1, y_2) = 1 \quad (2)$$

Substituting it into equation (1) we obtain

$$\frac{\partial C(x,y;t)}{\partial t} = -xyC(x,y;t) + 4\int_x^\infty\int_y^\infty C(x_1,y_1;t)dx_1dy_1. \quad (3)$$

The coefficient $xy$ of $C(x, y; t)$ of the loss term implies that seeds of cracks are nucleated on the blocks preferentially with respect to their areas which is consistent with the definition of our model.

Incorporating the 2-tuple Mellin transform given by

$$M(m,n;t) = \int_0^\infty\int_0^\infty x^{m-1}y^{n-1}C(x,y;t)dxdy, \quad (4)$$

in equation (3) we get

$$\frac{dM(m,n;t)}{dt} = \Big(\frac{4}{mn} - 1\Big)M(m+1,n+1;t). \quad (5)$$

Iterating equation (5) to get all the derivatives of $M(m, n; t)$ and then substituting them into the Taylor series expansion of $M(m, n; t)$ about $t = 0$ one can immediately write its solution in terms of generalized hypergeometric function [12]

$$M(m,n;t) = {}_2F_2\Big(a_+, a_-; m, n; -t\Big), \quad (6)$$

where $M(m, n; t) = M(n, m; t)$ for symmetry reason and

$$a_\pm = \frac{m+n}{2} \pm \Big[\Big(\frac{m-n}{2}\Big)^2 + 4\Big]^{\frac{1}{2}}. \quad (7)$$

One can see that (i) $M(1,1;t) = 1+3t$ is the total number of blocks $N(t)$ and (ii) $M(2,2;t) = 1$ is the sum of areas of all the blocks which is obviously a conserved quantity. Both properties are again consistent with the definition of the WPSL depicted in the algorithm. The behaviour of $M(m,n;t)$ in the long time limit is

$$M(m,n;t) \sim t^{-a_-}. \tag{8}$$

Thus, in addition to the conservation of total area the system is also governed by infinitely many non-trivial conservation laws as it implies

$$M(n,4/n;t) \sim \text{constant} \quad \forall n. \tag{9}$$

We used numerical simulation to verify equation (9) or its discrete counterpart $\sum_i^N x_i^{n-1} y_i^{4/n-1}$ if we label all the available blocks as $i = 1,2,...,N$. We found that the analytical solution is in perfect agreement with the numerical simulation which we performed based on the algorithm for the WPSL model (see figure 3).

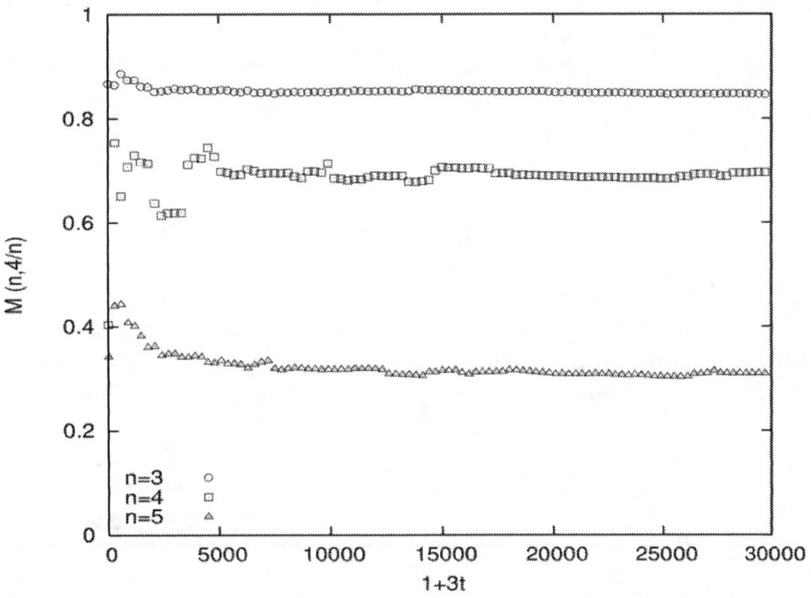


**Figure 3.** The plots of $\sum_i^N x_i^{n-1} y_i^{4/n-1}$ vs $N$ for $n = 3,4,5$ are drawn using data collected from one realization.

We now find it interesting to focus on the distribution function $n(x,t) = \int_0^\infty C(x,y,t)dy$ that describes the concentration of blocks which have length $x$ at time $t$ regardless of the size of their widths $y$. Then the $q$th moment of $n(x,t)$ is defined as

$$M_q(t) = \int_0^\infty x^q n(x,t)dx. \tag{10}$$

Appreciating the fact that $M_q(t) = M(q+1,1;t)$ and using equation (8) one can immediately write its solution

$$M_q(t) \sim t^{\{\sqrt{q^2+16}-(q+2)\}/2}. \tag{11}$$

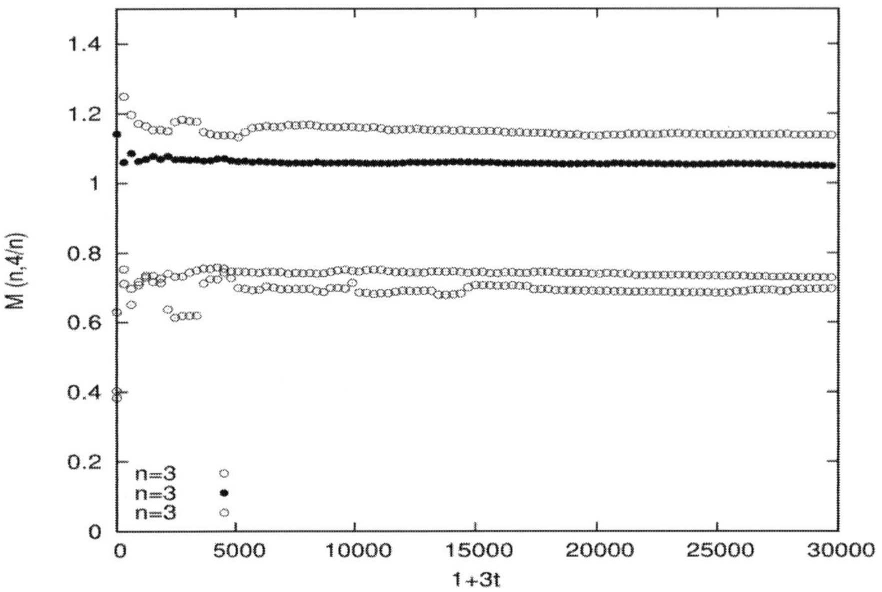


**Figure 4.** The plots of $\sum_i^N x_i^3$ vs $N$ for four different realizations shows that the numerical value is different at every independent realization.

Note that for symmetry reason it does not matter whether we consider the $q$th moments of $n(x,t)$ or that of $n(y,t)$ since we have $M(q+1,1;t) = M(1,q+1;t)$. According to equation (11) the quantity $M_3(t)$ and hence $\sum_i^N x_i^3$ or $\sum_i^N y_i^3$ is a conserved quantity. However, although $M_3(t)$ remains a constant against time in every independent realization the exact numerical value is found to be different in every different realization (see figure 4). It clearly indicates lack of self-averaging or wild fluctuation. This is supported by our analytical solution as we find that

$$< x^q >= \frac{\int_0^\infty x^q n(x,t)dx}{\int_0^\infty n(x,t)dx} \neq < x >^q = \Big(\frac{\int_0^\infty xn(x,t)dx}{\int_0^\infty n(x,t)dx}\Big)^q, \tag{12}$$

which suggest that a single length scale cannot characterize all the moments of the distribution function $n(x,t)$.

Of all the conservation laws we find that $M_3(t) = \sum_i x_i^3$ is a special one since we can use it as a multifractal measure consisting of members $p_i$, the fraction of the total measure $p_i = x_i^3/\sum_i x_i^3$, distributed on the geometric support WPSL. That is, we assume that the $i$th block is occupied with cubic power of its own length $x_i$. The corresponding "partition function" of multifractal formalism then is [13]

$$Z_q(t) = \sum_i p_i^q \sim M_{3q}(t). \tag{13}$$

Its solution can immediately be obtained from equation (11) to give

$$Z_q(t) \sim t^{\{\sqrt{9q^2+16}-(3q+2)\}/2}. \tag{14}$$

Using the square root of the mean area $\delta(t) = \sqrt{M(2,2;t)/M(1,1;t)} \sim t^{-1/2}$ as the yard-stick to express the partition function $Z_q$ gives the weighted number of squares $N(q,\delta)$ needed to cover the measure which we find decays following power-law

$$N(q,\delta) \sim \delta^{-\tau(q)}, \tag{15}$$

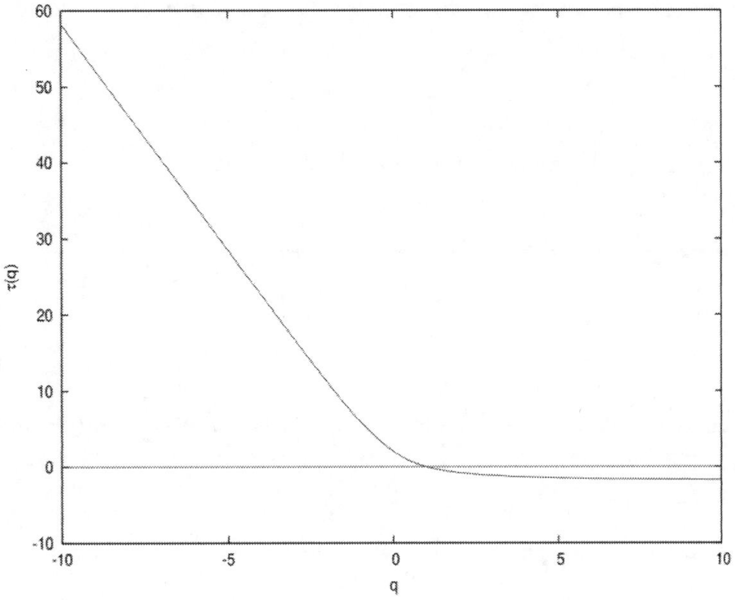


**Figure 5.** The plots of $\tau(q)$ vs $q$ to show its slope as a function of $q$ varies.

where the mass exponent

$$\tau(q) = \sqrt{9q^2 + 16} - (3q + 2). \tag{16}$$

The non-linear nature of $\tau(q)$, see figure 5 for instance, suggests that the gap exponent

$$\Delta = \tau(q) - \tau(q - 1) \tag{17}$$

is different for every $q$ value. It implies that we require an infinite hierarchy of exponents to specify how the moments of the probabilities $\{p\}$s scales with $\delta$. Now if we choose $q = 0$ then it gives an estimate of the number of squares $N(0, \delta = N(\delta)$ of sides $\delta$ we need to cover the support on which the members of the population is distributed. We find that $N(\delta)$ scales as

$$N(q = 0, \delta) \sim \delta^{-\tau(0)}, \tag{18}$$

where $\tau(0) = 2$ is the Hausdorff-Besicovitch dimension of the support [14]. On the other hand, if we choose $q = 1$ we have $Z_1 = \sum_i p_i = const.$ and hence we must have $\tau(1) = 0$. This is indeed the case according to equation (16). Therefore, $\tau(0) = 2$ (the dimension of the support) and $\tau(1) = 0$ (required by the normalization condition) are often considered as the first self-consistency check for the multifractal analysis.

The role of $q$ in the partition function $Z_q$ can be best understood by appreciating the fact that the large values of $q$ favours contributions in the $Z_q$ from the blocks with relatively high values of $p_i$ since $p_i^q >> p_j^q$ with $p_i > p_j$ provided $q >> 1$. On the other hand, $q << -1$ favours contributions in the $Z_q$ from those blocks which are occupied with relatively low values of the measure $p_i$. For further elucidation we find it worthwhile to consider the slope of the curve $\tau(q)$ vs $q$ which is given by

$$\frac{d\tau(q)}{dq} = -\lim_{\delta \to 0} \frac{\sum_i p_i^q \ln p_i}{(\sum_i p_i^q) \ln \delta}. \tag{19}$$

Now $\tau(q)$ vs $q$ curve has the maximum slope in the limit $q \to -\infty$ and hence we can write

$$\frac{d\tau(q)}{dq} = -\alpha_{\max}. \tag{20}$$

It implies that the right hand side of equation (19) is dominated by $p_{\min}$, the minimum of $p_i$, in the sum and hence we have

$$\left.\frac{d\tau(q)}{dq}\right|_{q\to-\infty} = -\lim_{\delta\to 0}\frac{\ln p_{\min}}{\ln\delta} = -\alpha_{\max}, \tag{21}$$

and hence we get

$$p_{\min} \sim \delta^{\alpha_{\max}}. \tag{22}$$

A similar argument in the limit $q \to \infty$ leads to the conclusion that the minimum slope of the $\tau(q)$ vs $q$ curve is given by

$$\left.\frac{d\tau(q)}{dq}\right|_{q\to+\infty} = -\lim_{\delta\to 0}\frac{\ln p_{\max}}{\ln\delta} = -\alpha_{\min}, \tag{23}$$

where $p_{\max}$ is the largest value of $p$ which gives the minimum value of $\alpha$ and hence

$$p_{\max} \sim \delta^{\alpha_{\min}}. \tag{24}$$

So, in general we can write

$$\frac{d\tau(q)}{dq} = -\alpha(q), \tag{25}$$

where the exponent $\alpha(q)$ is known as the Lipschitz-Hölder exponent and

$$p \sim \delta^{\alpha}. \tag{26}$$

We now perform the Legendre transformation of the mass exponent $\tau(q)$ by using the Lipschitz-Hölder exponent given equation (25) as an independent variable to obtain the new function

$$f(\alpha) = q\alpha + \tau(q). \tag{27}$$

Replacing $\tau(q)$ in equation (15) in favour of $f(\alpha)$ we find that

$$N(q(\alpha), \delta) \sim \lim_{\delta\to 0} \delta^{q\alpha - f(\alpha)}. \tag{28}$$

On the other hand, using $p \sim \delta^{\alpha}$ in the expression for partition function $Z_q = \sum_i p^q$ and replacing the sum by integral while indexing the blocks by continuous Lipschitz-Hölder exponent $\alpha$ as variable with a weight $\rho(\alpha)$ we obtain

$$N(q(\alpha), \delta) \sim \int \rho(\alpha) d\alpha N(\alpha, \delta)\delta^{q\alpha}, \tag{29}$$

where $N(\alpha, \delta)$ is the number of squares of side $\delta$ needed to cover the measure indexed by $\alpha$. Comparing Eqs. (28) and (29) we find

$$N(\alpha, \delta) \sim \delta^{-f(\alpha)}. \tag{30}$$

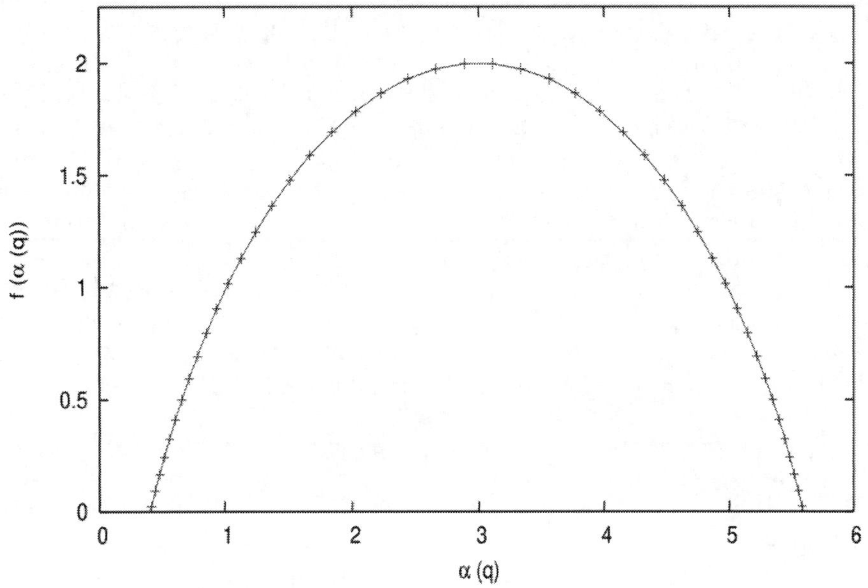


**Figure 6.** The $f(\alpha)$ spectrum.

It implies that we have a spectrum of spatially intertwined fractal dimensions

$$f(\alpha(q)) = \frac{16}{\sqrt{9q^2+16}} - 2, \tag{31}$$

are needed to characterize the measure. That is, the size disorder of the blocks are multifractal in character since the measure $\{p_\alpha\}$ is related to size of the blocks. That is, the distribution of $\{p_\alpha\}$ in WPSL can be subdivided into a union of fractal subsets each with fractal dimension $f(\alpha) \leq 2$ in which the measure $p_\alpha$ scales as $\delta^\alpha$. Note that $f(\alpha)$ is always concave in character (see figure 6) with a single maximum at $q = 0$ which corresponds to the dimension of the WPSL with empty blocks.

On the other hand, we find that the entropy $S(\delta) = -\sum_i p_i \ln p_i$ associated with the partition of the measure on the support (WPSL) by using the relation $\sum_i p_i^q \sim \delta^{-\tau(q)}$ in the definition of $S(\delta)$. Then a few steps of algebraic manipulation reveals that $S(\delta)$ exhibits scaling

$$S(\delta) = \ln \delta^{-\alpha(1)} \tag{32}$$

where the exponent $\alpha_1 = \frac{6}{5}$ obtained from

$$\alpha(q) = -\left. d\tau(q)/dq \right|_q . \tag{33}$$

It is interesting to note that $\alpha(1)$ is related to the generalized dimension $D_q$, also related to the Rényi entropy $H_q(p) = \frac{1}{q-1} \ln \sum_i p_i^q$ in the information theory, given by

$$D_q = \lim_{\delta \to 0} \left[ \frac{1}{q-1} \frac{\ln \sum_i p_i^q}{\ln \delta} \right] = \frac{\tau(q)}{1-q}, \tag{34}$$

which is often used in the multifractal formalism as it can also provide insightful interpretation. For instance, $D_0 = \tau(0)$ is the dimension of the support, $D_1 = \alpha_1$ is the Renyi information dimension and $D_2$ is known as the correlation dimension [15, 16]. Multifractal analysis was initially proposed to treat turbulence but later successfully applied in a wide range of exciting field of research. For instance, it has been found recently that the wild fluctuations of the wave functions at the Anderson and the quantum Hall transition can be best described by multifractality [17, 18]. Recently, though it has got renewed momentum as it has been found that the probability density function at the Anderson and the quantum Hall transition exhibits multifractality since in the vicinity of the transition point fluctuations are wild - a characteristic feature of multifractal behaviour [18].

In an attempt to understand the origin of multifractality we now consider the case where the generator divides the initiator into four equal blocks instead of randomly into four blocks. If the generator is applied over and over again thereafter to only one of the available squares by picking preferentially with respect to their areas then it results in the kinetic square lattice (KSL). Within the rate equation approach it can be described by the kernel

$$F(x_1, x_2, y_1, y_2) = (x_1 + x_2)(y_1 + y_2)\delta(x_1 - x_2)\delta(y_1 - y_2). \tag{35}$$

and hence the resulting rate equation can be obtained after substituting it in equation (1) to give

$$\frac{\partial C(x, y; t)}{\partial t} = -\frac{1}{4}xyC(x, y; t) + 4^2 xyC(2x, 2y; t). \tag{36}$$

Incorporating equation (4) in equation (36) yield

$$\frac{dM(m, n; t)}{dt} = -\Big(\frac{1}{4} - \frac{4}{2^{m+n}}\Big)M(m + 1, n + 1; t). \tag{37}$$

To obtain the solution of this equation in the long-time limit, we assume the following power-law asymptotic behaviour of $M(m, n; t)$ and write

$$M(m, n; t) \sim A(m, n)t^{\theta(m+n)}, \tag{38}$$

with $\theta(4) = 0$ since total area obtained by setting $m = n = 2$ is an obvious conserved quantity. Using it in equation (37) yields the following difference equation

$$\theta(m + n + 2) = \theta(m + n) - 1. \tag{39}$$

Iterating it subject to the condition that $\theta(4) = 0$ gives

$$M(m, n; t) \sim t^{-\frac{m+n-4}{2}}. \tag{40}$$

Apparently it appears that in addition to the conservation of the total area $M(2, 2; t)$, we find that the integrals $M(3, 1; t)$ and $M(1, 3; t)$) are also conserved. Interestingly, all the three integrals $M(2, 2; t)$, $M(3, 1; t)$ and $M(1, 3; t)$ effectively describe the same physical quantity since all the blocks are square in shape and hence

$$\sum_{i=1}^{N} x_i^2 = \sum_{i=1}^{N} y_i^2 = \sum_{i=1}^{N} x_i y_i. \tag{41}$$

Therefore, in reality the system obeys only one conservation law - conservation of total area.

We again look into the $q$th moment of $n(x, t)$ using equation (10) and appreciating the fact that $M_q(t)$ equal to $M(q + 1, 1; t)$ or $M(1, q + 1; t)$ we immediately find that

$$M_q(t) \sim t^{-\frac{q-2}{2}}. \tag{42}$$

Unlike in the previous case where the exponent of the power-law solution of the $M_q(t)$ is non-linear, here we have an exponent which is linear in $q$. It immediately implies that in the case kinetic square lattice

$$<x^q>=<x>^q, \tag{43}$$

and hence a single length-scale is enough to characterize all the moments of $n(x,t)$. That is, the system now exhibits simple scaling instead of multiscaling. Like before let us consider that each block is occupied with a fraction of the measure equal to square of its own length or area $p_i = \sum_i^N x_i^2$ and hence the corresponding partition function is

$$Z_q = \sum_i^N p_i^q = M_{2q}(t). \tag{44}$$

Using equation (42) we can immediately write its solution

$$Z_q(t) \sim t^{-\frac{2q-2}{2}}. \tag{45}$$

Expressing it in terms of the square root of the average area $\delta \sim t^{-1/2}$ gives the weighted number of square $N(q,\delta)$ of side $\delta$ needed to cover the measure which has the following power-law solution

$$N(q,\delta) \sim \delta^{-(2-2q)}, \tag{46}$$

where mass exponent $\tau(q)$ is

$$\tau(q) = 2 - 2q. \tag{47}$$

The Legendre transform of the mass exponent is a constant

$$f(\alpha) = 2, \tag{48}$$

and so is the generalized dimension

$$D_q = 2. \tag{49}$$

We thus find that if the generator divides the initiator into four equal square and we apply it thereafter sequentially then the resulting lattice is no longer a multifractal. The reason is that the distribution of the population in the resulting support in this case is uniform. The two model therefore provides a unique opportunity to look for possible origin of multifractality. The two models discussed in this talk differs only in the definition of the generator. In the case when the generator divides the initiator randomly into four blocks and we apply it over and over again sequentially then we have multifractality since the underlying mechanism in this case is governed by random multiplicative process. This is not, however, the case if the generator divides the initiator into equal four blocks and apply it over and over again sequentially since the resulting dynamics is governed by deterministic multiplicative process instead.

## 4. Scale-free properties of WPSL

Defining each step of the algorithm as one time unit and imposing periodic boundary condition in the simulation we can immediately write an exact expression for the total number of blocks as a function of time: $N(t) = 1 + 3t$. However, unlike the square lattice where every block shares its borders exactly with four neighbours, the blocks in the WPSL can have their common borders with at least four or more blocks albeit all the blocks have exactly four sides. In fact, we can characterize individual blocks of the WPSL by their coordination numbers, which is defined as the number of neighbours with whom they have common borders, which we find neither a constant like the square lattice nor has a typical mean value like the Voronoi lattice. Instead, the

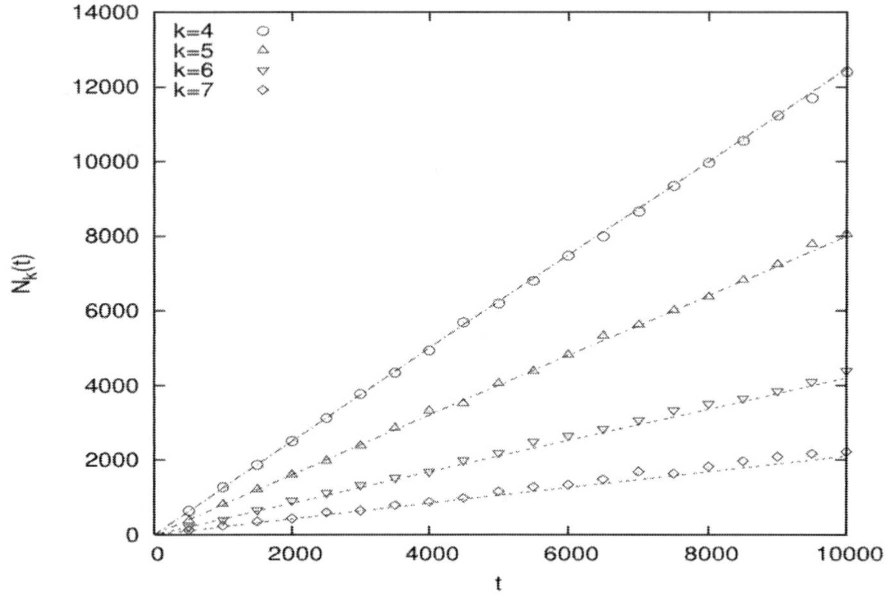


**Figure 7.** Shown is the growth of the number of blocks $N_k(t)$ with exactly $k = 4, 5, 6, 7$ neighbours as a function of time.

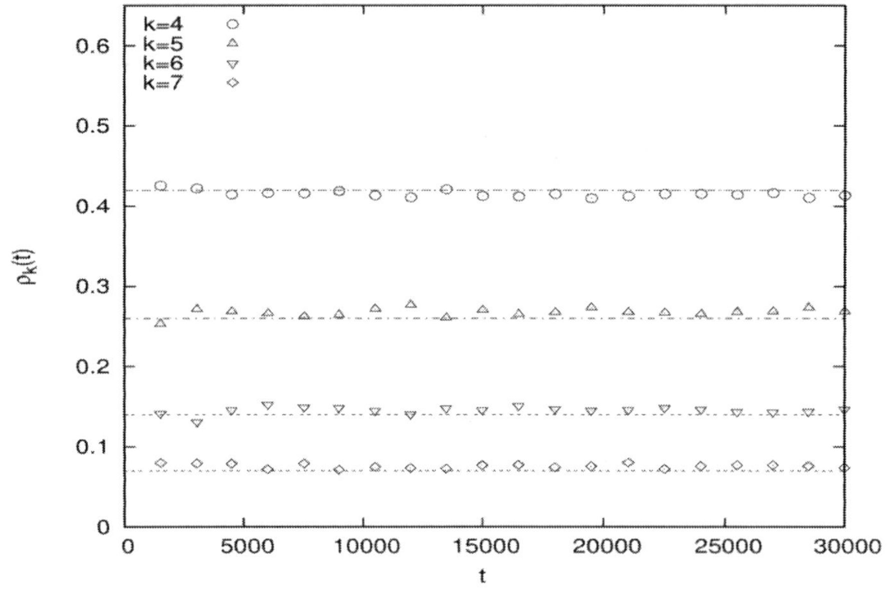


**Figure 8.** Fraction of the blocks $\rho_k(t)$ with $k = 4, 5, 6, 7$ neighbours are drawn as a function of time.

coordination number that each block assumes in the WPSL is random and evolves with time. The coordination number disorder in the WPSL can, therefore, be regarded as of annealed type. Our data extracted from numerical simulation suggest that the number of blocks $N_k(t)$ which have coordination number $k$ (or $k$ neighbours) continue to grow linearly with time $N_k(t) = m_k t$ (see figure 7). On the other hand, the number of total blocks $N(t) = \sum_k N_k(t)$ at time $t$ in the lattice also grow linearly with time $N(t) = 1 + 3t$ and hence $N(t) \sim 3t$ in the long-time limit. The ratio of the two quantities $\rho_k(t) = N_k(t)/N(t)$ that describes the fraction of the total blocks which have coordination number $k$ is $\rho_k(t) = m_k/3$ and it is clearly independent of time *vis-a-vis* of the size of the lattice (see figure 8). It implies that we can take a lattice of sufficiently large size and study its properties as these properties without being worried about its exact size.

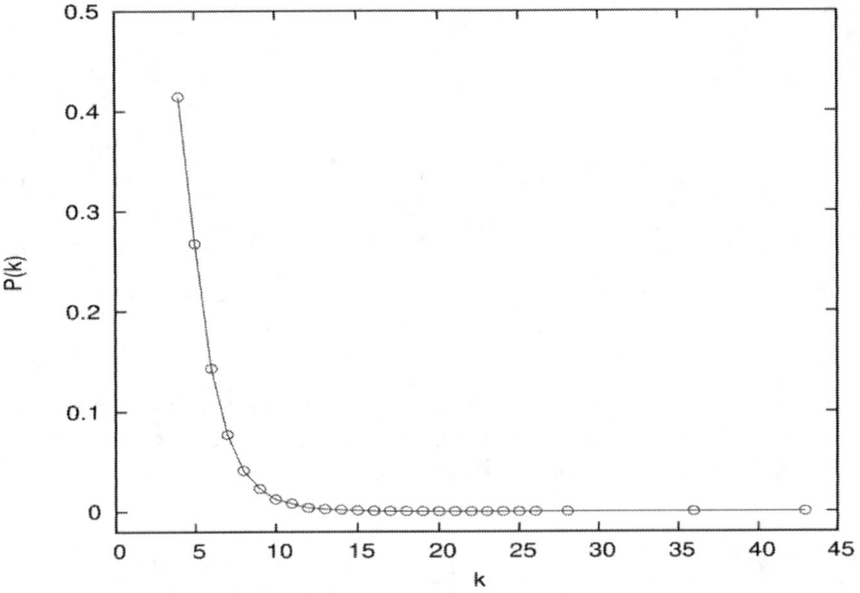


**Figure 9.** Shown is the coordination number distribution function $P(k)$ vs $k$ which is also equivalent to the degree distribution of the DWPSL.

We now assume that that blocks in the WPSL are equivalent regardless of their sizes and then ask: What is the probability that a block picked at random with equal *a priori* probability from all the available blocks of the lattice have coordination number $k$? Answer to this question lies in collecting data for $\rho_t$ as a function of $k$ and plotting a normalized histogram which effectively gives the coordination number distribution function $\rho_t(k) \equiv P(k)$ where subscript $t$ indicates fixed time. We first plot $P(k)$ in figure 9 to have a glimpse of its behaviour as a function of $k$. A closer look into the data reveals that there exist scarce data points near the tail. The long tail with scarce data points turns into a noisy or fat-tail when we plot $\ln P(k)$ vs $\ln(k)$. This is drawn in figure 10 where data points represents ensemble average over 50 independent realizations and it clearly suggests that fitting a straight line near the tail is possible. It implies that the coordination number distribution decays obeying power-law

$$P(k) \sim k^{-\gamma}, \tag{50}$$

with heavy or fat-tail reflecting scarce data points near the tail-end of figure 10. However, the noise in the tail-end complicate the process of identifying the range over which the power-law holds and hence of estimating the exponent $\gamma$. One way of reducing the noise at the tail-end is to plot cumulative distribution $P(k' \geq k)$ which is related to degree distribution $P(k)$ via

$$P(k) = -\frac{dP(k' \geq k)}{dk}. \tag{51}$$

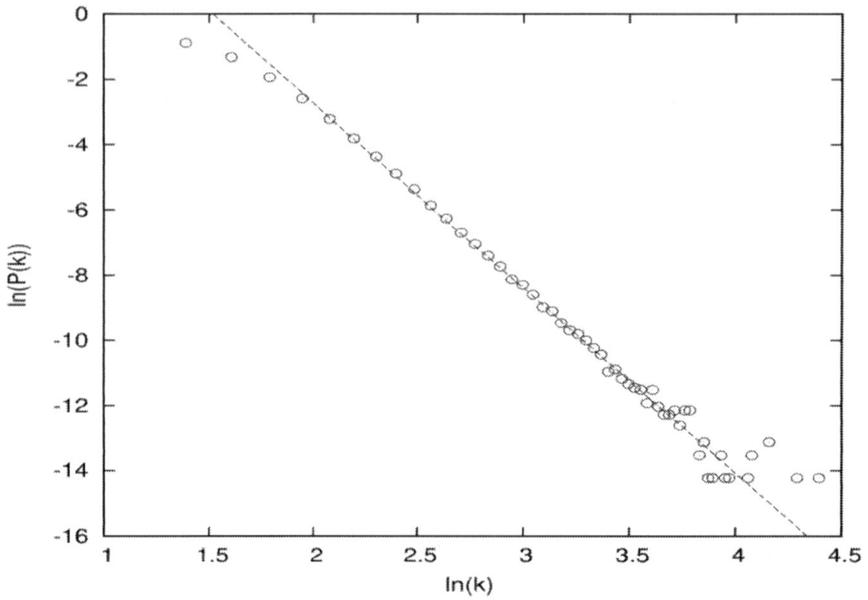


**Figure 10.** Drawn is plot of $\ln(P(k))$ against $\ln(k)$ for the DWPSL network where data points represent average of 50 independent realizations. The line have slope exactly equal to 5.66 revealing power-law degree distribution with exponent $\gamma = 5.66$.

We therefore plot $\ln(P(k' > k))$ vs $\ln(k)$ in figure 11 using the same data of figure 10 and find that the heavy tail smoothes out naturally where no data is obscured. The straight line fit of figure 11 has a slope $\gamma - 1 = 4.66$ which indicates that the degree distribution (figure 10) decays following power-law with exponent $\gamma = 5.66$. A power-law distribution function is often regarded as scale-free since it looks the same whatever scale we look at it.

We thus find that in the large-size limit the WPSL develops some order in the sense that its annealed coordination number disorder is scale-free in character. This is in sharp contrast to the quenched coordination number disorder found in the Voronoi lattice where it is almost impossible to find cells which have significantly higher or fewer neighbours than the mean value $k = 6$ [19]. In fact, it has been shown that the coordination number distribution of the Voronoi lattice is Gaussian in character instead. The power-law coordination number distribution in the WPSL is reminiscent of the scale-free degree distribution of complex network. The past decade have witnessed a surge of interest in the theory of complex network resulting in the dramatic advances in the field thanks to the seminal work of A.-L. Barabasi and his co-workers. It is worth mentioning that the dual of the WPSL, obtained replacing each block with a node at its centre and common border between blocks with an edge joining the two nodes (see figure 12 where we have illustrated the network aspect of the WPSL) is a network whose degree distribution has the same data as the coordination number distribution of the WPSL.

The square lattice, on the other hand, is self-dual and its degree distribution is $P(k) = \delta(k-4)$. Further, it is interesting to point out that the exponent $\gamma = 5.66$ is significantly higher than usually found in most real-life network which is typically $2 < \gamma \leq 3$. This suggests that in addition to the PA rule the network in question has to obey some constraints. For instance, nodes in the WPSL are spatially embedded in Euclidean space, links gained by the incoming nodes are constrained by the spatial location and the fitting parameter of the nodes. Owing to its unique dynamics this was not unexpected. Perhaps, it is noteworthy to mention that the

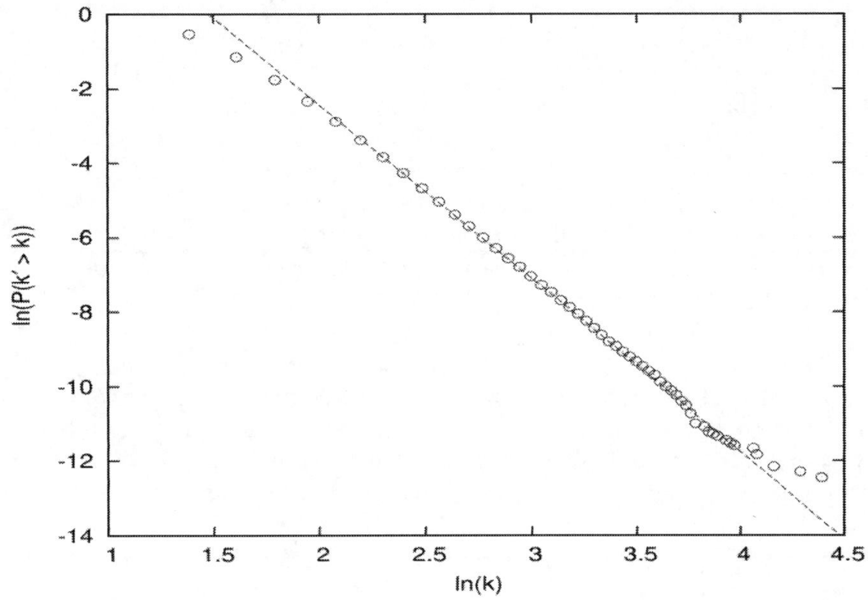


**Figure 11.** Cumulative degree distribution $P(k' > k)$ is shown using the same data as of figure 4. The dotted line with slope equal to $\gamma - 1 = 4.66$ is drawn to guide our eyes.

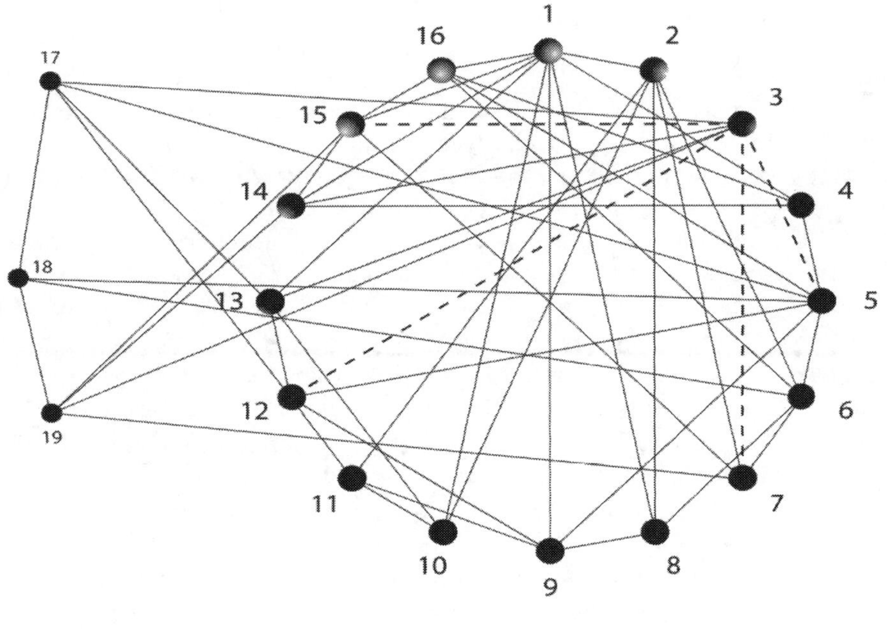


**Figure 12.** Nodes in the circle and their links with other nodes illustrate the network topology at step 5 of Fig. 2. New nodes (17, 18 and 19) and links to be added and links to be removed (broken lines) in step 6 again of Fig. 2 are also shown to illustrate the dynamics.

degree distribution of the electric power grid, whose nodes like WPSL are also embedded in the spatial position, is shown to exhibit power-law but with exponent $\gamma_{\text{power}} = 4$ [20].

The power-law degree distribution $P(k)$ has been found in many seemingly unrelated real life networks. It implies that there must exists some common underlying mechanisms for which disparate systems behave in such a remarkably similar fashion [21]. Barabasi and Albert argued that the growth and the PA rule are the main essence behind the emergence of such power-law. Indeed, the DWPSL network too grows with time but in sharp contrast to the BA model where network grows by addition of one single node with $m$ edges per unit time, the DWPSL network grows by addition of a group of three nodes which are already linked by two edges. It also differs in the way incoming nodes establish links with the existing nodes.

To understand the growth mechanism of the DWPSL network let us look into the $j$th step of the algorithm. First a node, say it is labeled as $a_p$, is picked from the $(3j-2)$ nodes preferentially with respect to the fitness parameter of a node (i.e., according to their respective areas). Secondly, connect the node $a_p$ with two new nodes $(3j-1)$ and $(3j+1)$ in order to establish their links with the existing network. At the same time, at least two or more links of $a_p$ with other nodes are removed (though the exact number depends on the number of neighbours $a_p$ already has) in favour of linking them among the three incoming nodes in a self-organized fashion. In the process, the degree $k_p$ of the node $a_p$ will either decrease (may turn into a node with marginally low degree in the case it is highly connected node) or at best remain the same but will never increase. It, therefore, may appear that the PA rule is not followed here. A closer look into the dynamics, however, reveals otherwise. It is interesting to note that an existing nodes during the process gain links only if one of its neighbour is picked not itself. It implies that the higher the links (or degree) a node has, the higher its chance of gaining more links since they can be reached in a larger number of ways. It essentially embodies the intuitive idea of PA rule. Therefore, the DWPSL network can be seen to follow preferential attachment rule but in disguise.

## 5. Summary

To summarize, we proposed a model that generates weighted planar stochastic lattice (WPSL) which emerges through evolution. Unlike the regular lattice where every block or cell has exactly the same coordination number, the WPSL is seemingly disordered both in terms of the coordination number and in terms of its block size distribution. One of our primary goal was to find some order in such seemingly disordered system since finding order is always an attractive proposition for physicists. To this end, our first attempt was to quantify the size disorder of the lattice and we found several interesting results. One of the most interesting results that we found is that $\sum_i^N x_i^{n-1} y_i^{4/n-1}$ remains a constant $\forall\, n$ regardless of the size of the lattice where blocks are labeled as $i = 1, 2, ..., N$ and $x_i$ and $y_i$ are their respective length and width. Yet another interesting fact is that the numerical values of the conserved quantities $\sum_i^N x_i^{n-1} y_i^{4/n-1}$ for a given value of $n$ except $n = 2$ are never the same in different realizations which is clearly an indication of wild fluctuation. Of the conservation laws, we found a special one obtained by choosing either $n = 1$ or $n = 3$ to give $\sum_{i=1}^N y_i^3$ or $\sum_{i=1}^N x_i^3$ which can be picked as a measure. It implies that if the blocks are populated with a fraction of the measure equal to cubic power of their respective length or width then its distribution on the WPSL is multifractal in nature - a property that quantifies the wild fluctuation of block size distribution of the WPSL. That is, the probability distribution of $x_i^3$ is multifractal in the sense that a single exponent is not enough to characterize its distribution on the support, namely the WPSL, instead we need a spectrum of exponent $f(\alpha)$. We derived an exact expression for the $f(\alpha)$ spectrum and obtained information dimension $D_1 = 6/5$ with which the entropy of the measure scales $S(\delta) \sim \ln \delta^{-D_1}$. To look for a possible origin of multifractality we also studied the kinetic square lattice which is actually the deterministic counterpart of the WPSL. It led to the conclusion that as soon as randomness is ceased from the definition of the generator the resulting lattice can no longer be quantified as multifractal rather it possesses all the properties of the square lattice but only in the long-time limit.

Our second attempt was to quantify the coordination number disorder. We have shown numerically that the coordination number disorder is scale-free in character since the degree distribution of its dual (DWPSL), which is topologically identical to the network obtained by considering blocks of the WPSL as nodes and the common border between blocks as links, exhibits power-law. In other words, the coordination number distribution function of the WPSL decays following power-law with exponent $\gamma = 5.66$. However, the novelty of this network is that it grows by addition of a group of already linked nodes which then establish links with the existing nodes following PA rule, though in disguise, in the sense that existing nodes gain links only if one of their neighbour is picked, not the node itself. Finally, such multifractal lattice with scale-free coordination disorder can be of great interest as it has the potential to mimic disordered medium on which one can study various physical phenomena like percolation and random walk problems etc. One may also study the phase transition and critical behaviour in such multifractal scale-free lattice. Indeed, phase transition and critical behaviour in complex network have attracted much of our recent attention. The advantage of the WPSL over many known complex networks is that its nodes or blocks are embedded in spatial position. That is, one may assume that interacting particles located on the sites with great many different neighbours and therefore its result must differ from that of a system where sites are located on a regular lattice. We intend to work in these directions in our future endeavour.

NIP gratefully acknowledges support from the Bose Centre for Advanced Study and Research in Natural Sciences.

[1] Rao M, Sengupta S, and Sahu H K 1995 *Phys. Rev. Lett.* **75**, 2164
[2] Atkinson H V 1998 *Acta Metall,* **36** 469
[3] J. C. M. Mombach J C M, Vasconcellos M A Z, and de Almeida R M C J. 1990 *Phys. D* **23**, 600
[4] Weaire D and Rivier N, 1984 *Contemp. Phys.* **25**, 59
[5] Okabe A, Boots B, Sugihara K, and Chiu S N 2000 *Spatial Tessellations - Concepts and Applications of Voronoi Diagrams* (Chicester: Wiley)
[6] Delaney G W, Hutzler S and Aste T 2008 *Phys. Rev. Lett.* **101** 120602
[7] Hassan M K, Hassan M Z, and Pavel N I 2010 *New J. Phys.* **12** 093045
[8] Hassan M K and Rodgers G J 1996 *Phys. Lett. A* **218**, 207
[9] Krapivsky P L and Ben-Naim E 1994 *Phys. Rev. E* **50** 3502
[10] Krapivsky P L and Ben-Naim E 1996 *Phys. Rev. Lett.* **76** 3234
[11] Majumdar S N, Dean D S and Krapivsky P L 2005 *Pramana - J. Phys.* **64**, 1187
[12] Luke Y L 1969 *The special functions and their approximations I* (New York: Academic Press)
[13] Stanley H E 1996 in *Fractals and disordered systems* eds. Bunde A and Havlin S (New York: Springer) pp. 1-49
[14] Feder J 1988 *Fractals* (New York: Plenum, New York)
[15] Hentschel H G E and Procaccia I 1983 *Physica* **8D**, 435
[16] Sporting J, Weickert J 1999 *IEEE Trans. Info. Theory* **45**, 1051
[17] Mandelbrot B B 1982 *Fractal geometry of nature* (New York: W. H. Freeman).
[18] Evers F and Mirlin A D 2008 *Rev. Mod. Phys.* **80**, 1355
[19] de Oliveira M M, Alves S G, Ferreira S C, and Dickman R 2008 *Phys. Rev. E* **78**, 031133
[20] Barabasi A L and R. Albert R 1999 *Science* **286**, 509
[21] Albert R and Barabasi A L 2002 *Rev. Mod. Phys.* **74**, 47

# Non-equilibrium Dynamics, Thermalization and Entropy Production

**Haye Hinrichsen**[1], **Christian Gogolin**[2], **and Peter Janotta**[1]

[1] Universität Würzburg, Fakultät für Physik und Astronomie 97074 Würzburg, Germany

[2] Institute for Physics and Astronomy, Potsdam University, 14476 Potsdam, Germany

E-mail: hinrichsen@physik.uni-wuerzburg.de

**Abstract.** This paper addresses fundamental aspects of statistical mechanics such as the motivation of a classical state space with spontaneous transitions, the meaning of non-equilibrium in the context of thermalization, and the justification of these concepts from the quantum-mechanical point of view. After an introductory part we focus on the problem of entropy production in non-equilibrium systems. In particular, the generally accepted formula for entropy production in the environment is analyzed from a critical perspective. It is shown that this formula is only valid in the limit of separated time scales of the system's and the environmental degrees of freedom. Finally, we present an alternative simple proof of the fluctuation theorem.

## 1. Introduction

Classical non-relativistic statistical physics is based on a certain set of postulates. Starting point is a physical entity, called *system*, which is characterized by a set $\Omega_{\text{sys}}$ of possible configurations. Although this configuration space could be continuous, it is useful to think of it as a countable set of discrete microstates $s \in \Omega_{\text{sys}}$. Being classical means that the actual configuration of the system is a matter of objective reality, i.e. at any time $t$ the system *is* in a well-defined configuration $s(t)$. The system is assumed to evolve in time by spontaneous transitions $s \to s'$ which occur randomly with certain transition rates $w_{ss'} > 0$. This commonly accepted framework is believed to subsume the emerging classical behavior of a complex quantum system subjected to decoherence.

Although in a classical system the trajectory of microscopic configurations is in principle measurable, an external observer is usually not able to access this information in detail. The observer would instead express his/her partial knowledge in terms of the probability $P_s(t)$ to find the system at time $t$ in the configuration $s$. In contrast to the unpredictable trajectory $s(t)$, the probability distribution $P_s(t)$ evolves deterministically according to the master equation $\dot{P}_s(t) = \sum_{s' \in \Omega_{\text{sys}}} (P_{s'}(t) w_{s's} - P_s(t) w_{ss'})$.

Journal of Physics: Conference Series **297** (2011) 012011 doi:10.1088/1742-6596/297/1/012011

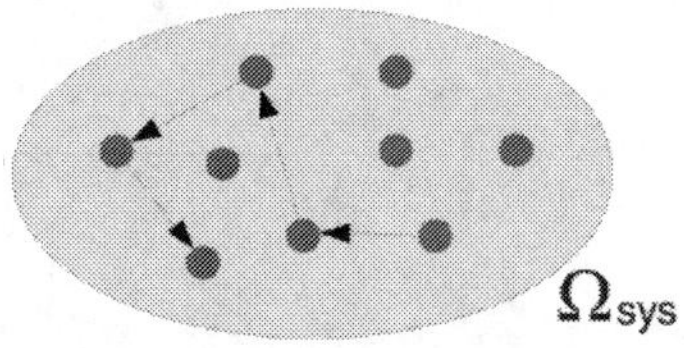


**Figure 1.** Cartoon of a complex statistical system as a space $\Omega_{sys}$ of configurations (red dots). At any time the system *is* in one of the configurations and evolves by spontaneous transitions (indicated by arrows) selected randomly according to specific transition rates.

A system is said to *equilibrate* if the probability distribution $P_s(t)$ becomes stationary in the limit $t \to \infty$. In addition, if the probability currents $J_{ss'} = P_s(t)w_{ss'}$ and $J_{s's} = P_{s'}(t)w_{s's}$ between all pairs of configurations $s, s'$ individually cancel each other in the stationary state, the system is said to *thermalize*, obeying *detailed balance*. The predictive power of equilibrium statistical mechanics relies on the fact that the stationary probability distribution of a thermalized system is universal and can be classified into a small number of thermodynamic ensembles. In particular, an isolated system thermalizes in such a way that each available configuration is visited with the same probability. This fundamental *equal a priori probability postulate* is at the core of equilibrium statistical mechanics, from which all other thermodynamic ensembles can be derived.

Classical thermodynamics is concerned with time-dependent phenomena close to equilibrium. Introducing notions such as thermodynamic forces, potentials and currents it makes statements how such systems relax towards thermal equilibrium. Contrarily, non-equilibrium statistical physics deals with systems that do not thermalize, meaning that their probability currents do not vanish even in the stationary state. Typically such systems need to be driven from the outside in order to prevent them from thermalizing.

The basic framework of statistical physics and thermodynamics sketched above is often taken for granted. However, one should be aware that the underlying postulates are highly non-trivial. In this paper we address some of these issues from a critical perspective. Where does the cartoon of spontaneous hopping between configurations come from? How can we be sure that the system will equilibrate into a stationary state? What is known about thermalization and the justification of the *equal a priori probability postulate*? What is the meaning of non-equilibrium in an environment that thermalizes? What is behind the commonly accepted formula for the entropy production of non-equilibrium systems? In the following we make an attempt to discuss some of these questions in a common context, giving partial answers and pointing out open questions. Moreover, we would like to draw the readers attention to the fascinating links between classical statistical physics and recent developments in quantum information science.

The paper is organized as follows. In the following section we summarize existing textbook knowledge on stochastic system in a compressed form suitable for beginners in the field. Sect. 3 deals with the question how we can justify the basic assumptions and postulates of classical statistical physics from the quantum perspective. In Sect. 4 we will discuss the problem of entropy production in non-equilibrium systems, analyzing the conditions under which commonly accepted formulas are valid. A discussion of the fluctuation theorem together with an alternative compact proof is presented in Sect. 5. The paper closes with concluding remarks in Sect. 6.

## 2. Setup of classical statistical physics

### Configuration space and dynamics

As outlined in the introduction, classical statistical physics is mainly concerned with models of complex systems having the following properties:

(i) The system is characterized by a certain set $\Omega_{\rm sys}$ of configurations $s \in \Omega_{\rm sys}$, also called microstates. Usually the configuration space is implicitly specified by the definition of a model. For example, in a reaction-diffusion model this space is the set of all possible particle configurations while in a growth model the microstates are identified with the possible configurations of an interface.

(ii) The states are classical, i.e., at any time the system *is* in one particular configuration $s(t)$.

(iii) The system evolves randomly by instantaneous transitions $s \to s'$ occurring spontaneously with certain transition rates $w_{ss'} \geq 0$. In numerical simulations, this dynamics is approximated by random-sequential update algorithms.

Starting with an initial configuration $s_0$ the system evolves randomly through an unpredictable sequence of configurations $s_0 \to s_1 \to s_2 \to \ldots$ by instantaneous transitions. These transition take place at certain points of time $t_1, t_2, \ldots$ which are distributed according to a Poisson distribution like shot noise. Such a sequence of transitions is called a *stochastic path.*

Although the actual stochastic path of the system is unpredictable, the probability $P_s(t)$ to find the system in configuration $s$ at time $t$ evolves deterministically according to the master equation

$$\frac{\mathrm{d}}{\mathrm{d}t} P_s(t) = \sum_{s' \in \Omega} \Big( J_{s's}(t) - J_{ss'}(t) \Big) , \tag{1}$$

where

$$J_{ss'}(t) = P_s(t) w_{ss'} \tag{2}$$

is the probability current flowing from configuration $s$ to configuration $s'$. The system is said

- to be *stationary* or *equilibrated* if the probability distribution $P_s(t)$ is time-independent, meaning that for a given configuration $s$ all incoming and outgoing probability currents cancel.
- to *equilibrate* if the master equation evolves into a stationary distribution in the limit $t \to \infty$, denoted as $P_s = P_s(\infty)$.

For simplicity we will assume that this stationary state is unique and independent of the initial state. Note that systems with a finite configuration space always relax into a stationary state while for systems with an infinite or continuous configuration space a stationary state may not exist. Moreover, we will assume that the dynamics of the system under consideration is ergodic, i.e., the network of transitions is connected so that each configuration can be reached.

### Detailed balance

A stationary system is said to *thermalize* if it evolves into a stationary state which obeys *detailed balance*. This means that the probability currents between all pairs of configurations cancel, i.e.

$$J_{ss'} = J_{s's} \qquad \forall s, s' , \tag{3}$$

or equivalently

$$\frac{w_{ss'}}{w_{s's}} = \frac{P_{s'}}{P_s} \qquad \forall s, s' , \tag{4}$$

where $P_s$ is the stationary probability distribution. It is worth being noted that detailed balance can be defined in an alternative way without knowing the stationary probability distribution. To see this let us consider a closed loop of three transitions $s_1 \to s_2 \to s_3 \to s_1$. For these transitions Eq. (4) provides a system of three equations. By multiplying all equations one can eliminate the probabilities $P_{s_i}$, arriving at the condition $w_{s_1s_2}w_{s_2s_3}w_{s_3s_1} = w_{s_1s_3}w_{s_3s_2}w_{s_2s_1}$. A similar result is obtained for any closed loop of transitions, hence the condition of detailed balance can be recast as

$$\prod_i w_{s_is_{i+1}} = \prod_i w_{s_{i+1}s_i} \tag{5}$$

for all closed loops in $\Omega_{\text{sys}}$. A system with this property is said to have a *balanced dynamics*.

Note that in a system with balanced dynamics we may rescale a pair of opposite rates by $w_{ss'} \to \Lambda w_{ss'}$ and $w_{s's} \to \Lambda w_{s's}$ without breaking the detailed balance condition. This intervention changes the dynamics of the model and therewith its relaxation, but the stationary state (if existent and unique) remains the same. This statement holds even if pairs of opposite rates are set to zero as long as this manipulation does not break ergodicity.

### Equal a priori postulate

Classical statistical physics is based on the so-called *equal a priori postulate*. This postulate states that an isolated system, which does not interact physically with the outside world, will thermalize into a stationary state where all accessible configurations occur with the same probability, i.e.

$$P_s = const = 1/|\Omega_{\text{sys}}| \,. \qquad \forall s \in \Omega_{\text{sys}} \tag{6}$$

It is assumed that this state obeys detailed balance since otherwise one could use the non-vanishing probability currents to construct a *perpetuum mobile*. The immediate consequence would be that all transitions are reversible and that opposite rates coincide, i.e. $w_{ss'} = w_{s's}$.

### Entropy

Entropy is probably the most fundamental concept of statistical physics. From the information-theoretic point of view, the entropy of a system is defined as the amount of information (measured in bits) which is necessary to describe the configuration of the system. Since the description of a highly ordered configuration requires less information than a disordered one, entropy can be viewed as a measure of disorder.

The amount of information which is necessary to describe a configuration depends on the already existing partial knowledge of the observer at a given time. For example, deterministic systems with a given initial configuration have no entropy because the observer can compute the entire trajectory in advance, having complete knowledge of the configuration as a function of time even without measuring it. Contrarily, in stochastic systems the observer has only a

Journal of Physics: Conference Series **297** (2011) 012011 doi:10.1088/1742-6596/297/1/012011

partial knowledge about the system expressed in terms of the probability distribution $P_s(t)$. In this situation the amount of information needed to specify a particular configuration $s \in \Omega_{\rm sys}$ is $-\log_2 P_s(t)$ bits, meaning that rare configurations have more entropy than frequent ones.

Different scientific communities define entropy with different prefactors. In information science one uses the logarithm to base 2 so that entropy is directly measured in bits. Mathematicians instead prefer a natural logarithm while physicists are accustomed to put an historically motivated prefactor $k_B$ in front, giving entropy the unit of an energy. In what follows we set $k_B = 1$, defining the entropy of an *individual* configuration $s$ as

$$S_{\rm sys}(t,s) = -\ln P_s(t)\,. \tag{7}$$

Since this entropy depends on the actual configuration $s$, it will fluctuate along the stochastic path. However, its expectation value, expressing the observer's average lack of information, evolves deterministically and is given by

$$S_{\rm sys}(t) = \langle S_{\rm sys}(t,s)\rangle_s = -\sum_{s\in\Omega_{\rm sys}} P_s(t)\ln P_s(t)\,, \tag{8}$$

where $\langle\ldots\rangle$ denotes the ensemble average over independent realizations of randomness. Apart from the prefactor, this is just the usual definition of Shannon's entropy [1, 2].

Up to this point entropy is just an information-theoretic concept for the description of configurations. The point where entropy takes on a *physical* meaning is the *equal a priori postulate*, stating that an isolated system thermalizes in such a way that the entropy takes the largest possible value $S_{\rm sys} = \ln|\Omega_{\rm sys}|$. As is well-known, all other thermodynamic ensembles can be derived from this postulate.

The numerical determination of entropies is a nontrivial task because of the highly non-linear influence of the logarithm. To measure an entropy numerically, one first has to estimate the probabilities $P_s(t)$. The resulting symmetrically distributed sampling errors in finite data sets are amplified by the logarithm, leading to a considerable systematic bias in entropy estimates. Various methods have been suggested to reduce this bias on the expense of the statistical error, see e.g. [3, 4].

**Subsystems**

In most physical situations the system under consideration is not isolated, instead it interacts with the environment. In this case the usual approach of statistical physics is to consider the system combined with the environment as a composite system. This superordinate *total system* is then assumed to be isolated, following the same rules as outlined above. To distinguish the total system from its parts, we will use the suffixes 'tot' for the total system while 'sys' and 'env' refer to the embedded subsystem and its environment, respectively.

The total system is characterized by a certain space $\Omega_{\rm tot}$ of classical configurations $c \in \Omega_{\rm tot}$ (not to be confused with system configurations $s \in \Omega_{\rm sys}$). The number of these configurations may be enormous and they are usually not accessible in experiments, but in principle there should be a corresponding probability distribution $P_c(t)$ evolving by a master equation

$$\frac{\rm d}{{\rm d}t}P_c(t) = \sum_{c'\in\Omega_{\rm tot}} \Bigl(J_{c'c}(t) - J_{cc'}(t)\Bigr)\,, \qquad J_{cc'}(t) = P_c(t)w_{cc'} \tag{9}$$

with certain time-independent transition rates $w_{cc'} \geq 0$.

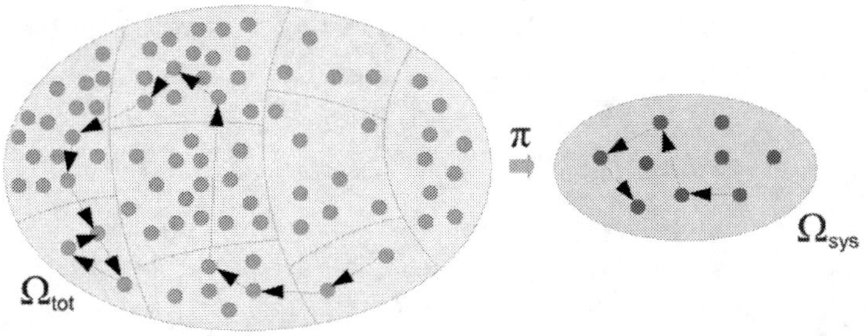


**Figure 2.** A subsystem is defined by a projection $\pi$ which maps each state $s \in \Omega_{\rm tot}$ of the total system (left) onto a particular state $c \in \Omega_{\rm sys}$ of the subsystem (right), dividing the state space of the total system into sectors. The figure shows a stochastic path in the total system together with the corresponding stochastic path in the subsystem.

Let us now consider an embedded subsystem. Obviously, for every classical configuration $c \in \Omega_{\rm tot}$ of the total system we will find the subsystem in a well-defined unique configuration $s \in \Omega_{\rm sys}$. Conversely, for a given configuration of the subsystem $s \in \Omega_{\rm sys}$ the environment (and therewith the total system) can be in many different states. This relationship can be expressed in terms of a surjective map $\pi : \Omega_{\rm tot} \to \Omega_{\rm sys}$ which projects every configuration $c$ of the total system onto the corresponding configuration $s$ of the subsystem, as sketched in schematically Fig. 2.

The projection $\pi$ divides the space $\Omega_{\rm tot}$ into sectors $\pi^{-1}(c) \subset \Omega_{\rm tot}$ which consist of all configurations which are mapped onto the same $s$. Therefore, the probability to find the subsystem in configuration $s \in \Omega_{\rm sys}$ is the sum over all probabilities in the corresponding sector, i.e.

$$P_s(t) \;=\; \sum_{c(s)} P_c(t)\,, \tag{10}$$

where the sum runs over all configurations $c \in \Omega_{\rm tot}$ with $\pi(c) = s$. Likewise, the projected probability current $J_{ss'}$ in the subsystem flowing from configuration $s$ to configuration $s'$ is the sum of all corresponding probability currents in the total system:

$$J_{ss'}(t) \;=\; \sum_{c(s)} \sum_{c'(s')} J_{cc'}(t) \;=\; \sum_{c(s)} P_c(t) \sum_{c'(s')} w_{cc'}\,. \tag{11}$$

This allows us to define effective transition rates in the subsystem by

$$w_{ss'}(t) \;=\; \frac{J_{ss'}(t)}{P_s(t)} \;=\; \frac{\sum_{c(s)} P_c(t) \sum_{c'(s')} w_{cc'}}{\sum_{c(s)} P_c(t)}\,. \tag{12}$$

In contrast to the transition rates of the total system, which are usually assumed to be constant, the effective transition rates in the subsystem may depend on time. With these time-dependent rates the subsystem evolves according to the master equation

$$\frac{\mathrm{d}}{\mathrm{d}t} P_s(t) \;=\; \sum_{s' \in \Omega_{\rm sys}} \Bigl( J_{s's}(t) - J_{ss'}(t) \Bigr) \qquad J_{ss'}(t) = P_s(t) w_{ss'}(t)\,. \tag{13}$$

From the subsystems point of view this time dependence reflects the unknown dynamics in the environment. Moreover, ergodicity plays a subtle role: Even if the dynamics of the total system was ergodic, the dynamics *within* the sectors $\pi^{-1}(c)$ is generally non-ergodic and may decompose into several ergodic subsectors. As we will see in Sect. 4, this allows the environmental entropy to increase even if the subsystem is stationary.

**Systems far from thermal equilibrium**

In Nature many systems are not thermalized but out of equilibrium. For this reason the study of non-equilibrium systems plays an increasingly important role in statistical physics. It should be noted that different communities use term 'non-equilibrium' in a different way. In the context of thermodynamics, non-equilibrium refers to non-stationary situations close to thermal equilibrium, while statistical physicists use this term for systems violating detailed balance. In the following we use the nomenclature that a system is

- in *thermal equilibrium* if its probability distribution is stationary obeying detailed balance.
- *thermalizing* if it relaxes towards thermal equilibrium with balanced rates obeying Eq. (5).
- in a *non-thermal equilibrium* if its probability distribution is in a stationary state without satisfying detailed balance.
- in *out of thermal equilibrium* if it is non-stationary violating Eq. (5).

Since in Nature isolated systems are expected to thermalize, we can conclude that conversely a non-thermalizing system must always interact with the environment. This means that an external drive is needed to prevent the system from thermalizing, maintaining its non-vanishing probability currents. On the other hand, the total system composed of laboratory system and environment should thermalize. This raises the question how a thermalizing total system can contain a non-thermalizing subsystem?

The answer to this question is given in Eq. (12). Even if the total system was predetermined to thermalize, meaning that the rates $w_{cc'}$ obey Eq. (5), one can easily show that the effective rates $w_{ss'}$ of transitions in the subsystem are generally not balanced. Therefore, a thermalizing 'Universe' may in fact contain subsystem out of thermal equilibrium. The apparent contradiction is resolved by the observation that the projected rates $w_{ss'}(t)$ depend on time: Although these rates may initially violate detailed balance, they will slowly change as the 'Universe' continues to thermalize, eventually converging to values where they do obey detailed balance. This process reflects our everyday experience that any non-thermalizing system will eventually thermalize when the external drive runs out of power.

## 3. Justification from the quantum perspective

The standard setup of statistical mechanics as described in section 2 is amazingly successful in explaining a wide range of physical processes. In stark contrast to this strong *justification by corroboration* the question of whether and how it can be justified *microscopically* is still open to a great extent. Over the course of the last century many famous physicists, such as Ludwig Boltzmann, have tried to derive statistical physics from Newtonian mechanics. Despite major efforts no fully convincing and commonly accepted microscopic foundation for statistical physics could be found [5]. Nowadays, quantum mechanics is the commonly accepted microscopic theory but a complete justification of statistical mechanics from quantum mechanics is yet to be achieved. This is surprising as already the founding fathers of quantum theory have worked on this problem [6].

Recently, stimulated by new experiments [7,8] and novel methods from quantum information theory, a renewed interest in such fundamental questions ignited a flurry of activity in this field with significant new insights, leading to a reconsideration of the axiomatic system of statistical

Journal of Physics: Conference Series **297** (2011) 012011 doi:10.1088/1742-6596/297/1/012011

mechanics. As we shall see below, quantum mechanics and statistical physics apparently contradict each other but, very surprisingly, at the same time some genuine features of quantum mechanics can be used to justify essential assumptions of statistical mechanics. In this Section we discuss some of these recent developments, which we think are of interest to a wider audience, in a non-technical fashion. Readers not interested in the relationship of quantum mechanics and statistical physics can safely skip this section.

**Emergence of the classical state space**

One aspect of the quantum mechanical foundations of statistical mechanics is to explain why macroscopic systems can be well described without considering the quantum mechanical nature at the microscopic level. In the following we discuss a general mechanism that explains how classical ensembles can emerge from the microscopic quantum dynamics.

As an example for an intrinsically quantum mechanical process that can be successfully described within the framework of statistical mechanics, let us consider the emission and absorption of light (photons) by atoms. This example includes both equilibrium situations, like the interaction of atoms with black body ration, and extreme nonequilibrium situation, like the pumping of a laser. The statement that emission and absorption of light can be described within the framework of statistical mechanics is to be understood in the following sense: quantum mechanics enters the description only in so far as it defines the energy levels of the atom. The relevant energy eigenstates of the quantum Hamiltonian are then taken to be the configurations $s$ that constitute the classical state space $\Omega_{\text{sys}}$ of the atom while the dynamics is entirely described in terms of transition rates between these levels, corresponding to absorption, spontaneous emission, and stimulated emission. Although this description turned out to be useful, it is not at all clear why this simplified, classical treatment of-light matter interaction is eligible. As an essential feature, quantum mechanics allows systems to be in *coherent superpositions* of energy eigenstates rather than *probabilistic superpositions*. Why is it possible to simply ignore this fundamental feature of quantum mechanics and work with a classical description that only includes incoherent, *probabilistic superpositions* of energy eigenstates and transitions between them?

The first step in resolving this problem is to realize that it is in general impossible to completely isolate a system from its environment. By tracing out the degrees of freedom in the environment it is easy to show that the time evolution of such an interacting system is not unitary anymore. In particular, the interaction with the environment can suppress coherent superpositions in the system, a process called *decoherence* [9–11]. The insights obtained in the field of decoherence theory have proved to be extremely valuable, for both applications and from a fundamental perspective. However, most of the results are restricted to specific models or rely on assumptions such as a special form of the interaction or on approximations that are hard to control. Given the broad applicability of statistical mechanics and thermodynamics one would rather want to have an explanation for the emergence of classical state spaces from quantum theory that does not rely on such details.

Recently, building on earlier works [12, 13], a result in this spirit was obtained in Ref. [14]. It can be summarized informally as follows. Whenever a system interacts *weakly* with an environment its quantum state (density operator) is close to being in a purely probabilistic superposition of energy eigenstates for most times. The result shows that quantum mechanics implies a generic mechanism leading to a suppression of coherent superpositions in such a way that an observer measuring the system at an arbitrary point in time is most likely to find the system in a state where only classical probabilistic superpositions contribute significantly. In other words, the coherent quantum dynamics of system plus environment suppresses coherence

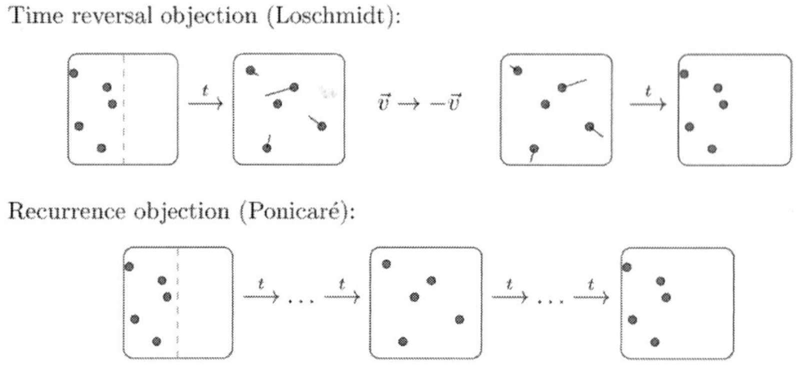


**Figure 3.** Schematic depiction of the two objections against irreversibility in systems based on Newtonian mechanics. According to Loschmidt's *time reversal objection*, for every process that increases the entropy of a system, say the spreading of particles in a container after removal of a wall, there exists a corresponding time-reversed process, which would reduce the entropy of the system. According to Ponicaré's *recurrence objection*, whenever the entropy of a system increased, the recurrent nature of Newtonian evolution implies that the entropy has to decrease after a sufficiently long period of time. The two objections illustrate that the Second Law in its usual form is incompatible with Newtonian mechanics. Equivalent objections apply to quantum mechanics which is also time reversal invariant and recurrent.

in the subsystem. The result can be expressed as a rigorous inequality bounding the off-diagonal entries in the density matrix of the subsystem. Its proof does not rely on any special properties of the interaction, it rather follows directly from the full unitary evolution of the total system without any approximations. The inequality is meaningful only if the interaction with the environment is *weak*, meaning that it must be smaller (or at least not significantly larger) than the gaps between the energy levels of the system.

Can this result explain why the energy eigenbasis is the correct choice for the state space in the classical description of an atom absorbing and emitting light? Thinking of the atom plus the surrounding light field as a quantum system subjected to decoherence by interaction with other atoms of the gas it does. In fact, the energy gaps of a few eV are several orders of magnitude larger than the thermal energy of the atoms (about 25 meV at room temperature), which sets the relevant energy scale for the coupling to other atoms of the gas. The weak perturbation caused by thermal scattering in the gas naturally leads to a decoherence into the energy eigenbasis, meaning that the classical description in terms of a probability $P_s(t)$ to be in the energy eigenstate $s$ at time $t$ describes the state of the system almost completely.

### Equilibration and apparent irreversibility

Another important issue is an apparent incompatibility between quantum mechanics and statistical physics: While the Second Law, a corner stone of statistical physics, postulates the irreversible convergence of $P_s(t)$ to a stable equilibrium distribution in the limit $t \to \infty$, the unitary dynamics of isolated quantum systems is *time reversal invariant* and *recurrent*; meaning that for every initial state the system will eventually return to an almost indistinguishable state after a certain (possibly extremely large) recurrence time. The famous objections brought forward by Loschmidt and Ponicaré against Boltzmann's H-Theorem, which was intended to be a proof of the Second Law for Newtonian mechanics, the so called *time reversal objection* and the *recurrence objection* (see Fig. 3) equally apply if quantum mechanics is taken to be the underlying microscopic theory.

These two objections show that true equilibration in the usual sense of statistical mechanics

Journal of Physics: Conference Series **297** (2011) 012011 doi:10.1088/1742-6596/297/1/012011

is impossible in quantum theory. The origin of the tendency to evolve towards equilibration, which constitutes a crucial part of the framework of statistical physics, on first sight, appears to be miraculous from the quantum perspective. However, recent studies show a way how it is possible to resolve this apparent contradiction. All we can hope for is to find equilibration in a weaker sense, meaning that the system is *almost equilibrated for most times*. That is, the expectation values of a set of relevant observables could *evolve towards* and then *stay close* to a certain *equilibrium value* for *most times* during the evolution, even if the system started far from equilibrium. The notion of equilibration in classical statistical mechanics that comes closest to the quantum version of equilibrium seems to be what is called a *non-thermal equilibrium* above.

Of particular interest are observables acting on a small part of a larger composite system. Surprisingly, as shown in a series of recent papers [12, 15, 16], it is possible to rigorously prove equilibration for almost all times for such observables from the unitary, time reversal invariant, and recurrent time evolution of quantum mechanics. More specifically, whenever the dimension of the Hilbert space of a subsystem is much smaller than a quantity called the effective dimension of the initial state, it is shown that the unitary dynamics of the full system is such that for most times the states of all small subsystems are practically indistinguishable from apparent equilibrium states. The proof requires that the Hamiltonian is fully interactive and non-degenerate, i.e., it does not decompose into non-interacting subsystems, – a very weak and quite natural assumption.

The effective dimension, which is defined as $d_{\text{eff}}(\rho) = 1/\text{Tr}\rho^2$, measures how many eigenstates of the Hamiltonian contribute significantly to the initial state. Can we expect it to be large in realistic situations? In realistic many particle systems energy is approximately extensive. Even if we assume it to grow at most polynomially with the number of constituents, then an energy interval of fixed width will usually contain exponentially many energy eigenstates. This implies that even states with a very small energy uncertainty will usually consist of exponentially many energy eigenstates and thus it is safe to assume that the effective dimension will be very large in realistic situations. Thus we can conclude that equilibration of small subsystems in large, interacting quantum systems is a generic property. Similar results can be obtained for certain sets of global observables [15, 16].

### Maximum entropy principle

In his seminal papers [1, 2] E. T. Jaynes suggested a *maximum entropy principle* as a possible foundation for statistical mechanics. In short, Jaynes argues that the correct method to calculate expectation values of observables that give only limited knowledge about a system is to take the state with maximum entropy among the configurations that are compatible with our partial knowledge. His argument is based on information theoretic considerations of the method of statistical inference.

Surprisingly, the unitary dynamics of quantum mechanics does also naturally imply a maximum entropy principle that is quite similar in style. Recently it was shown in Ref. [17] that whenever an expectation value of some observable equilibrates in the sense defined above, it equilibrates towards the expectation value it would have in the state that maximizes the von Neumann entropy among the states that have the same expectation values for all conserved quantities, i.e., of all observables that commute with the Hamiltonian. Again the result is derived without making any special assumptions and without approximations. In contrast to the common approach in which the maximum entropy principle appears as an axiom, this result follows directly from first principles when we chose a microscopic quantum description. What makes the maximum entropy principle from quantum dynamics different form the usual Jaynes' principle is that the number of conserved quantities of a quantum system grows with the dimension of the Hilbert space and thus exponentially with the system size. Contrarily, Jaynes formulated his maximum entropy principle having situations with partial knowledge about a handful of natural physical observables in mind. How exactly these two maximum entropy principles are related is yet to be explored in more detail.

### Thermalization

One of the most important applications of equilibrium statistical physics is to calculate the properties of systems at a well defined temperature.The standard assumption going into these calculations is that the state of such a system is described by a Gibbs state. The Gibbs state and the canonical ensemble can be derived from the equal a priory probability postulate under certain assumptions about the density of states of a bath with which the system can exchange energy. Alternatively it is possible to justify the Gibbs state by using Jaynes' maximum entropy principle, showing that the Gibbs state is the state that maximizes the conditional entropy given a fixed energy expectation value. However, it remains unclear how, and under which conditions, subsystems of quantum systems actually thermalize, by which, in this section, we mean that it equilibrates towards a Gibbs states with a well defined temperature. Note that it is not easy to relate the notion of thermalization we use in this section to the detailed balance condition used throughout the rest of this article.

Earlier works attempting to solve this problem [18–20] either rely on certain unproven hypotheses such as the so-called *eigenstate thermalization hypothesis*, or they are restricted to quite special situations such as coupling Hamiltonians of a special form, or they merely prove typicality arguments instead of dynamical relaxation towards a Gibbs state. Although the results obtained in these papers are very useful and have significantly improved our understanding of the process of thermalization, they do not yet draw a complete and coherent picture.

An attempt to settle the question of thermalization will be made in a forthcoming paper [21]. As discussed above we already know conditions under which we can rigorously guarantee equilibration [12, 15, 16]. What remains to be done is to identify a set of conditions under which one can guarantee that the equilibrium state of a subsystem is close to a Gibbs state. By using a novel perturbation theory argument and carefully bounding all the errors in an approximation similar to that of [20] one can indeed identify such a set of sufficient (and more or less necessary) conditions, that can be summarized in a non-technical way as follows:

(i) The energy content and the Hilbert space dimension of the bath must be much larger than the respective quantities of the system.

(ii) The coupling between them must be strong enough, in particular much stronger than the gaps of the decoupled Hamiltonian. This ensures that the eigenbasis of the full Hamiltonian is sufficiently entangled (a lack of entanglement provably prevents thermalization [17]).

(iii) At the same time, the coupling must be weak in the sense that it is much smaller than the energy uncertainty of the initial state. This is a natural counterpart to the weak coupling assumption known from classical statistical mechanics. item[(iv)] The energy uncertainty of the initial state must be small compared to the energy content and at the same time large compared to the level spacing. Moreover, the energy distribution must satisfy certain technical smoothness conditions.

(v) The spectrum of the bath must be well approximable by an exponential on the scale of the energy uncertainty and the density of states must grow faster than exponential. This property of the bath is ultimately the reason for the exponential form of the Gibbs state and is also required in the classical derivation of the canonical ensemble. Most natural many particle systems have this property.

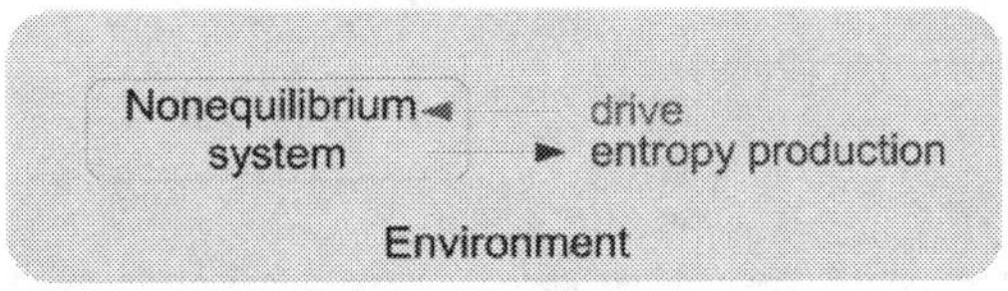


**Figure 4.** Entropy production: A non-thermalizing system cannot exist on its own but must be driven from the outside. The external drive, that keeps the system away from thermal equilibrium, inevitably increases the entropy in the environment.

In summary, one can say that more or less the same conditions that are used in the classical derivation of the canonical ensemble appear naturally in the proof of dynamical thermalization.

### Time scales

The most important open problem for the approach described above is that rigorous bound on the time scales for decoherence/equilibration/thermalization are not yet known. The results derived in [12–17] only tell us that decoherence/equilibration/thermalization must eventually happen under the given conditions, but they do not tell us how long it takes. In general this seems to be tough question, but for exactly solvable models the time scales can be derived [22].

## 4. Entropy production

Returning to the classical framework, let us now study the problem of entropy production. As outlined in Sect. 2, thermalizing systems (i.e. systems with balanced rates relaxing into thermal equilibrium) can contain subsystems which are out of thermal equilibrium in the sense that the transition rates $w_{ss'}$ do not obey detailed balance. The apparent contradiction is resolved by observing that the effective rates in the subsystem are generally time-dependent and will eventually adjust in such a way that the subsystem thermalizes as well. However, for a limited time it is possible to keep them constant in such a way that they violate detailed balance. This is exactly what happens in experiments far from equilibrium – typically they rely on external power and will quickly thermalize as soon as power is turned off.

The external drive which is necessary to keep a subsystem away from thermal equilibrium will on average increase the entropy in the environment, as sketched in Fig. 4. In the following we discuss various attempts to quantify this entropy production.

### Entropy changes

For a subsystem embedded in an environment we distinguish three types of configurational entropies, namely, the configurational entropy of the total system ('Universe'), the entropy of the subsystem (experiment) and the entropy in its environment:

$$S_{\text{tot}}(t,c) = -\ln P_c(t)\,, \qquad (14)$$
$$S_{\text{sys}}(t,s) = -\ln P_s(t)\,, \qquad (15)$$
$$S_{\text{env}}(t,c) = S_{\text{tot}}(t,c) - S_{\text{sys}}(t,\pi(c))\,. \qquad (16)$$

Averaging over many realizations the corresponding mean entropies are given by

$$\bar{S}_{\rm tot}(t) \;=\; \langle S_{\rm tot}(t,c)\rangle_c \;=\; -\sum_{c\in\Omega_{\rm tot}} P_c(t)\ln P_c(t)\,, \tag{17}$$

$$\bar{S}_{\rm sys}(t) \;=\; \langle S_{\rm sys}(t,s)\rangle_s \;=\; -\sum_{s\in\Omega_{\rm sys}} P_s(t)\ln P_s(t)\,, \tag{18}$$

$$\bar{S}_{\rm env}(t) \;=\; \langle S_{\rm env}(t,c)\rangle_s \;=\; \bar{S}_{\rm tot}(t) - \bar{S}_{\rm sys}(t)\,. \tag{19}$$

The time derivative of these averages is given by

$$\frac{\mathrm{d}}{\mathrm{d}t}\bar{S}_{\rm tot}(t) \;=\; \sum_{c,c'\in\Omega_{\rm tot}} J_{cc'}(t)\ln\frac{P_c(t)}{P_{c'}(t)}\,, \tag{20}$$

$$\frac{\mathrm{d}}{\mathrm{d}t}\bar{S}_{\rm sys}(t) \;=\; \sum_{s,s'\in\Omega_{\rm sys}} J_{ss'}(t)\ln\frac{P_s(t)}{P_{s'}(t)}\,, \tag{21}$$

where we used the master equations (13) and (9).

Let us now consider a temporal regime in which the rates of the subsystem can be considered as constant. In this case the subsystem, which is often small compared to the environment, may quickly reach a non-thermalized stationary state, which is often referred to as *non-equilibrium steady state* (NESS) in the literature. In this case the average entropy of the subsystem will saturate whereas the total system thermalizes according to the Second Law, meaning that the average entropy in the environment increases. This environmental entropy production is the price Nature has to pay for keeping a subsystem away from thermal equilibrium.

To be more specific, let us now assume that the total system follows a particular stochastic path $\Gamma : t \mapsto c(t)$

$$\Gamma:\; c_0 \to c_1 \to c_2 \to \dots \qquad \text{at times} \qquad t_0, t_1, t_2, \dots \tag{22}$$

Whenever $\pi(c_i) \neq \pi(c_{i+1})$ a transition in the total system implies a transition in the subsystem, as sketched in Fig. 2. Denoting the corresponding transition times by $t_{n_i}$, the projected stochastic path of the subsystem $\gamma = \pi[\Gamma]$ reads

$$\gamma:\; s_0 \to s_1 \to s_2 \to \dots \qquad \text{at times} \qquad t_{n_0}, t_{n_1}, t_{n_2}, \dots \tag{23}$$

where $s_i = \pi(c_{n_i})$. Along their respective stochastic paths the configurational entropies of the total system and the subsystem are given by

$$S^{\Gamma}_{\rm tot}(t) \;=\; -\ln P_{c(t)}(t)\,, \tag{24}$$

$$S^{\gamma}_{\rm sys}(t) \;=\; -\ln P_{s(t)}(t)\,. \tag{25}$$

How do these quantities change with time? Following Ref. [23] the temporal evolution of the configurational entropy is made up of a continuous contribution caused by the deterministic evolution of the master equation and a discontinuous contribution occurring whenever the system hops to a different configuration. This means that the time derivative of the systems entropy is given by

$$\frac{\mathrm{d}}{\mathrm{d}t}S^{\gamma}_{\rm sys}(t) \;=\; -\frac{\dot{P}_{s(t)}(t)}{P_{s(t)}(t)} - \sum_j \delta(t-t_{n_j})\ln\frac{P_{s_j}(t)}{P_{s_{j-1}}(t)}\,. \tag{26}$$

Similarly, the total entropy of the 'Universe' is expected to change as

$$\frac{\mathrm{d}}{\mathrm{d}t}S^{\Gamma}_{\mathrm{tot}}(t) \;=\; -\frac{\dot{P}_{c(t)}(t)}{P_{c(t)}(t)} - \sum_n \delta(t-t_n)\ln\frac{P_{c_n}(t)}{P_{c_{n-1}}(t)} \tag{27}$$

so that the environmental entropy production is given by their difference:

$$\frac{\mathrm{d}}{\mathrm{d}t}S^{\Gamma}_{\mathrm{env}}(t) \;=\; \frac{\mathrm{d}}{\mathrm{d}t}S^{\gamma}_{\mathrm{sys}}(t) - \frac{\mathrm{d}}{\mathrm{d}t}S^{\Gamma}_{\mathrm{tot}}(t)\,. \tag{28}$$

This formula is exact but useless from a practical point of view because the actual stochastic trajectory $\Gamma$ of the total system (the whole 'Universe') is generally not known.

**Effective environmental entropy production**

Based on previous work by Andrieux and Gaspard [24], Seifert suggested a very compact formula for the effective entropy production in the environment caused by the embedded subsystem [23]:

$$\boxed{\frac{\mathrm{d}}{\mathrm{d}t}S^{\gamma}_{\mathrm{env}}(t) \;=\; -\sum_j \delta(t-t_{n_j})\ln\frac{w_{s_j s_{j+1}}(t)}{w_{s_{j+1} s_j}(t)}\,.} \tag{29}$$

This formula tells us that each transition $s \to s'$ in the subsystem causes an instantaneous change of the environmental entropy by the log ratio of the forward rate $w_{ss'}$ divided by the backward rate $w_{s's}$. Together with Eq. (26) this formula would imply that the *total* entropy changes according to

$$\frac{\mathrm{d}}{\mathrm{d}t}S^{\gamma}_{\mathrm{tot}}(t) \;=\; -\frac{\dot{P}_{s(t)}(t)}{P_{s(t)}(t)} - \sum_j \delta(t-t_{n_j})\ln\frac{P_{s_j}(t)w_{s_j s_{j+1}}(t)}{P_{s_{j-1}}(t)w_{s_{j+1} s_j}(t)}\,. \tag{30}$$

This expression differs significantly from the exact formula (27) so that it can be only meaningful in an effective sense under certain conditions or in a particular limit.

Before discussing these underlying assumptions in detail, we like to note that Eq. (29) is indeed very elegant. It does not require any knowledge about the nature of the environment, instead it depends exclusively on the stochastic path of the subsystem and the corresponding transition rates. Moreover, this quantity can be computed very easily in numerical simulations: Whenever the program selects the move $s \to s'$, all what has to be done is to increase the environmental entropy variable by $\ln(w_{ss'}/w_{s's})$[1]. Note that the logarithmic ratio of the rates requires each transition to be reversible.

To motivate formula (29) heuristically, Seifert argues the corresponding *averages* of the entropy production reproduce a well-known result in the literature. More specifically, he shows that Eq. (29) averaged over many possible paths gives the expression.

$$\frac{\mathrm{d}}{\mathrm{d}t}\bar{S}_{\mathrm{env}}(t) \;=\; \sum_{s,s'\in\Omega_{\mathrm{sys}}} J_{ss'}(t)\ln\frac{w_{ss'}(t)}{w_{s's}(t)}\,. \tag{31}$$

Combined with Eq. (21) one obtains the *average* entropy production in the total system

$$\frac{\mathrm{d}}{\mathrm{d}t}\bar{S}_{\mathrm{tot}}(t) \;=\; \sum_{s,s'\in\Omega_{\mathrm{sys}}} J_{ss'}(t)\ln\frac{P_s(t)w_{ss'}(t)}{P_{s'}(t)w_{s's}(t)}\,. \tag{32}$$

[1] In order to avoid unnecessary floating point operations, it is useful to store all possible log ratios of the rates in an array.

This formula was first introduced by Schnakenberg [25] and has been frequently used in chemistry and physics [26]. It is in fact very interesting to see how Schnakenberg derived this formula. As described in detail in Appendix A, he considered a fictitious chemical system of homogenized interacting substances which resemble the dynamics of the master equation in terms of particle concentrations. Applying standard methods of thermodynamics, he was able to prove Eq. (32). The rational behind this derivation is to assume that the environment is always close to thermal equilibrium.

**Limit of fast thermalization in the environment**

In the following we show that the entropy production formula is correct in the limit where the environment equilibrates immediately whenever a transition occurs in the subsystem. This requires a separation of time scales of the internal dynamics of the subsystem on the one hand and the relaxation in the environment on the other.

As shown in Sect. 2, the projection $\pi : \Omega_{\text{tot}} \to \Omega_{\text{sys}}$ divides the configuration space of the total system into sectors of configurations $c$ mapped onto the same $s$ (see Fig. 2). It is important to note that even if the dynamics of the total system was ergodic, the dynamics *within* these sectors is generally non-ergodic, meaning that they may split up into various ergodic subsectors. For example, if the subsystem starting in a certain configuration returns to the same configuration after some time, the corresponding subsector may have changed, reflecting the change of entropy in the environment. For a given stochastic path of the small system, we will denote the corresponding subsector as $\Omega^{s(t)}_{\text{tot}} \subset \Omega_{\text{tot}}$.

Let us now assume that the environmental degrees of freedom thermalize almost instantaneously, reaching maximal entropy within the actual subsector. This means that the system quickly reaches a state where the probabilities

$$P_c(t) = \begin{cases} P_s(t)/N_s(t) & \text{if } c \in \Omega^{s(t)}_{\text{tot}} \\ 0 & \text{otherwise.} \end{cases} \tag{33}$$

are constant on $\Omega^{s(t)}_{\text{tot}}$. This implies that in the formula (28) for the environmental entropy production, namely

$$\frac{\mathrm{d}}{\mathrm{d}t} S^{\Gamma}_{\text{env}}(t) = \frac{\dot{P}_{c(t)}(t)}{P_{c(t)}(t)} - \frac{\dot{P}_{s(t)}(t)}{P_{s(t)}(t)} + \sum_j \delta(t - t_{n_j}) \ln \frac{P_{s_j}(t)}{P_{s_{j-1}}(t)} - \sum_n \delta(t - t_n) \ln \frac{P_{c_n}(t)}{P_{c_{n-1}}(t)}, \tag{34}$$

the first two terms cancel. Therefore, this expression reduces to

$$\frac{\mathrm{d}}{\mathrm{d}t} S^{\Gamma}_{\text{env}}(t) = \sum_j \delta(t - t_{n_j}) \ln \frac{P_{s_j}(t)}{P_{s_{j-1}}(t)} - \sum_n \delta(t - t_n) \ln \frac{P_{\pi(c_n)}(t) N_{\pi(c_{n-1})}(t)}{P_{\pi(c_{n-1})}(t) N_{\pi(c_n)}(t)}. \tag{35}$$

Obviously, only those terms in the second sum will contribute where $\pi(c_{n-1}) \neq \pi(c_n)$, i.e. where $n = n_j$, hence the sum can be reorganized as

$$\frac{\mathrm{d}}{\mathrm{d}t} S^{\gamma}_{\text{env}}(t) = \sum_j \delta(t - t_{n_j}) \left[ \frac{P_{s_j}(t)}{P_{s_{j-1}}(t)} - \ln \frac{P_{s_j}(t) N_{s_{j-1}}(t)}{P_{s_{j-1}}(t) N_{s_j}(t)} \right] = \sum_j \delta(t - t_{n_j}) \ln \frac{N_{s_j}(t)}{N_{s_{j-1}}(t)} \tag{36}$$

which now depends only on the stochastic path $\gamma$ of the subsystem. This result, saying that the entropy increase is given by the logarithmic ratio of the number of available configurations, is very plausible under the assumption of instantaneous thermalization of the environment. It

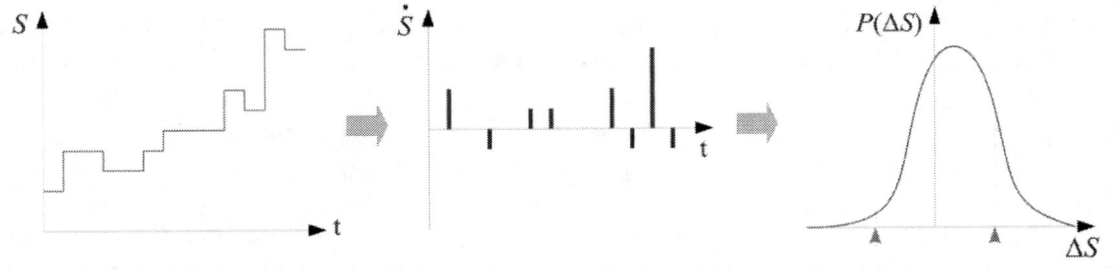


**Figure 5.** Configurational entropy of an isolated system, its temporal derivative and the corresponding probability distribution of entropy differences.

remains to be shown that this ratio is related to the ratio of the effective rates. In fact, inserting (33) into (12) we obtain

$$w_{ss'}(t) = \frac{\sum_{c(s)} P_c(t) \sum_{c'(s')} w_{cc'}}{\sum_{c(s)} P_c(t)} = \frac{\sum_{c(s)} \sum_{c'(s')} w_{cc'}}{N_s(t)}, \tag{37}$$

hence $w_{ss'}/w_{s's} = N_{s'}/N_s$. Inserting this relationship into Eq. (36) we arrive at the formula for the effective entropy production (29). This proves that this formula is valid under the assumption that the environmental degrees of freedom thermalize instantaneously after each transition of the subsystem.

### Conjecture for systems in a non-thermalized environment

What happens if the environment does not thermalize immediately after each transition of the subsystem? To answer this question we performed numerical simulations of a three-state system in an environment with 50 configurations and randomly chosen transition rates. The results (not shown here) suggest that the entropy production predicted by formula (29) deviates from the true entropy production in both directions. However, averaging over many independent stochastic paths while keeping the rates fixed, the formula seems to systematically underestimate the actual entropy production. This leads us to the conjecture that formula (29) may serve as a lower bound for the expectation value of the entropy production.

## 5. Fluctuation theorem revisited

The Second Law of thermodynamics tells us that the entropy $S_{\text{tot}}$ of an isolated system increases on average during thermalization, approaching the maximal value $S_{\text{tot}} = \ln|\Omega_{\text{tot}}|$. The fluctuation theorem generalizes this statement by studying the actual fluctuations of the total entropy, including the Second Law as a special case [27–35]. In this section we want to suggest an alternative particularly transparent proof of the fluctuation theorem.

The generic situation is sketched in Fig. 5. Although the average entropy of an isolated system will increase, the actual configurational entropy will not grow monotonously, instead it jumps discontinuously by finite differences $\Delta S_{\text{tot}}$. As the average entropy increases, these differences will be preferentially positive, but sometimes fluctuations in opposite direction may occur. Depending on the system under consideration, these differences will be distributed according to a certain asymmetric distribution. The fluctuation theorem states that this distribution is constrained by the condition

$$\frac{P(\Delta S_{\text{tot}})}{P(-\Delta S_{\text{tot}})} = e^{\Delta S_{\text{tot}}}. \tag{38}$$

Journal of Physics: Conference Series **297** (2011) 012011 doi:10.1088/1742-6596/297/1/012011

It relates pairs of values at opposite locations on the abscissa, as marked by the red arrows in Fig. 5. In other words, given one half of the distribution, the fluctuation theorem predicts the other half.

The fluctuation theorem holds for any stochastic system and is usually proved by comparing the entropy production along a given stochastic path with the entropy production along the reverse path. Here we suggest an alternative simple proof. Starting point is the observation that the fluctuation relation is invariant under convolution, i.e. if two function $f, g$ satisfy the property $f(x) = e^x f(-x)$ and $g(x) = e^x g(-x)$ their convolution product will satisfy the same property:

$$\begin{aligned}(f * g)(x) &= \int \mathrm{d}y\, f(y)g(x-y) = \int \mathrm{d}y\, e^y f(-y)e^{x-y}g(-x+y) \\ &= e^x \int \mathrm{d}y\, f(-y)g(y-x) = e^x \int \mathrm{d}y\, f(y)g(-y+x) = e^x (f * g)(-x)\,.\end{aligned} \tag{39}$$

This means that the sum of random variables obeying the fluctuation relation will again obey the fluctuation relation. The remaining proof consists of two steps:

(i) First we prove the fluctuation theorem for a *single* transition. According to Eq. (27), an *individual* transition $c \to c'$ changes the configurational entropy of an isolated system by $\Delta S_{\mathrm{tot}} = \ln J_{cc'} - \ln J_{c'c}$. Since transition occurs with frequency $P_c w_{cc'} = J_{cc'}$, we have $P(\Delta S_{\mathrm{tot}}) = J_{cc'}$ and similarly $P(-\Delta S_{\mathrm{tot}}) = J_{c'c}$ for the reverse transition. Therefore the fluctuation relation
$$\Delta S_{\mathrm{tot}} = \ln\Big(P(\Delta S_{\mathrm{tot}})/P(-\Delta S_{\mathrm{tot}})\Big) \tag{40}$$
holds trivially for a single transition.

(ii) The entropy change $\Delta S$ over a finite time is the sum of entropy changes caused by individual transition. Summing random variables means to convolve their probability distributions. Since the fluctuation theorem is invariant under convolution, it follows that this sum will automatically obey the fluctuation relation as well.

As well-known, the fluctuation theorem implies the nonequilibrium partition identity
$$\langle \exp(-\Delta S_{\mathrm{tot}}) \rangle = 1 \tag{41}$$
as well as the Second Law of thermodynamics
$$\langle \Delta S_{\mathrm{tot}} \rangle \geq 0. \tag{42}$$
Note that the fluctuation theorem holds exactly only in isolated systems. However, it may also hold approximately in the limit $t \to \infty$ for the environmental entropy under certain assumptions if the subsystem is stationary.

## 6. Concluding remarks

In this paper we have addressed several aspects of classical non-equilibrium statistical physics, describing its general setup and its justification from the quantum perspective. In particular, we have focused on the problem of entropy production. As we have pointed out, the commonly accepted formula for entropy production in the environment $\Delta S_{\mathrm{env}} = \ln(w_{ss'}/w_{s's})$ holds only in situations where the environment thermalizes almost immediately after each transition of the subsystem. Whether this separation of time scales is valid in realistic situations remains to be seen. Moreover, we have suggested a conjecture that this formula gives a lower bound to the average entropy production in the environment.

Journal of Physics: Conference Series **297** (2011) 012011 doi:10.1088/1742-6596/297/1/012011

## Appendix A. Tracing the historical route to entropy production

It is instructive to retrace how the formula for entropy production was derived by Schnakenberg in 1976 [25]. To quantify entropy production, Schnakenberg considers a fictitious chemical system that mimics the dynamics of the master equation. This fictitious system is based on the following assumptions:

- Each configuration $c$ of the original system is associated with a chemical species $X_c$ in an ideal homogeneous mixture of molecules.
- The molecules react under isothermal and isochoric conditions by $X_c \rightleftharpoons X_{c'}$ in such a way that their concentrations $[X_c] = N_c/V$ evolve in the same way as the probabilities in the master equation, i.e.
$$\frac{\mathrm{d}}{\mathrm{d}t}[X_c] = \sum_{c'}\Big([X_{c'}]w_{c'c} - [X_c]w_{cc'}\Big). \tag{A.1}$$
- The reactions are so slow that standard methods of thermodynamics can be applied.

Under isothermal and isochoric conditions the chemical reactions change the particle numbers $N_i$ in such a way that the Helmholtz free energy $F$ is maximized. In chemistry the corresponding thermodynamic current is called the *extent of reaction* $\xi_{cc'}$, which is defined as the expectation value of the accumulated number of forward reactions $X_c \to X_{c'}$ minus the number of backward reactions $X_c \leftarrow X_{c'}$. Note that $\xi_{cc'}$ does not account for fluctuations, instead it is understood as a macroscopic deterministic quantity that grows continuously as

$$\dot{\xi}_{cc'} = N_c w_{cc'} - N_{c'} w_{c'c}\,. \tag{A.2}$$

According to conventional thermodynamics, a thermodynamic flux is caused by a conjugate thermodynamic force which is the partial derivative of the thermodynamic potential with respect to the flux. In chemistry the thermodynamic force conjugate to the extent of reaction $\dot{\xi}_{cc'}$ is the so-called *chemical affinity*

$$A_{cc'} = \left.\frac{\partial F}{\partial \xi_{cc'}}\right|_{V,T}. \tag{A.3}$$

With this definition the temporal change of the free energy is given by

$$\dot{F} = \sum_{cc'} A_{cc'}\dot{\xi}_{cc'}\,. \tag{A.4}$$

The affinity is related to the chemical potential of the involved substances as follows. On the one hand, the reaction changes the particle number by $\dot{N}_c = -\dot{\xi}_{cc'}$ and $\dot{N}_{c'} = +\dot{\xi}_{cc'}$. On the other hand, the change of the free energy can be expressed as $\dot{F} = \sum_c \frac{\partial F}{\partial N_c}\dot{N}_c = \sum_c \mu_c \dot{N}_c$. Comparing this expression with Eq. (A.4) the affinity can be expressed as

$$A_{cc'} = \mu_{c'} - \mu_c\,. \tag{A.5}$$

For an ideal mixture the Helmholtz free energy is given by

$$F = \sum_c N_c(q_c + k_B T \ln N_c)\,, \tag{A.6}$$

where $q_c$ is a temperature-dependent constant. The chemical potential of species $X_c$ is

$$\mu_c = \frac{\partial F}{\partial N_c} = \mu_c^0 + k_B T \ln N_c \tag{A.7}$$

Journal of Physics: Conference Series **297** (2011) 012011 doi:10.1088/1742-6596/297/1/012011

with $\mu_c^0 = q_c + k_B T$, so that the affinity is given by

$$A_{cc'} = \mu_c^0 - \mu_{c'}^0 + k_B T \ln \frac{N_{c'}}{N_c}\,. \tag{A.8}$$

The fictitious chemical system relaxes towards an equilibrium state that corresponds to the stationary state of the original master equation. In this state the particle numbers $N_c$ attain certain stationary equilibrium values $N_c^{\rm eq}$. Moreover, the thermodynamic flux and its conjugate force vanish in equilibrium:

$$A_{cc'}^{\rm eq} = \xi_{cc'}^{\rm eq} = 0\,. \tag{A.9}$$

Because of $A_{cc'}^{\rm eq} = 0$ we have

$$\mu_c^0 - \mu_{c'}^0 = k_B T \ln \frac{N_{c'}^{\rm eq}}{N_c^{\rm eq}}\,, \tag{A.10}$$

which allows one to express the affinity as

$$A_{cc'} = k_B T \ln \frac{N_{c'} N_c^{\rm eq}}{N_c N_{c'}^{\rm eq}}\,. \tag{A.11}$$

On the other hand, $\xi_{cc'}^{\rm eq} = 0$ implies that

$$\frac{N_c^{\rm eq}}{N_{c'}^{\rm eq}} = \frac{w_{c'c}}{w_{cc'}}\,. \tag{A.12}$$

Inserting this relation into Eqs. (A.11) and (A.4) the change of the free energy (caused by all reactions $X_c \rightleftharpoons X_{c'}$) is given by

$$\dot{F} = k_B T \sum_{cc'} \xi_{cc'} \ln \frac{N_{c'} w_{c'c}}{N_c w_{cc'}}\,. \tag{A.13}$$

Since temperature $T$ and internal energy $U$ of the mixture remain constant the variation of the free energy $F = U - TS$ is fully absorbed in a change of the entropy, i.e. $\dot{F} = -T\dot{S}$. This allows one to derive a formula for the entropy production

$$\dot{S} = -k_B \sum_{cc'} \xi_{cc'} \ln \frac{[X_c] w_{cc'}}{[X_{c'}] w_{c'c}}\,, \tag{A.14}$$

where we used the fact that $N_c/N_{c'} = [X_c]/[X_{c'}]$.

Since the extent of reaction $\xi_{cc'}$ just counts the number of reactions from $c$ to $c'$, this formula is just the continuum limit of Eq. (30), proving the formula for entropy production (29). However, it is important to recall the underlying assumptions. Each component of the fictitious chemical systems was assumed to be internally equilibrated and methods of ordinary thermodynamics were used to derive Eq. (A.14). This means that the formula will only be meaningful in the limit where the environment equilibrates almost immediately.

## Acknowledgments

This work was initiated at conference STATPHYS-VII in Kolkata, India. HH would like to thank the organizers for the warm hospitality and for the lively and productive atmosphere. CG and PJ thank the Centre for Quantum Technologies, Singapore, where part of this work was done.

## References

[1] Jaynes E, 1957, *Information theory and statistical mechanics I*, Phys. Rev. **106** 620.
[2] Jaynes E, 1957, *Information theory and statistical mechanics II*, Phys. Rev. **108** 171.
[3] Schürmann T and Grassberger P, 1996 Chaos **6** 414.
[4] Bonachela JA, Hinrichsen H and Muñoz MA, 2008 J. Phys. A: Math. Theor. **41** 202001.
[5] Uffink J, *Compendium of the foundations of classical statistical physics*, `http://philsci-archive.pitt.edu/2691/`.
[6] Schrödinger E, 1927 Ann. Phys. **388**, 956; von Neumann J, 1929 Z. Phys. **57**, 30.
[7] Bloch I, Dalibard J and Zwerger W, 2008 Rev. Mod. Phys. **80** 885.
[8] Trotzky S, Chen Y-A, Flesch A, McCulloch IP, Schollwöck U, Eisert J, and Bloch I, 2011 preprint arXiv:1101.2659.
[9] Zurek WH, 2003 Rev. Mod. Phys. 715.
[10] Joos E, Zeh H, Kiefer C, Giulini D, Kupsch J, Stamatescu IO, 1996 *Decoherence and the Appearance of a Classical World in Quantum Theory* (Springer, Berlin).
[11] Schlosshauer M, 2007 *Decoherence and the Quantum to Classical Transition* (Springer, Berlin).
[12] Linden N, Popescu S, Short AJ, and Winter A, 2009 Phys. Rev. E **79**, 061103.
[13] Linden N, Popescu S, Short AJ, and Winter A, 2010 New J. Phys. **12**, 055021.
[14] Gogolin C, 2010 Phys. Rev. E **81**, 051127.
[15] Reimann P, 2008 Phys. Rev. Lett. **101**, 190403.
[16] Short AJ, 2010 preprint arXiv:1012.4622.
[17] Gogolin C, Müller MP, and Eisert J, 2011 Phys. Rev. Lett. **106**, 040401.
[18] Srednicki M, 1994 Phys. Rev. E **50** 888. .
[19] Tasaki H, 1998 Phys. Rev. Lett **80** 1373.
[20] Goldstein S, 2006 Phys. Rev. Lett. **96** 050403.
[21] Riera A, Gogolin C, and Eisert J, in preparation.
[22] Cramer M, Dawson CM, Eisert J, and Osborne TJ, 2008 Phys. Rev. Lett. **100**, 030602; Cramer M and Eisert J, 2010 New J. Phys. **12**, 055020.
[23] Seifert U 2005 Phys. Rev. Lett **95**, 040602.
[24] Andrieux D and Gaspard P 2004 J. Chem. Phys. **121** 6167
[25] Schnakenberg J 1976 Rev. Mod. Phys. **48** 571
[26] Jiu-li L, Van den Broeck C, and Nicolis G, 1984 Z. Phys. B: Condens. Matter **56**, 165.
[27] Evans D J, Cohen E G D and Morriss GP, 1993 Phys. Rev. Lett. **71** 2401.
[28] Evans D J and Searles D J, 1994 Phys. Rev. E **50** 1645.
[29] Gallavotti G and Cohen E G D, 1995 Phys. Rev. Lett. **74** 2694.
[30] Kurchan J, 1998 J. Phys. A: Math. Gen. **31** 3719.
[31] Lebowitz J L and Spohn H, 1999 J. Stat. Phys. **95** 333.
[32] Maes C, 1999 J. Stat. Phys. **95** 367.
[33] Jiang D-Q, Qian M and Qian M-P, 2004 *Mathematical Theory of Nonequilibrium Steady States* (Berlin: Springer)
[34] Harris R J and Schütz G M, 2007 J. Stat. Mech. P07020.
[35] Kurchan J, 2007 J. Stat. Mech. P07005.

Journal of Physics: Conference Series **297** (2011) 012012 doi:10.1088/1742-6596/297/1/012012

# Pattern-recalling processes in quantum Hopfield networks far from saturation

**Jun-ichi Inoue**

Graduate School of Information Science and Technology, Hokkaido University, N14-W9, Kita-ku, Sapporo 060-0814, Japan

E-mail: j_inoue@complex.ist.hokudai.ac.jp

**Abstract.** As a mathematical model of associative memories, the Hopfield model was now well-established and a lot of studies to reveal the pattern-recalling process have been done from various different approaches. As well-known, a single neuron is itself an uncertain, noisy unit with a finite unnegligible error in the input-output relation. To model the situation artificially, a kind of 'heat bath' that surrounds neurons is introduced. The heat bath, which is a source of noise, is specified by the 'temperature'. Several studies concerning the pattern-recalling processes of the Hopfield model governed by the Glauber-dynamics at finite temperature were already reported. However, we might extend the 'thermal noise' to the quantum-mechanical variant. In this paper, in terms of the stochastic process of quantum-mechanical Markov chain Monte Carlo method (the quantum MCMC), we analytically derive macroscopically deterministic equations of order parameters such as 'overlap' in a quantum-mechanical variant of the Hopfield neural networks (let us call *quantum Hopfield model* or *quantum Hopfield networks*). For the case in which non-extensive number $p$ of patterns are embedded via asymmetric Hebbian connections, namely, $p/N \to 0$ for the number of neuron $N \to \infty$ ('far from saturation'), we evaluate the recalling processes for one of the built-in patterns under the influence of quantum-mechanical noise.

## 1. Introduction

Basic concept of associative memories in artificial neural networks was already proposed in early 70' s by a Japanese engineer Kaoru Nakano [1]. Unfortunately, in that time, nobody interested in his model, however in 80's, J.J. Hopfield [2, 3] pointed out that there exists an energy function (Lyapunov function) in the so-called *associatron* (the Nakano model) and the system can be treated as a kind of spin glasses. After his study, a lot of researchers who were working in the research field of condensed matter physics picked the so-called *Hopfield model* up for their brand-new 'target materials'.

Among these studies, a remarkable progress has been done by three theoretical physicists Amit, Gutfreund and Sompolinsky [4] who clearly (mathematically) defined the concept of *storage capacity* in the Hopfield model as a critical point at which system undergoes phase transitions from ferromagnetic retrieval to spin glass phases by utilizing the replica method. They also introduced a noise to prevent the network from retrieving one of the built-in patterns as a 'heat bath' which surrounds the neurons. They draw the phase diagram which is composed of three distinct phases, namely, ferromagnetic-retrieval, paramagnetic and spin glass phases. These phase boundaries are specified by two control parameters, namely, storage capacity and temperature of the heat bath.

To evaluate the storage capacity of the Hopfield model without energy function (for instance, Hopfield model having a non-monotonic input-output function [5]), Shiino and Fukai [6] proposed the so-called Self-Consistent Signal-to-Noise Analysis (SCSNA) which enables us to derive a couple of self-consistent macroscopic equations of state by making use of the concept of the TAP equations [3].

As we mentioned, these theoretical arguments are constructed for the case in which each neuron is surrounded by a heat bath at finite temperature. In this sense, the above studies revealed the robustness of the associative memories against *thermal noises* in the artificial brain.

However, we might consider a different kind of such noises, that is, *quantum-mechanical noise*. As such successful attempts, Ma and Gong [7], Nishimori and Nonomura [8] independently introduced the quantum-mechanical noise to the conventional Hopfield model by adding the transverse field to the classical Hamiltonian (from now on, we shall refer the model as *quantum Hopfield model*). Especially, Nishimori and Nonomura [8] investigated the structure of retrieval phase diagrams by using of the standard replica method with the assistance of the static approximation. Therefore, we might say that the equilibrium properties of the Hopfield model were now well-understood and the methodology to investigate the model was already established.

On the other hand, theory of the dynamics to evaluate the pattern-recalling processes is not yet well-established. However, up to now, several powerful approaches were proposed. For instance, Amari and Maginu [9] pointed out that the relevant macroscopic quantities in the synchronous neuro-dynamics are the overlap (the direction cosine) and the noise variance. They derived the update equations with respect to these quantities. After their study, the so-called Amari-Maginu theory was improved by taking into account the correlations in the noise variances by Okada [10].

Whereas for the asynchronous dynamics, Coolen and his co-authors established a general approach, so-called *dynamical replica theory* [11, 12]. They utilized two assumptions, namely, equipartitioning in the sub-shells and self-averaging of the intrinsic noise distribution to derive the deterministic flow equations for relevant macroscopic quantities. However, there is no such theoretical framework so far to investigate the pattern-recalling process of the quantum Hopfield model systematically.

In this paper, we propose such a candidate of dynamical theory to deal with the pattern-recalling processes in the quantum Hopfield model. We shall consider the stochastic process of quantum Monte Carlo method which is applied for the quantum Hopfield model and investigate the quantum neuro-dynamics through the differential equations with respect to the macroscopic quantities such as the overlap.

This paper is organized as follows. In the next section 2, we explain the basics of the conventional Hopfield model and its generic properties. Then, we categorize the model into two distinct classes according to the origin of noises in artificial brain. The quantum Hopfield model is clearly defined. In section 3, we explain the quantum Monte Carlo method based on the Suzuki-Trotter decomposition [13] and consider the stochastic process in order to investigate the pattern-recalling dynamics of the quantum Hopfield model. In section 4, we derive the macroscopic deterministic flow of the overlap between the neuronal state and one of the built-in patterns from the microscopic master equation which describes the stochastic processes in the quantum Monte Calro method for the quantum Hopfield model [14]. The general solution of the dynamics is obtained under the so-called static approximation. In the next section 5, we apply our general solution to a special case, namely, sequential recalling the built-in two patterns via asymmetric Hebb connections. The effect of quantum-mechanical noise is compared with that of the conventional thermal noise. The last section is summary.

Journal of Physics: Conference Series **297** (2011) 012012 doi:10.1088/1742-6596/297/1/012012

## 2. The Hopfield model

In this section, we briefly explain the basics of the conventional Hopfield model. Then, we shall divide the model into two classes, namely, the Hopfield model put in thermal noises (the model is referred to as *classical systems*) and the same model in the quantum-mechanical noise (the model is referred to as *quantum systems*). In following, we define each of the models explicitly.

### *2.1. The classical system*

Let us consider the network having $N$-neurons. Each neuron $S_i$ takes two states, namely, $S_i = +1$ (fire) and $S_i = -1$ (stationary). Neuronal states are given by the set of variables $S_i$, that is, $\boldsymbol{S} = (S_1, \cdots, S_N)$, $S_i \in \{+1, -1\}$. Each neuron is located on a complete graph, namely, graph topology of the network is 'fully-connected'. The synaptic connection between arbitrary two neurons, say, $S_i$ and $S_j$ is defined by the following Hebb rule:

$$J_{ij} = \frac{1}{N} \sum_{\mu,\nu} \xi_i^{\mu} A_{\mu\nu} \xi_j^{\nu} \tag{1}$$

where $\boldsymbol{\xi}^{\mu} = (\xi_1, \cdots, \xi_N), \xi_i^{\mu} \in \{+1, -1\}$ denote the embedded patterns and each of them is specified by a label $\mu = 1, \cdots, P$. $A_{\mu\nu}$ denotes $(P \times P)$-size matrix and $P$ stands for the number of built-in patterns. We should keep in mind that there exists an energy function (a Lyapunov function) in the system if the matrix $A_{\mu\nu}$ is symmetric.

Then, the output of the neuron $i$, that is, $S_i$ is determined by the sign of the local field $h_i$ as

$$h_i = \sum_{\mu,\nu=1}^{p} \xi_i^{\mu} A_{\mu\nu} m^{\nu} + \frac{1}{N} \sum_{\mu',\nu'=p+1}^{P} \xi_i^{\mu'} A_{\mu'\nu'} \sum_j \xi_j^{\nu'} S_j \tag{2}$$

where $A_{\mu\nu}$ and $A_{\mu'\nu'}$ are elements of $p \times p, (P-p) \times (P-p)$-size matrices, respectively. We also defined the overlap (the direction cosine) between the state of neurons $\boldsymbol{S}$ and one of the built-in patterns $\boldsymbol{\xi}^{\nu}$ by

$$m^{\nu} \equiv \frac{1}{N} (\boldsymbol{S} \cdot \boldsymbol{\xi}^{\nu}) = \frac{1}{N} \sum_i \xi_i^{\nu} S_i. \tag{3}$$

Here we should notice that the Hamiltonian of the system is given by $-\sum_i h_i S_i$. The first term appearing in the left hand side of equation (2) is a contribution from $p \sim \mathcal{O}(1)$ what we call 'condensed patterns', whereas the second term stands for the so-called 'cross-talk noise'. In this paper, we shall concentrate ourselves to the case in which the second term is negligibly small in comparison with the first term, namely, the case of $P = p \sim \mathcal{O}(1)$. In this sense, we can say that the network is 'far from its saturation'.

### *2.2. The quantum system*

To extend the classical system to the quantum-mechanical variant, we rewrite the local field $h_i$ as follows.

$$\boldsymbol{\phi}_i = \sum_{\mu,\nu=1}^{p} \xi_i^{\mu} A_{\mu\nu} \left( \frac{1}{N} \sum_i \xi_i^{\nu} \boldsymbol{\sigma}_i^z \right) \tag{4}$$

where $\boldsymbol{\sigma}_i^z$ $(i = 1, \cdots, N)$ stands for the $z$-component of the Pauli matrix. Thus, the Hamiltonian $\boldsymbol{H}_0 \equiv -\sum_i \boldsymbol{\phi}_i \boldsymbol{\sigma}_i^z$ is a diagonalized $(2^N \times 2^N)$-size matrix and the lowest eigenvalue is identical to the ground state of the classical Hamiltonian $-\sum_i \phi_i S_i$ ($S_i$ is an eigenvalue of the matrix $\boldsymbol{\sigma}_i^z$).

Journal of Physics: Conference Series **297** (2011) 012012 doi:10.1088/1742-6596/297/1/012012

Then, we introduce quantum-mechanical noise into the Hopfield neural network by adding the transverse field to the Hamiltonian as follows.

$$\boldsymbol{H} = \boldsymbol{H}_0 - \Gamma \sum_{i=1}^{N} \boldsymbol{\sigma}_i^x \tag{5}$$

where $\boldsymbol{\sigma}_i^x$ is the $x$-component of the Pauli matrix and transitions between eigenvectors of the classical Hamiltonian $\boldsymbol{H}_0$ are induced due to the off-diagonal elements of the matrix $\boldsymbol{H}$ for $\Gamma \neq 0$. In this paper, we mainly consider the system described by (5).

## 3. Quantum Monte Carlo method

The dynamics of the quantum model (5) follows Schrödinger equation. Thus, we should solve it or investigate the time dependence of the state $|\psi(t)\rangle$ by using the time-evolutionary operator $\mathrm{e}^{-i\boldsymbol{H}\Delta t/\hbar}$ defined for infinitesimal time $\Delta t$ as

$$|\psi(t+\Delta t)\rangle \;=\; \mathrm{e}^{-i\boldsymbol{H}\Delta t/\hbar}|\psi(t)\rangle. \tag{6}$$

However, even if we carry it out numerically, it is very hard for us to do it with reliable precision because $(2^N \times 2^N)$-size Hamilton matrix becomes huge for the number of neurons $N \gg 1$ as in a realistic brain. Hence, here we use the quantum Monte Carlo method to simulate the quantum system in our personal computer and consider the stochastic processes of Glauber-type to discuss the pattern-recalling dynamics of the quantum Hopfield model.

### *3.1. The Suzuki-Trotter decomposition*

The difficulty to carry out algebraic calculations in the model system is due to the non-commutation operators appearing in the Hamiltonian (5), namely, $\boldsymbol{H}_0, \boldsymbol{H}_1 \equiv -\Gamma \sum_i \boldsymbol{\sigma}_i^x$. Thus, we use the following Suzuki-Trotter decomposition [13] in order to deal with the system as a classical spin system.

$$\mathrm{tr}\,\mathrm{e}^{\beta(\boldsymbol{H}_0+\boldsymbol{H}_1)} = \lim_{M\to\infty} \mathrm{tr}\left(\exp\left(\frac{\beta \boldsymbol{H}_0}{M}\right)\exp\left(\frac{\beta \boldsymbol{H}_1}{M}\right)\right)^M \tag{7}$$

where $\beta$ denotes the 'inverse temperature' and $M$ is the number of the Trotter slices, for which the limit $M \to \infty$ should be taken. Thus, one can deal with $d$-dimensional quantum system as the corresponding $(d+1)$-dimensional classical system.

## 4. Derivation of the deterministic flows

In the previous section, we mentioned that we should simulate the quantum Hopfield model by means of the quantum Monte Carlo method to reveal the quantum neuro-dynamics through the time-dependence of the macroscopic quantities such as the overlap. However, in general, it is also very difficult to simulate the quantum-mechanical properties at the ground state by a personal computer even for finite size systems $(N, M < \infty)$.

With this fact in mind, in this section, we attempt to derive the macroscopic flow equations from the microscopic master equation for the classical system regarded as the quantum system in terms of the Suzuki-Trotter decomposition. This approach is efficiently possible because the Hopfield model is a fully-connected mean-field model such as the Sherrington-Kirkpatrick model [15] for spin glasses and its equilibrium properties are completely determined by several order parameters.

### 4.1. The master equation

After the Suzuki-Trotter decomposition (7), we obtain the local field for the neuron $i$ located on the $k$-th Trotter slice as follows.

$$\beta\phi_i(\boldsymbol{\sigma}_k : \sigma_i(k \pm 1)) = \frac{\beta}{M}\sum_{\mu,\nu}\xi_i^\nu A_{\mu\nu}\left\{\frac{1}{N}\sum_j \xi_j^\nu \sigma_j(k)\right\} + \frac{B}{2}\{\sigma_i(k-1) + \sigma_i(k+1)\} \quad (8)$$

where parameter $B$ is related to the amplitude of the transverse field (the strength of the quantum-mechanical noise) $\Gamma$ by

$$B = \frac{1}{2}\log\coth\left(\frac{\beta\Gamma}{M}\right). \quad (9)$$

In the classical limit $\Gamma \to 0$, the parameter $B$ goes to infinity. For the symmetric matrix $A_{\mu\nu}$, the Hamiltonian (scaled by $\beta$) of the system is given by $-\sum_i \beta\phi_i(\boldsymbol{\sigma}_k : \sigma(k \pm 1))\sigma_i(k)$.

Then, the transition probability which specifies the Glauber dynamics of the system is given by $w_i(\boldsymbol{\sigma}_k) = (1/2)[1 - \sigma_i(k)\tanh(\beta\phi_i(\boldsymbol{\sigma}_k : \sigma(k \pm 1)))]$. More explicitly, $w_i(\boldsymbol{\sigma}_k)$ denotes the probability that an arbitrary neuron $\sigma_i(k)$ changes its state as $\sigma_i(k) \to -\sigma_i(k)$ within the time unit. Therefore, the probability that the neuron $\sigma_i(k)$ takes $+1$ is obtained by setting $\sigma_i(k) = -1$ in the above $w_i(\boldsymbol{\sigma}_k)$ and we immediately find $\sigma_i(k) = \sigma_i(k-1) = \sigma_i(k+1)$ with probability 1 in the limit of $B \to \infty$ which implies the classical limit $\Gamma \to 0$.

Hence, the probability that a microscopic state including the $M$-Trotter slices $\{\boldsymbol{\sigma}_k\} \equiv (\boldsymbol{\sigma}_1, \cdots, \boldsymbol{\sigma}_M), \boldsymbol{\sigma}_k \equiv (\sigma_1(k), \cdots, \sigma_N(k))$ obeys the following master equation:

$$\frac{dp_t(\{\boldsymbol{\sigma}_k\})}{dt} = \sum_{k=1}^{M}\sum_{i=1}^{N}[p_t(F_i^{(k)}(\boldsymbol{\sigma}_k))w_i(F_i^{(k)}(\boldsymbol{\sigma}_k)) - p_t(\boldsymbol{\sigma}_k)w_i(\boldsymbol{\sigma}_k)] \quad (10)$$

where $F_i^{(k)}(\cdot)$ denotes a single spin flip operator for neuron $i$ on the Trotter slice $k$ as $\sigma_i(k) \to -\sigma_i(k)$. When we pick up the overlap between neuronal state $\boldsymbol{\sigma}_k$ and one of the built-in patterns $\boldsymbol{\xi}^\nu$, namely,

$$m_k \equiv \frac{1}{N}(\boldsymbol{\sigma}_k \cdot \boldsymbol{\xi}^\nu) = \frac{1}{N}\sum_i \xi_i^\nu \sigma_i(k) \quad (11)$$

as a relevant macroscopic quantity, the joint distribution of the set of the overlaps $\{m_1, \cdots, m_M\}$ at time $t$ is written in terms of the probability for realizations of microscopic states $p_t(\{\boldsymbol{\sigma}_k\})$ at the same time $t$ as

$$P_t(m_1^\nu, \cdots, m_M^\nu) = \sum_{\{\boldsymbol{\sigma}_k\}} p_t(\{\boldsymbol{\sigma}_k\})\prod_{k=1}^{M}\delta(m_k^\nu - m_k^\nu(\boldsymbol{\sigma}_k)) \quad (12)$$

where we defined the sums by

$$\sum_{\{\boldsymbol{\sigma}_k\}}(\cdots) \equiv \sum_{\boldsymbol{\sigma}_1}\cdots\sum_{\boldsymbol{\sigma}_M}(\cdots), \quad \sum_{\boldsymbol{\sigma}_k}(\cdots) \equiv \sum_{\sigma_1(k)=\pm 1}\cdots\sum_{\sigma_N(k)=\pm 1}(\cdots). \quad (13)$$

Taking the derivative of equation (12) with respect to $t$ and substituting (10) into the result, we have the following differential equations for the joint distribution

$$\begin{aligned}\frac{dP_t(m_1^\nu,\cdots,m_M^\nu)}{dt} &= \sum_k \frac{\partial}{\partial m_k^\nu}\{m_k^\nu P_t(m_1^\nu,\cdots,m_k^\nu,\cdots,m_M^\nu)\} \\ &- \sum_k \frac{\partial}{\partial m_k^\nu}\Bigg\{P_t(m_1^\nu,\cdots,m_k^\nu,\cdots,m_M^\nu)\int_{-\infty}^{\infty} D[\xi^\nu]d\xi^\nu \\ &\times \frac{\sum_{\{\boldsymbol{\sigma}_k\}} p_t(\{\boldsymbol{\sigma}_k\})\xi^\nu \tanh[\beta\phi(k)]\prod_{k,i}\delta(m_k^\nu - m_k^\nu(\boldsymbol{\sigma}_k))}{\sum_{\{\boldsymbol{\sigma}_k\}} p_t(\{\boldsymbol{\sigma}_k\})\prod_k \delta(m_k^\nu - m_k^\nu(\boldsymbol{\sigma}_k))}\Bigg\} \\ &\times \delta(\sigma(k+1)-\sigma_i(k+1))\delta(\sigma(k-1)-\sigma_i(k-1)) \qquad (14)\end{aligned}$$

where we introduced several notations

$$D[\xi^\nu] \equiv \frac{1}{N}\sum_i \delta(\xi^\nu - \xi_i^\nu) \qquad (15)$$

$$\beta\phi(k) \equiv \frac{\beta\sum_{\mu\nu}\xi^\mu A_{\mu\nu}}{M} m_k^\nu + \frac{B}{2}\sigma(k-1) + \frac{B}{2}\sigma(k+1) \qquad (16)$$

for simplicity.

Here we should notice that if the local field $\beta\phi(k)$ is independent of the microscopic variable $\{\boldsymbol{\sigma}_k\}$, one can get around the complicated expectation of the quantity $\tanh[\beta\phi(k)]$ over the time-dependent Gibbs measurement which is defined in the sub-shell: $\prod_k \delta(m_k^\nu - m_k^\nu(\boldsymbol{\sigma}_k))$. As the result, only procedure we should carry out to get the deterministic flow is to calculate the data average (the average over the built-in patterns). However, unfortunately, we clearly find from equation (16) that the local field depends on the $\{\boldsymbol{\sigma}_k\}$. To overcome the difficulty and to carry out the calculation, we assume that the probability $p_t(\{\boldsymbol{\sigma}_k\})$ of realizations for microscopic states during the dynamics is independent of $t$, namely,

$$p_t(\{\boldsymbol{\sigma}_k\}) = p(\{\boldsymbol{\sigma}_k\}). \qquad (17)$$

Then, our average over the time-dependent Gibbs measurement in the sub-shell is rewritten as

$$\begin{aligned}&\frac{\sum_{\{\boldsymbol{\sigma}_k\}} p_t(\{\boldsymbol{\sigma}_k\})\xi^\nu \tanh[\beta\phi(k)]\prod_{k,i}\delta(m_k^\nu - m_k^\nu(\boldsymbol{\sigma}_k))}{\sum_{\{\boldsymbol{\sigma}_k\}} p_t(\{\boldsymbol{\sigma}_k\})\prod_k \delta(m_k^\nu - m_k^\nu(\boldsymbol{\sigma}_k))} \\ &\times \delta(\sigma(k+1)-\sigma_i(k+1))\delta(\sigma(k-1)-\sigma_i(k-1)) \\ &\equiv \langle \xi^\nu \tanh[\beta\phi(k)]\prod_i \delta(\sigma(k+1)-\sigma_i(k+1))\delta(\sigma(k-1)-\sigma_i(k-1))\rangle_* \qquad (18)\end{aligned}$$

where $\langle\cdots\rangle_*$ stands for the average in the sub-shell defined by $m_k^\nu = m_k^\nu(\boldsymbol{\sigma}_k)\,(\forall_k)$:

$$\langle\cdots\rangle_* \equiv \frac{\sum_{\{\boldsymbol{\sigma}_k\}} p(\{\boldsymbol{\sigma}_k\})(\cdots)\prod_k \delta(m_k^\nu - m_k^\nu(\boldsymbol{\sigma}_k))}{\sum_{\{\boldsymbol{\sigma}_k\}} p(\{\boldsymbol{\sigma}_k\})\prod_k \delta(m_k^\nu - m_k^\nu(\boldsymbol{\sigma}_k))} \qquad (19)$$

If we notice that the Gibbs measurement in the sub-shell is rewritten as

$$\sum_{\{\boldsymbol{\sigma}_k\}} p(\{\boldsymbol{\sigma}_k\})\prod_k \delta(m_k^\nu - m_k^\nu(\boldsymbol{\sigma}_k)) = \mathrm{tr}_{\{\sigma\}}\exp\left[\beta\sum_{l=1}^{M}\phi(l)\sigma(l)\right] \qquad (20)$$

Journal of Physics: Conference Series **297** (2011) 012012 doi:10.1088/1742-6596/297/1/012012

$(\mathrm{tr}_{\{\sigma\}}(\cdots) \equiv \prod_k \sum_{\boldsymbol{\sigma}_k}(\cdots))$, and the quantity

$$\tanh\left[\beta\phi(k)\right] = \frac{\sum_{\sigma(k)=\pm 1} \sigma(k) \exp[\beta\phi(k)\sigma(k)]}{\sum_{\sigma(k)=\pm 1} \exp[\beta\phi(k)\sigma(k)]} \tag{21}$$

is independent of $\sigma(k)$, the average appearing in (18) leads to

$$\begin{aligned} \langle \xi^\nu \tanh[\beta\phi(k)] \prod_i \delta(\sigma(k\pm 1) - \sigma_i(k\pm 1)) \rangle_* &= \frac{\mathrm{tr}_{\{\sigma\}} \xi^\nu \{\frac{1}{M}\sum_{l=1}^M \sigma(l)\} \exp[\beta\phi(k)\sigma(k)]}{\mathrm{tr}_{\{\sigma\}} \exp[\beta\phi(k)\sigma(k)]} \\ &\equiv \xi^\nu \langle \sigma \rangle_{path}^{(\xi^\nu)} \end{aligned} \tag{22}$$

in the limit of $M \to \infty$. This is nothing but a path integral for the effective single neuron problem in which the neuron updates its state along the imaginary time axis: $\mathrm{tr}_{\{\sigma\}}(\cdots) \equiv \sum_{\sigma(1)=\pm 1} \cdots \sum_{\sigma(M)=\pm 1}(\cdots)$ with weights $\exp[\beta\phi(k)\sigma(k)]$, $(k = 1, \cdots, M)$.

Then, the differential equation (14) leads to

$$\begin{aligned} \frac{dP_t(m_1^\nu, \cdots, m_M^\nu)}{dt} &= \sum_k \frac{\partial}{\partial m_k^\nu} \{m_k^\nu P_t(m_1^\nu, \cdots, m_k^\nu, \cdots, m_M^\nu)\} \\ &- \sum_k \frac{\partial}{\partial m_k^\nu} \left\{ P_t(m_1^\nu, \cdots, m_k^\nu, \cdots, m_M^\nu) \int_{-\infty}^{\infty} D[\xi^\nu] d\xi^\nu \xi^\nu \langle \sigma \rangle_{path}^{(\xi^\nu)} \right\}. \end{aligned} \tag{23}$$

In order to derive the compact form of the differential equations with respect to the overlaps, we substitute $P_t(m_1^\nu, \cdots, m_M^\nu) = \prod_{k=1}^M \delta(m_k^\nu - m_k^\nu(t))$ into the above (23) and multiplying $m_l^\nu$ by both sides of the equation and carrying out the integral with respect to $dm_1^\nu \cdots dm_M^\nu$ by part, we have for $l = 1, \cdots, M$ as

$$\frac{dm_l^\nu}{dt} = -m_l^\nu + \int_{-\infty}^{\infty} D[\xi^\nu] d\xi^\nu \xi^\nu \langle \sigma \rangle_{path}^{(\xi^\nu)}. \tag{24}$$

Here we should notice that the path integral $\xi^\nu \langle \sigma \rangle_{path}^{(\xi^\nu)}$ depends on the embedded patterns $\boldsymbol{\xi}^\nu$. In the next subsection, we carry out the quenched average explicitly under the so-called static approximation.

### *4.2. Static approximation*

In order to obtain the final form of the deterministic flow, we assume that macroscopic quantities such as the overlap are independent of the Trotter slices $k$ during the dynamics. Namely, we must use the so-called static approximation:

$$m_k^\nu = m^\nu \, (\forall_k). \tag{25}$$

Under the static approximation, let us use the following inverse process of the Suzuki-Trotter decomposition (7):

$$\lim_{M\to\infty} Z_M = \mathrm{tr} \exp\left[\beta \sum_{\mu\nu} \xi^\mu A_{\mu\nu} m^\nu \sigma_z + \beta\Gamma\sigma_x\right] \tag{26}$$

$$Z_M \equiv \mathrm{tr}_{\{\sigma\}} \exp\left[\frac{\beta \sum_{\mu\nu} \xi^\mu A_{\mu\nu} m^\nu}{M} \sum_k \sigma(k) + B \sum_k \sigma(k)\sigma(k+1)\right] \tag{27}$$

In our previous study [14], we numerically checked the validity of static approximation by computer simulations and found that the approximation is successfully valid for the pure-ferromagnetic system but it is deviated from the good approximation for the disordered systems. The validity of the static approximation was recently argued by Takahashi and Matsuda [16] from the different perspective.

Then, one can calculate the path integral immediately as

$$\langle\sigma\rangle_{path}^{(\xi^\nu)} = \frac{\sum_{\mu\nu}\xi^\mu A_{\mu\nu}m^\nu}{\sqrt{(\sum_{\mu\nu}\xi^\mu A_{\mu\nu}m^\nu)^2+\Gamma^2}}\tanh\beta\sqrt{\left(\sum_{\mu\nu}\xi^\mu A_{\mu\nu}m^\nu\right)^2+\Gamma^2}. \tag{28}$$

Inserting this result into (24), we obtain

$$\frac{dm^\nu}{dt} = -m^\nu + \mathbb{E}_{\boldsymbol{\xi}}\left[\frac{\xi^\nu\sum_{\mu\nu}\xi^\mu A_{\mu\nu}m^\nu}{\sqrt{(\sum_{\mu\nu}\xi^\mu A_{\mu\nu}m^\nu)^2+\Gamma^2}}\tanh\beta\sqrt{\left(\sum_{\mu\nu}\xi^\mu A_{\mu\nu}m^\nu\right)^2+\Gamma^2}\right] \tag{29}$$

where we should bear in mind that the empirical distribution $D[\xi^\nu]$ in (24) was replaced by the built-in pattern distribution $\mathcal{P}(\xi^\nu)$ as

$$\lim_{N\to\infty}\frac{1}{N}\sum_i\delta(\xi_i^\nu-\xi^\nu)=\mathcal{P}(\xi^\nu) \tag{30}$$

in the limit of $N\to\infty$ and the average is now carried out explicitly as

$$\int D[\xi^\nu]d\xi^\nu(\cdots)=\int\mathcal{P}(\xi^\nu)d\xi^\nu(\cdots)\equiv\mathbb{E}_{\boldsymbol{\xi}}[\cdots]. \tag{31}$$

Equation (29) is a general solution for the problem in this paper.

*4.3. The classical and zero-temperature limits*

It is easy for us to take the classical limit $\Gamma\to 0$ in the result (29). Actually, we have immediately

$$\frac{dm^\nu}{dt} = -m^\nu + \mathbb{E}_{\boldsymbol{\xi}}\left[\xi^\nu\tanh\left(\beta\sum_{\mu\nu}\xi^\mu A_{\mu\nu}m^\nu\right)\right]. \tag{32}$$

The above equation is identical to the result by Coolen and Ruijgrok [17] who considered the retrieval process of the conventional Hopfield model under thermal noise.

We can also take the zero-temperature limit $\beta\to\infty$ in (29) as

$$\frac{dm^\nu}{dt} = -m^\nu + \mathbb{E}_{\boldsymbol{\xi}}\left[\frac{\xi^\nu\sum_{\mu\nu}\xi^\mu A_{\mu\nu}m^\nu}{\sqrt{(\sum_{\mu\nu}\xi^\mu A_{\mu\nu}m^\nu)^2+\Gamma^2}}\right]. \tag{33}$$

Thus, the equation (29) including the above two limiting cases is our general solution for the neuro-dynamics of the quantum Hopfield model in which $\mathcal{O}(1)$ patterns are embedded. Thus, we can discuss any kind of situations for such pattern-recalling processes and the solution is always derived from (29) explicitly.

## 5. Limit cycle solution for asymmetric connections

In this section, we discuss a special case of the general solution (29). Namely, we investigate the pattern-recalling processes of the quantum Hopfield model with asymmetric connections $\boldsymbol{A}\equiv\{A_{\mu\nu}\}$.

*5.1. Result for two-patterns*

Let us consider the case in which just only two patterns are embedded via the following matrix:

$$A = \begin{pmatrix} 1 & -1 \\ 1 & 1 \end{pmatrix} \tag{34}$$

Then, from the general solution (29), the differential equations with respect to the two overlaps $m_1$ and $m_2$ are written as

$$\frac{dm_1}{dt} = -m_1 + \frac{m_1}{\sqrt{(2m_1)^2 + \Gamma^2}} - \frac{m_2}{\sqrt{(2m_2)^2 + \Gamma^2}}$$
$$\frac{dm_2}{dt} = -m_2 + \frac{m_1}{\sqrt{(2m_1)^2 + \Gamma^2}} + \frac{m_2}{\sqrt{(2m_2)^2 + \Gamma^2}}.$$

In Figure 1, we show the time evolutions of the overlaps $m_1$ and $m_2$ for the case of the amplitude $\Gamma = 0.01$. From this figure, we clearly find that the neuronal state evolves as

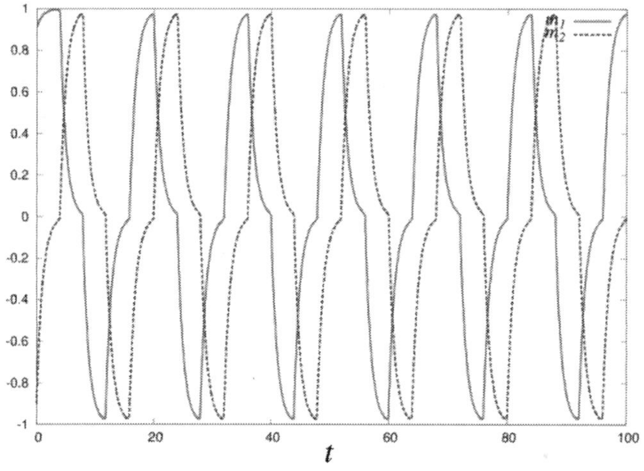


**Figure 1.** Time evolutions of $m_1$ and $m_2$ for the case of $\Gamma = 0.01$.

$A \to B \to \overline{A} \to \overline{B} \to A \to B \to \cdots$ ($\overline{A}, \overline{B}$ denote the 'mirror images' of $A$ and $B$, respectively), namely, the network behaves as a limit cycle.

To compare the effects of thermal and quantum noises on the pattern-recalling processes, we plot the trajectories $m_1$-$m_2$ for $(T \equiv \beta^{-1}, \Gamma) = (0, 0.01), (0.01, 0)$ (left panel), $(T, \Gamma) = (0, 0.8), (0.8, 0)$ (right panel) in Figure 2. From these panels, we find that the limit cycles are getting collapsed as the strength of the noise level is increasing for both thermal and quantum-mechanical noises, and eventually the trajectories shrink to the origin $(m_1, m_2) = (0, 0)$ in the limit of $T, \Gamma \to \infty$.

## 6. Summary

In this paper, we considered the stochastic process of quantum Monte Carlo method applied for the quantum Hopfield model and investigated the quantum neuro-dynamics through the differential equations with respect to the macroscopic quantities such as the overlap. By using the present approach, one can evaluate the 'inhomogeneous' Markovian stochastic process of quantum Monte Carlo method (in which amplitude $\Gamma$ is time-dependent [18, 19]) such as

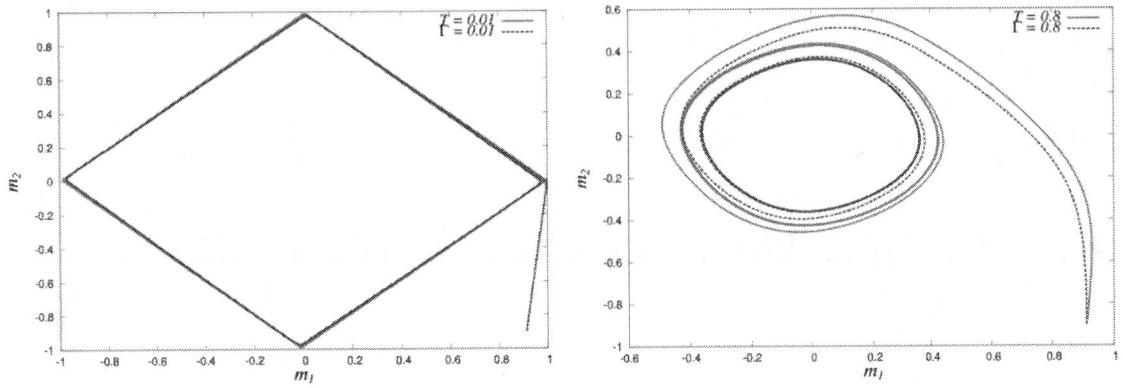


**Figure 2.** Trajectories $m_1$-$m_2$ for $(T, \Gamma) = (0, 0.01), (0.01, 0)$ (left panel), $(T, \Gamma) = (0, 0.8), (0.8, 0)$ (right panel).

quantum annealing [20, 21, 22, 23, 24, 25]. In the next step of the present study, we are planning to extend this formulation to the probabilistic information processing described by spin glasses including a peculiar type of antiferromagnet [26].

## Acknowledgments

We thank B.K. Chakrabarti, A.K. Chandra, P. Sen and S. Dasgupta for fruitful discussion. We also thank local organizers of Statphys-Kolkata VII for their warm hospitality. This work was financially supported by Grant-in-Aid for Scientific Research (C) of Japan Society for the Promotion of Science, No. 22500195.

## References

[1] Nakano K 1972 *IEEE Trans. on Systems, Man, and Sybernetics* **SMC-2** 380
[2] Hopfield J J 1982 *PNAS* **79** 2554
[3] Mézard M, Parisi G and Virasoro M A 1987 *Spin Glass Theory and Beyond* (Singapore: World Scientific)
[4] Amit D, Gutfreund H and Somplolinsky H 1985 *Phys. Rev. Lett.* **55** 1530
[5] Inoue J 1996 *J. Phys. A* **29** 4815
[6] Shiino M and Fukai T *J. Phys. A* **25** L375
[7] Ma Y Q and Gong C D 1992 *Phys. Rev. B* **45** 793
[8] Nishimori H and Nonomura Y 1996 *J. Phys. Soc. Japan* **65** 3780
[9] Amari S and Maginu K 1988 *Neural Networks* **1** 63
[10] Okada M 1995 *Neural Networks* **8** 833
[11] Coolen A C C and Sherrington D 1994 *Phys. Rev. Let.* **49** 1921
[12] Coolen A C C, Laughton S N and Sherrington D 1996 *Phys. Rev. B* **53** 8184
[13] Suzuki M 1976 *Prog. Theor. Phys.* **56** 1454
[14] Inoue J 2010 *Journal of Physics: Conference Series* **233** 012010
[15] Sherrington D and Kirkpatrick S 1975 *Phys. Rev. Lett.* **35** 1792
[16] Takahashi K and Matsuda Y 2010 *J. Phys. Soc. Japan* **79** 043712
[17] Coolen A C C and Ruijgrok Th W 1988 *Phys. Rev. A* **38** 4253
[18] Das A, Sengupta K, Sen D and Chakrabarti B K 2006 *Phys. Rev. B* **74** 144423
[19] Das A 2010 *Phys. Rev. B* **82** 172402
[20] Kadowaki T and Nishimori H 1998 *Physical Review E* **58** 5355
[21] Farhi E, Goldstone J, Gutmann S, Lapan J, Lundgren A and Preda P 2001 *Science* **292** 472
[22] Morita S and Nishimori H 2006 *J. Phys. A* **39** 13903
[23] Suzuki S and Okada M 2005 *J. Phys. Soc. Jpn.* **74** 1649
[24] Santoro G E and Tosatti 2006 *J. Phys. A* **41** 209801
[25] Das A and Chakrabarti B K 2008 *Rev. Mod. Phys.* **80** 1061
[26] Chandra A K, Inoue J and Chakrabarti B K 2010 *Phys. Rev. E* **81** 021101

Journal of Physics: Conference Series **297** (2011) 012013 doi:10.1088/1742-6596/297/1/012013

# Physics of non-Abelian vortices in Bose-Einstein condensates

**Michikazu Kobayashi**

Department of Basics Sciences, The University of Tokyo, Bldg. 16 #728, Kobama 3-8-1, Meguro-ku, Tokyo 153-8902, Japan

E-mail: michikaz@cat.phys.s.u-tokyo.ac.jp

**Abstract.** A wide order-parameter manifolds of Bose-Einstein condensates (BECs) with spin degrees of freedom enables various kinds of topological defects which never appear in conventional scalar BECs. In this paper, we focus on the characteristic example of them; non-Abelian vortices and their dynamics in the cyclic phase of a spin-2 BEC. Non-Abelian vortices shows the unique effect in their collision dynamics, i.e. unlike Abelian vortices, they do neither reconnect themselves nor pass through each other but create a rung vortex between them.

## 1. Introduction

Quantized vortices have been one of the main topics in superfluid systems, and had a long research history since their theoretical prediction [1] and experimental observations [2]. Quantized vortices have quantized circulations of the superfluid velocity, and a vortex in a scalar Bose-Einstein condensate, for example, has the circulation being integer multiple of $h/M$. Here $h$ is the Planck's constant and $M$ is the mass of atoms. The quantized circulation has been observed in superfluid $^4$He [2], superfluid $^3$He [3], and atomic BECs [4, 5].

Quantized vortices are the topological defects of order parameters of superfluid, and can be classified as topological invariants defined by the topological structures of the order parameters. Topological invariants of quantized vortices (line defects) are defined by how the order parameters change along the closed path encircling the vortices. In the case of scalar BECs, for example, the order-parameter manifold is $U(1)$, and the $U(1)$ gauge changes by integer multiple of $2\pi$ along the closed path. As a result, topological invariants of vortices in scalar BECs are classified as the additive group of integers, which corresponds to the circulations of vortices being integer multiple of $h/M$.

For the case of multi-component BECs, a discrete symmetry enables vortices to have fractional circulations. Characteristic examples are half-quantized vortices in superfluid $^3$He-A [6], vortex cores in superfluid $^3$He-B, and the polar phase of a spin-1 spinor BEC . As other kind of vortices having interesting topological invariants, we discuss non-Abelian vortices here. For BECs with spin degrees of freedom, besides the $U(1)$ gauge, the direction of the $SO(3)$ spin rotates along the closed path encircling the vortices. Because $SO(3)$ is the non-Abelian group, topological invariants can be classified as discrete non-Abelian subgroup of $SO(3)$, defining non-Abelian vortices.

The non-Abelian property of non-Abelian vortices becomes remarkable in their collision dynamics. When two $U(1)$ vortices collide, they reconnect themselves. This reconnection of vortices has been studied theoretically [10, 11, 12], and observed in superfluid $^4$He [13]. As other abnormal examples, collision of two $U(1)$ vortices in the attractive BEC [14, 15] or two $U(1) \times U(1)$ vortices in the attractive two-component BEC [16] results in the formation of a new vortex bridging two colliding vortices, like a rung of a ladder. We define this vortex as a rung vortex.

When the vortices are non-Abelian, the situation changes. It was predicted that the collision of two non-Abelian vortices produces a rung vortex [17, 18, 19]. In this paper, furthermore, we show that for two vortices with non-commutative topological invariants, both reconnection and passing through are topologically forbidden and only the formation of a rung vortex is allowed. As the other dynamics demonstrating the genuine non-Abelian character, we show that the two twisted vortices with commutative topological invariants can unravel, whereas those with non-commutative ones cannot.

Non-Abelian vortices and their collision dynamics can be realized in the cyclic phase of the spin-2 BEC which has the non-Abelian tetrahedral symmetry.

In this paper, we briefly overview our study of non-Abelian vortices. In Sec. 2, we define non-Abelian vortices based on homotopy theory and discuss their collision dynamics algebraically. In Sec. 3, we review the spin-2 BECs and non-Abelian vortices, and show our recent results for several collision dynamics of non-Abelian vortices.

## 2. Non-Abelian vortices and their collision dynamics

In this section, we define topological invariants of quantized vortices based on homotopy theory and non-Abelian vortices. After that, we discuss collision dynamics of vortices.

### 2.1. Order-parameter manifold

Quantized vortices appear in the symmetry broken system such as superfluid phase of liquid helium and ultracold atomic BECs. Let the Lie group $G$ be a set of transformations which do not change the free energy of the system, and the subgroup $H$ be a set of transformations which do not change the order parameter $\psi$ of the symmetry broken system:

$$H = \{h \in G | h\psi = \psi\}. \tag{1}$$

In cases of superfluid $^4$He, $s$-wave superconductors, and scalar BECs, the order parameter can be described by $\psi = |\psi|e^{i\phi}$, where $\phi$ is the phase of the order parameter. Because the free energy remains invariant under the $U(1)$ gauge transformation: $\phi \to \phi + \delta\phi$, then $G \simeq U(1)$ and only the identity transformations remains $\psi$ unchanged: $H \simeq \{1\}$. The degrees of freedom of the order parameter can be described as the element of $G$, whereas several elements in $G$ are equivalent due to Eq. (1). Then, the exact degrees of freedom of the order parameter become the coset space: $G/H$, which is called order-parameter manifold.

### 2.2. Topological invariants of quantized vortices

Let us consider the order parameter of the system $\psi(\mathbf{x})$ as a continuous function of the coordinate $\mathbf{x}$ in real space. A closed loop with a fixed base point in real space can be mapped into a loop in the order-parameter manifold (Fig. 1). If two loops in the order-parameter manifold can be transformed to each other through a continuous function, those loops are homotopic. If a loop in real space does not encircle a vortex, the corresponding loop in the order-parameter manifold is homotopic to the trivial loop: the base point. In contrast, a loop in real space encircles a vortex, the mapped loop in the order-parameter manifold is not homotopic to the base point. The topological invariant of a vortex can be defined by the set of homotopic loops in the order parameter manifold.

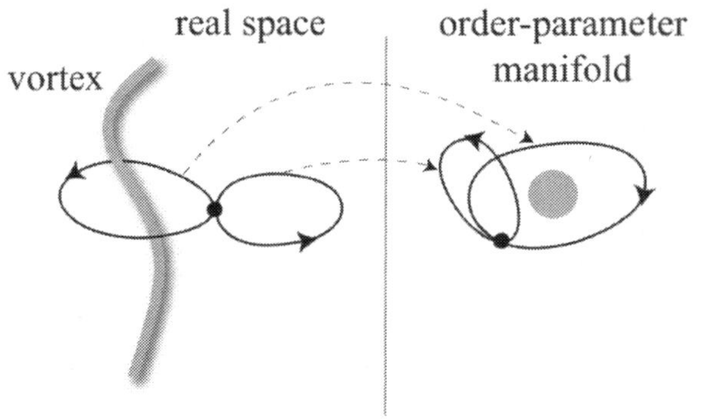


**Figure 1.** Loops in real space (left) and mapped loops in the order-parameter manifold (right).

When the order parameter can be described as a complex scalar function $\psi(\mathbf{x}) = |\psi(\mathbf{x})|e^{i\phi(\mathbf{x})}$, for example, the order-parameter manifold $G/H \simeq U(1)$ is isomorphic to a $S^1$ circle. Because loops encircling the $S^1$ circle $n$ times for the nonzero integer $n$ are not homotopic to a point, there exist vortices for such loops. Therefore, topological invariants of vortices can be classified as the integer $n$. When a loop encircle two vortices in real space, the topological invariants of which are classified as $n_1$ and $n_2$ respectively, the mapped loop in the order-parameter manifold encircle the $S^1$ circle $n_1 + n_2$ times, which means that two vortices with topological invariants $n_1$ and $n_2$ can combine to one vortex with the topological invariant $n_1 + n_2$.

The topological invariant of a vortex can be defined by the map from the loop in real space to that in the order-parameter manifold, namely the fundamental group $\pi_1[G/H]$. For $G/H \simeq U(1)$ case, the above discussion is summarized as

$$\pi_1[G/H \simeq U(1)] \simeq \mathbb{Z}. \tag{2}$$

When $G$ is connected ($\pi_0[G] \simeq 1$), and simply connected ($\pi_1[G] \simeq 1$), then

$$\pi_1[G/H] \simeq \pi_0[H]. \tag{3}$$

Here, the 0th homotopy set $\pi_0[H]$ is the set of homotopic points. When $H$ is discrete group, it satisfies that $\pi_0[H] \simeq H$.

### *2.3. Definition of non-Abelian vortices*

For the term of non-Abelian vortices, there are three definitions:

(i) Vortices for non-Abelian $G$,
(ii) Vortices for non-Abelian $H$,
(iii) Vortices for non-Abelian $\pi_1[G/H]$.

Definition (i) is the largest and includes definition (ii) which includes definition (iii). Although definition (ii) is widely used for non-Abelian vortices, we here adopt definition (iii).

When $\pi_1[G/H]$ is non-Abelian, the topological invariant of a vortex depends on the base point of the loop in real space among the conjugacy class. To see this, we consider a vortex and paths as shown in Fig. 2. A loop $P_1 : u \to v \to u' \to u$ with the base point $u$ in real space encircles the vortex and is mapped to the loop in the order-parameter manifold which defines the topological invariant $a \in \pi_1[G/H]$. Suppose the different point $s$ from $u$ in real space which is mapped to the same point as that mapped from $u$ in the order-parameter manifold, and an unclosed path $s \to u$ in real space which is mapped to the loop in the order-parameter manifold which defines

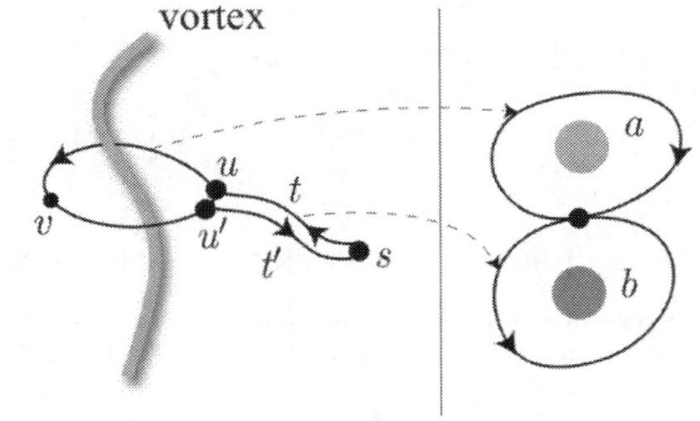


**Figure 2.** Paths encircling a non-Abelian vortex in real space (left) and mapped paths in the order-parameter manifold (right). A loop $u \to v \to u' \to u$ is mapped to the loop giving the topological invariant $a$, and the path $s \to u$ (or $u' \to s$) is mapped to the loop giving the topological invariant $b$.

the topological invariant $b \in \pi_1[G/H]$. Then the unclosed path $u' \to s$ gives the topological invariant $b^{-1}$ and the loop $P_2 : s \to u \to v \to u' \to s$ gives the topological invariant $b^{-1}ab$. When $\pi_1[G/H]$ is Abelian, paths $P_1$ and $P_2$ give the same topological invariant, i.e. $a = b^{-1}ab$. However, non-commutative $a$ and $b$ for non-Abelian $\pi_1[G/H]$ give different topological invariant for the two paths within the same conjugacy class.

### 2.4. Collision of non-Abelian vortices

Here, we consider algebraic and geometric structures of collisions of vortices. Let us consider two colliding vortices and four paths $a$, $b$, $c$, and $d$, as shown in Fig. 3 (a), and assume that paths $a$ and $b$ define the topological invariants of vortices as $A$ and $B$, respectively. With the fixed base

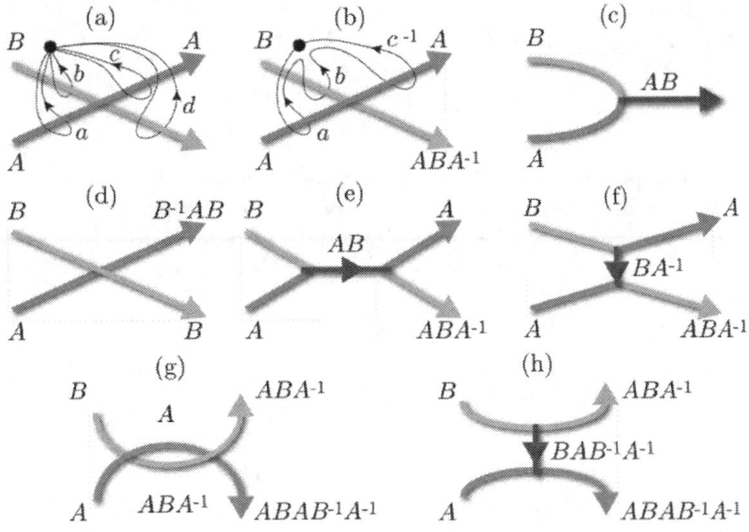


**Figure 3.** (a) Four closed paths for two colliding vortices, where $A$ and $B$ denote the corresponding topological invariants. (b) Path homotopic to path $d$ in (a), and topological invariants of vortices. (c) $Y$-shaped junction. (d)-(f) Collision patterns from two vortices starting from the configuration in (a). (d) Passing through. (e) Rung vortex $AB$ between two vortices. (f) Rung vortex $BA^{-1}$ between two vortices. (g) Twisted vortices. (h) Collision of twisted vortices and rung vortex $BAB^{-1}A^{-1}$.

point, paths $a$ and $c$ are topologically equivalent. On the other hand, path $d$ is topologically different from path $b$, but it can be continuously deformed to the path shown in Fig. 3 (b). As a result, path $d$ defines the topological invariant as $ABA^{-1}$. When the invariants $A$ and $B$ are non-commutative for non-Abelian vortices, paths $b$ and $d$ give different topological invariants for the same vortex. Similarly, we can define the topological invariants of vortices for a $Y$-shaped junction as shown in Fig. 3 (c). Upon the collision of two vortices [Fig. 3(a)], three possible consequences [Figs. 3 (d) - 3 (f)] follow, where Fig. 3 (d) shows "passing through", and Fig. 3 (e) and 3 (f) show formation of "rung vortices" that bridge the two vortices.

*2.4.1. Same topological invariants:* $A = B$ In this case, $A$ and $B$ are commutative, and all three cases are topologically allowed. In particular, Fig. 3 (f) reduces to reconnection because the rung vortex vanishes identically ($BA^{-1} = 1$). As other dynamics, Fig. 3 (e) shows the formation of the doubly quantized rung vortex $AB = A^2$ which is energetically unfavorable. Passing through in Fig. 3 (d) is also energetically unfavorable because the doubly quantized vortex is formed at just the moment of passing through. Therefore, reconnection is the most favorable dynamics in this case, except for specific situations such as the collision of attractive vortices as in type-I superconductors, where whether a rung vortex is formed or not is strongly dependent on the kinematic parameters of the collision [14, 15].

*2.4.2. Different and commutative topological invariants:* $A \neq B$ *and* $AB = BA$ In this case, both Figs. 3 (e) and 3 (f) reduce to the formation of a rung vortex which increases the total vortex line length and is energetically unfavorable. Therefore, passing through in Fig. 3 (d) is the most favorable dynamics. One exception is the formation of a rung vortex $AB$ for the attractive $U(1) \times U(1)$ BEC where a vortex $A$ and the other vortex $B$ are attractive each other and the rung vortex $AB$ (Fig. 3 (e)) or $BA^{-1}$ (Fig. 3 (f)) is formed [16].

*2.4.3. Non-commutative topological invariants:* $AB \neq BA$ When $A$ and $B$ are non-commutative for non-Abelian vortices, the transition from Fig. 3 (a) to 3 (e) is topologically forbidden because they are topologically distinct. Therefore, a rung vortex with the topological invariant of $AB$ or $BA^{-1}$ must be formed after the collision, regardless of the kinematic parameters such as the collision angles and the initial relative speed.

*2.4.4. Collision of two twisted vortices* Collision of two twisted vortices as shown in Fig. 3 (g) shows the genuine non-Abelian character. Figure 3 (h) shows the formation of the rung vortex $BAB^{-1}A^{-1}$ which always vanishes for Abelian vortices. Therefore, twisted vortices with commutative topological invariants can unravel, whereas twisted non-Abelian vortices with non-commutative ones cannot.

## 3. Non-Abelian vortices in spin-2 Bose-Einstein condensates

Spinor BECs with spin degrees of freedom admit various kind of topological excitations such as not only quantized vortices but also monopoles and skyrmions, reflecting their rich variety of order-parameter manifolds [20, 21, 22].

In this section, we show that non-Abelian vortices can exist in the cyclic phase of the spin-2 BEC. To do this, we first discuss the theory of the spin-2 BEC [23, 24, 25, 26, 27, 28, 29, 30, 31, 34, 35] and the cyclic phase which is one of possible ground states.

*3.1. Spin-2 BEC*

*3.1.1. Hamiltonian* The system of spin-2 bosons can be described as 5-components Bose field $\hat{\Psi} = (\Psi_2, \Psi_1, \Psi_0, \Psi_{-1}, \Psi_{-2})^T$. The Hamiltonian becomes

$$\begin{aligned}\hat{H} &= \int d\mathbf{x}\, \hat{\Psi}^\dagger_m(\mathbf{x}) \left[ -\frac{\hbar^2}{2M}\nabla^2 + V^{\rm ext}(\mathbf{x}) \right] \hat{\Psi}_m(\mathbf{x}) \\ &+ \frac{1}{2}\int d\mathbf{x}_1 d\mathbf{x}_2\, \hat{\Psi}^\dagger_{m_1}(\mathbf{x}_1)\hat{\Psi}^\dagger_{m_2}(\mathbf{x}_2) V^{\rm int}_{m_1,m_2,m_1',m_2'}(\mathbf{x}_1 - \mathbf{x}_2)\hat{\Psi}_{m_2'}(\mathbf{x}_2)\hat{\Psi}_{m_1'}(\mathbf{x}_1),\end{aligned} \tag{4}$$

where $M$ is the particle's mass, $V^{\rm ext}$ is the external potential, and $V^{\rm int}$ is the particle interaction, and repeated indices are assumed to be summed over $2, 1, \cdots, -2$.

As an actual system, we consider spinor BECs of dilute cold atomic gases. In this system, the main inter-particle interaction is the two body $s$-wave scattering. When the dipole-dipole interaction is neglected, we obtain

$$V^{\rm int}_{m_1,m_2,m_1',m_2'}(\mathbf{x}_1 - \mathbf{x}_2) = \frac{4\pi\hbar^2}{M}\delta(\mathbf{x}_1 - \mathbf{x}_2) \sum_{S=0,2,4} \sum_{M_s=-S}^{S} a_S\, C^{S,M_s}_{m_1,m_2} \left( C^{S,M_s}_{m_1',m_2'} \right)^*, \tag{5}$$

Here, $a_S$ is the two body $s$-wave scattering length for the total spin $S$ channel and $C^{S,M_s}_{m_1,m_2}$ is the Clebsch-Gordan coefficient. From Eqs. (4) and (5), we obtain

$$\begin{aligned}\hat{H} &= \int d\mathbf{x}\, \hat{\Psi}^\dagger_m(\mathbf{x}) \left[ -\frac{\hbar^2}{2M}\nabla^2 + V^{\rm ext}(\mathbf{x}) \right] \hat{\Psi}_m(\mathbf{x}) \\ &+ \frac{1}{2}\int d\mathbf{x}\, \hat{N}\left[ c_0\hat{n}^2(\mathbf{x}) + c_1\hat{\mathbf{F}}^2(\mathbf{x}) + c_2\hat{A}^\dagger_{20}(\mathbf{x})\hat{A}_{20}(\mathbf{x}) \right],\end{aligned} \tag{6}$$

where,

$$\begin{aligned}\hat{n}(\mathbf{x}) &= \sum_{m=-2}^{2} \hat{\Psi}^\dagger_m(\mathbf{x})\hat{\Psi}_m(\mathbf{x}), \\ \hat{\mathbf{F}}(\mathbf{x}) &= \sum_{m,n=-2}^{2} \mathbf{f}_{m,n}\hat{\Psi}^\dagger_m(\mathbf{x})\hat{\Psi}_n(\mathbf{x}), \\ \hat{A}_{20}(\mathbf{x}) &= \left[ 2\hat{\Psi}_2(\mathbf{x})\hat{\Psi}_{-2}(\mathbf{x}) + \hat{\Psi}_1(\mathbf{x})\hat{\Psi}_{-1}(\mathbf{x}) + \hat{\Psi}_0^2(\mathbf{x}) \right],\end{aligned} \tag{7}$$

are the density, spin density, and the pair-singlet operators, and $\hat{N}[\cdots]$ is the operator for the normal ordering. $\mathbf{f}_{m,n} = (f^x_{m,n}, f^y_{m,n}, f^z_{m,n})$ is the $5 \times 5$ spin matrix:

$$\begin{aligned} f^+ &= \begin{pmatrix} 0 & 2 & 0 & 0 & 0 \\ 0 & 0 & \sqrt{6} & 0 & 0 \\ 0 & 0 & 0 & \sqrt{6} & 0 \\ 0 & 0 & 0 & 0 & 2 \\ 0 & 0 & 0 & 0 & 0 \end{pmatrix}, \quad f^z = \begin{pmatrix} 2 & 0 & 0 & 0 & 0 \\ 0 & 1 & 0 & 0 & 0 \\ 0 & 0 & 0 & 0 & 0 \\ 0 & 0 & 0 & -1 & 0 \\ 0 & 0 & 0 & 0 & -2 \end{pmatrix}, \\ f^x &= \frac{f^+ + (f^+)^T}{2}, \quad f^y = \frac{f^+ - (f^+)^T}{2i}. \end{aligned} \tag{8}$$

Coefficients $c_0$, $c_1$, and $c_2$ satisfy

$$c_0 = \frac{4\pi\hbar^2}{M}\frac{4a_2 + 3a_4}{7}, \quad c_1 = \frac{4\pi\hbar^2}{M}\frac{a_4 - a_2}{7}, \quad c_2 = \frac{4\pi\hbar^2}{M}\frac{7a_0 - 10a_2 + 3a_4}{35}. \tag{9}$$

Here, we apply the mean-field approximation. Assuming that all particles occupy the single-particle ground state, we obtain [27]

$$H = \int d\mathbf{x} \left[ \frac{\hbar^2}{2M} |\nabla \Psi_m(\mathbf{x})|^2 + V^{\text{ext}}(\mathbf{x}) n(\mathbf{x}) \right] + \frac{1}{2} \int d\mathbf{x} \left[ c_0 n^2(\mathbf{x}) + c_1 \mathbf{F}^2(\mathbf{x}) + c_2 |A_{20}(\mathbf{x})|^2 \right], \tag{10}$$

where,

$$\begin{aligned} n(\mathbf{x}) &= \sum_{m=-2}^{2} |\Psi_m(\mathbf{x})|^2, \\ \mathbf{F}(\mathbf{x}) &= \sum_{m,n=-2}^{2} \mathbf{f}_{m,n} \Psi_m^*(\mathbf{x}) \Psi_n(\mathbf{x}), \\ A_{20}(\mathbf{x}) &= \sum_{m=-2}^{2} (-1)^m \Psi_m(\mathbf{x}) \Psi_{-m}(\mathbf{x}), \end{aligned} \tag{11}$$

are the density, spin density, and singlet-pair amplitude. Although, for actual cold atomic BECs, we have to consider the optical trapping potential and the linear and quadratic Zeeman effects due to the external magnetic field as $V^{\text{ext}}(\mathbf{x})$, we neglect these effects for simplicity and obtain the Hamiltonian density:

$$\frac{H}{V} = \frac{1}{2} \left( c_0 n^2 + c_1 \mathbf{F}^2 + c_2 |A_{20}|^2 \right), \tag{12}$$

where $V$ is the volume of the system.

*3.1.2. Order-parameter manifold* Here, we consider the transformation $G$ which keeps the Hamiltonian in Eq. (12) invariant. For dynamical stability of BECs, $c_0 > 0$ is required. Because the interactions proportional to $c_0$, $c_1$, and $c_2$ have $SU(5)$, $SO(3)$, and $SO(5)$ symmetries respectively [35], in addition to the $U(1)$ gauge symmetry, $G$ becomes

$$\begin{aligned} &G \simeq U(5)_{g+s} \simeq \frac{U(1)_g \times SU(5)_s}{\mathbb{Z}_{g+s}} \quad \text{for } c_1 = c_2 = 0, \\ &G \simeq U(1)_g \times SO(5)_s \quad \text{for } c_1 = 0 \text{ and } c_2 \neq 0, \\ &G \simeq U(1)_g \times SO(3)_s \quad \text{for } c_1 \neq 0. \end{aligned} \tag{13}$$

Here, $g$ and $s$ denote the $U(1)$ gauge and spin parts respectively.

To see the symmetry of the order parameter in detail, we define the density renormalized spinor order parameter $\zeta = (\zeta_2, \zeta_1, \zeta_0, \zeta_{-1}, \zeta_{-2})^T$:

$$\Psi_m = \sqrt{n} \zeta_m, \quad \sum_{m=-2}^{2} |\zeta_m|^2 = 1, \tag{14}$$

and the following nematic tensor [23, 29]. Expanding $\zeta$ by the rank-2 spherical harmonics $Y_{2,m}(\mathbf{e})$ ($\mathbf{e}$ is the unit vector in spin space), we obtain

$$\zeta^{\Sigma}(\mathbf{e}) \equiv \sum_{m=-2}^{2} \zeta_m Y_{2,m}(\mathbf{e}) = \frac{1}{2} \sqrt{\frac{15}{8\pi}} \mathbf{e}^T Q(\zeta) \mathbf{e}, \tag{15}$$

Journal of Physics: Conference Series **297** (2011) 012013 doi:10.1088/1742-6596/297/1/012013

where

$$Q(\zeta) = \begin{pmatrix} \zeta_2 + \zeta_{-2} - \sqrt{\frac{2}{3}}\zeta_0 & i(\zeta_2 - \zeta_{-2}) & -\zeta_1 + \zeta{-1} \\ i(\zeta_2 - \zeta_{-2}) & -\zeta_2 - \zeta_{-2} - \sqrt{\frac{2}{3}}\zeta_0 & -i(\zeta_1 + \zeta_{-1}) \\ -\zeta_1 + \zeta{-1} & -i(\zeta_1 + \zeta_{-1}) & 2\sqrt{\frac{2}{3}}\zeta_0 \end{pmatrix} \tag{16}$$

is the traceless symmetric tensor. The spin rotation of the spinor $\zeta$ and the $SO(3)$ rotation of $Q(\zeta)$ have 1 to 1 correspondence:

$$Q(e^{-i\theta\mathbf{f}\cdot\mathbf{e}}\zeta) = R(\mathbf{e},\theta)Q(\zeta)R^T(\mathbf{e},\theta). \tag{17}$$

Here, $R(\mathbf{e},\theta)$ is the 3-dimensional rotation matrix with the rotation axis $\mathbf{e}$ and the rotation angle $\theta$. Including the $U(1)$ gauge transformation, we can show that the Hamiltonian is invariant when the arbitrary $Q(\zeta)$ is transformed to

$$\hat{T}[\phi, R(\mathbf{e},\theta)]Q(\zeta) \equiv e^{i\phi}R(\mathbf{e},\theta)Q(\zeta)R^T(\mathbf{e},\theta). \tag{18}$$

*3.1.3. Ground state* The ground state can be obtained by finding the $\zeta$ which minimize the Hamiltonian [23, 24, 27, 34].

(i) ferromagnetic-1 phase: $c_1 < 0$ and $c_2 > 4c_1$.

$$\zeta_{\rm F1} = (1,0,0,0,0)^T, \quad Q(\zeta_{\rm F1}) = \begin{pmatrix} 1 & i & 0 \\ i & -1 & 0 \\ 0 & 0 & 0 \end{pmatrix}. \tag{19}$$

There is a continuous transformation $\hat{T}[2\phi, R(\hat{\mathbf{z}},\phi)]$ keeping $Q(\zeta_{\rm F1})$ invariant, which reveals that the spin rotation (along the $z$ axis) and the gauge transformation is equivalent, namely $H_{\rm F1} \simeq U(1)_{2g+s}$. The order-parameter manifold becomes

$$\left(\frac{G}{H}\right)_{\rm F1} \simeq \frac{U(1)_g \times SO(3)_s}{U(1)_{2g+s}} \simeq \frac{SO(3)_{g+s}}{(\mathbb{Z}_2)_{g+s}}. \tag{20}$$

(ii) uniaxial nematic phase: $c_1 > 0$ and $c_2 < 0$.

$$\zeta_{\rm UN} = (0,0,1,0,0)^T, \quad Q(\zeta_{\rm UN}) = \sqrt{\frac{2}{3}}\begin{pmatrix} -1 & 0 & 0 \\ 0 & -1 & 0 \\ 0 & 0 & 2 \end{pmatrix}, \tag{21}$$

which remains invariant under a continuous transformation $\hat{T}[0, R(\hat{\mathbf{z}},\phi)]$ and a discrete transformation $\hat{T}[0, R(a\hat{\mathbf{x}} + b\hat{\mathbf{y}}, \pi)]$. These two transformations form the 2nd orthogonal group $H_{\rm UN} \simeq O(2)_s$ and the order-parameter manifold becomes

$$\left(\frac{G}{H}\right)_{\rm UN} \simeq \frac{U(1)_g \times SO(3)_s}{O(2)_s} \simeq U(1)_g \times \mathbb{RP}^2_s, \tag{22}$$

where $\mathbb{RP}^2$ is the real projective plane.

Journal of Physics: Conference Series **297** (2011) 012013 doi:10.1088/1742-6596/297/1/012013

(iii) biaxial nematic phase: $c_1 < 0$ and $c_2 < 4c_1$.

$$\zeta_{\rm BN} = \sqrt{\frac{1}{2}}(1,0,0,0,1)^T, \quad Q(\zeta_{\rm BN}) = \sqrt{2}\begin{pmatrix} 1 & 0 & 0 \\ 0 & -1 & 0 \\ 0 & 0 & 0 \end{pmatrix}. \tag{23}$$

There are seven discrete transformations $\hat{T}[0, R(\{\hat{\mathbf{x}}, \hat{\mathbf{y}}, \hat{\mathbf{z}}\}, \pi)]$, $\hat{T}[\pi, R(\hat{\mathbf{y}}, \{\pi/4, 3\pi/4\})]$, and $\hat{T}[\pi, R(\{\hat{\mathbf{x}} \pm \hat{\mathbf{y}}\}, \pi)]$ keeping $Q(\zeta_{\rm BN})$ invariant. These transformations form the 4th dihedral group $H_{\rm BN} \simeq (D_4)_{g+s}$ with the identity transformation and the order-parameter manifold becomes

$$\left(\frac{G}{H}\right)_{\rm BN} \simeq \frac{U(1)_g \times SO(3)_s}{(D_4)_{g+s}}. \tag{24}$$

(iv) cyclic phase: $c_1 > 0$ and $c_2 > 0$.

$$\zeta_{\rm C} = \frac{1}{2}(i, 0, \sqrt{2}, 0, i)^T, \quad Q(\zeta_{\rm C}) = \frac{\sqrt{3}}{2}\begin{pmatrix} e^{2\pi i/3} & 0 & 0 \\ 0 & e^{-2\pi i/3} & 0 \\ 0 & 0 & 1 \end{pmatrix}. \tag{25}$$

There are eleven discrete transformations $\hat{T}[0, R(\{\hat{\mathbf{x}}, \hat{\mathbf{y}}, \hat{\mathbf{z}}\}, \pi)]$, $\hat{T}[2\pi/3, R(\{\mathbf{e}_{1,2,3,4}\}, 2\pi/3)]$, and $\hat{T}[-2\pi/3, R(\{\mathbf{e}_{1,2,3,4}\}, -2\pi/3)]$ keeping $Q(\zeta_{\rm C})$ invariant. Here $\mathbf{e}_1 = (\hat{\mathbf{x}} + \hat{\mathbf{y}} + \hat{\mathbf{z}})/\sqrt{3}$, $\mathbf{e}_2 = (\hat{\mathbf{x}} - \hat{\mathbf{y}} - \hat{\mathbf{z}})/\sqrt{3}$, $\mathbf{e}_3 = (-\hat{\mathbf{x}} + \hat{\mathbf{y}} - \hat{\mathbf{z}})/\sqrt{3}$, $\mathbf{e}_4 = (-\hat{\mathbf{x}} - \hat{\mathbf{y}} + \hat{\mathbf{z}})/\sqrt{3}$. These transformations form the tetrahedral group $H_{\rm C} \simeq T_{g+s}$ with the identity transformation [19, 26, 30] and the order-parameter manifold becomes

$$\left(\frac{G}{H}\right)_{\rm C} \simeq \frac{U(1)_g \times SO(3)_s}{T_{g+s}}. \tag{26}$$

(v) ferromagnetic-2 phase (This phase cannot exist as the ground state).

$$\zeta_{\rm F2} = (0,1,0,0,0)^T, \quad Q(\zeta_{\rm F2}) = \begin{pmatrix} 0 & 0 & -1 \\ 0 & 0 & -i \\ -1 & -i & 0 \end{pmatrix}. \tag{27}$$

There is a continuous transformation $\hat{T}[\phi, R(\hat{\mathbf{z}}, \phi)]$ keeping $Q(\zeta_{\rm F2})$ invariant, namely $H_{\rm F2} \simeq U(1)_{g+s}$. The order-parameter manifold becomes

$$\left(\frac{G}{H}\right)_{\rm F2} \simeq \frac{U(1)_g \times SO(3)_s}{U(1)_{g+s}} \simeq SO(3)_{g+s}. \tag{28}$$

This state appear in the vortex core for example [19].

Each phase can also be characterized by the spin density $\mathbf{F}^2$, the singlet-pair amplitude $|A_{20}|^2$, and the singlet-trio amplitude $|A_{30}|^2$ defined as

$$A_{30} = \frac{3\sqrt{6}}{2}\left(\Psi_1^2\Psi_{-2} + \Psi_{-1}^2\Psi_2\right) + \Psi_0\left(\Psi_0^2 - 3\Psi_1\Psi_{-1} - 6\Psi_2\Psi_{-2}\right). \tag{29}$$

These values for each phase is summarized in TABLE 1 [32, 34].

The symmetry of each phase can be visualized by $\zeta^\Sigma$. Figure 4 (a) shows the $c_1 - c_2$ phase diagram and the shape of $\zeta^\Sigma(\mathbf{e})$. In the ferromagnetic-1 phase, $\zeta^\Sigma(\mathbf{e})$ shows the disk shape and

| | order parameter | $H$ | $\mathbf{F}^2$ | $\lvert A_{20}\rvert^2$ | $\lvert A_{30}\rvert^3$ |
|---|---|---|---|---|---|
| ferromagnetic-1 | $(1,0,0,0,0)^T$ | $U(1)_{2g+s}$ | $4n^2$ | 0 | 0 |
| ferromagnetic-2 | $(0,1,0,0,0)^T$ | $U(1)_{g+s}$ | $n^2$ | 0 | 0 |
| uniaxial nematic | $(0,0,1,0,0)^T$ | $O(2)_s$ | 0 | $n^2$ | $n^3$ |
| biaxial nematic | $(1,0,0,0,1)^T/\sqrt{2}$ | $(D_4)_{g+s}$ | 0 | $n^2$ | 0 |
| cyclic | $(i,0,\sqrt{2},0,i)^T/2$ | $T_{g+s}$ | 0 | 0 | $2n^3$ |

**Table 1.** Order parameter, $H$, $\mathbf{F}^2$, $|A_{20}|^2$, $|A_{30}|^3$ for each phase.

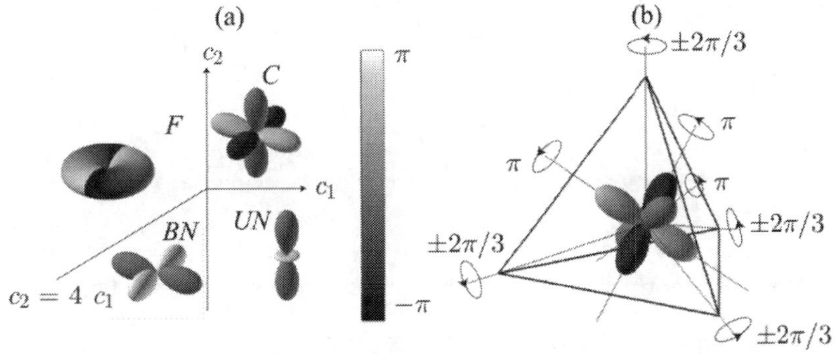


**Figure 4.** (a) $c_1 - c_2$ phase diagram and the shape of $\zeta^\Sigma(\mathbf{e})$ (The shape is calculated as $|\zeta^\Sigma(\mathbf{e})|^2$, and the color represents the phase of $\zeta^\Sigma(\mathbf{e})$). Here $F$, $UN$, $BN$, and $C$ represent the ferromagnetic-1, uniaxial nematic, biaxial nematic, and cyclic phases respectively. (b) Schematic illustration showing that the symmetry axes of the cyclic order parameter constitute a tetrahedron.

the phase changes by $4\pi$ around the disk. The rotation along the axis perpendicular to the disk and the $U(1)$ gauge transformation are equivalent which corresponds to $H_{\rm f1} \simeq U(1)_{2g+s}$. From the dumbbell and cloverleaf shapes of $\zeta^\Sigma(\mathbf{e})$ for the uniaxial nematic and biaxial nematic phases, it is easy to see that these phases have cylindrical symmetry $H_{\rm un} \simeq O(2)_s$ and square symmetry $H_{\rm bn} \simeq (D_4)_{g+s}$ respectively. In the cyclic phase, $\zeta^\Sigma(\mathbf{e})$ takes the triad shape which has the tetrahedral symmetry as shown in Fig. 4 (b).

Among four ground states, the biaxial nematic and cyclic phases gives the non-Abelian $\pi_1[G/H]$. However, the biaxial nematic and uniaxial nematic phases are energetically degenerate in the mean-field framework. The zero-point fluctuation lifts this degeneracy [32, 33, 34, 35]. In the mean-field theory, therefore, the biaxial nematic phase is difficult to study non-Abelian vortices and the cyclic phase is the best theoretical model to study those.

### *3.2. Non-Abelian vortices in the cyclic phase*

We calculate the topological invariants of vortices in the cyclic phase from the order-parameter manifold of Eq. (26) [19, 26, 30]. Because $SO(3)$ is not simply connected, lets consider the mapping from $SO(3)$ to simply connected $SU(2)$:

$$\frac{U(1)_g \times SO(3)_s}{T_{g+s}} \simeq \frac{U(1)_g \times SU(2)_s}{T^*_{g+s}} \tag{30}$$

Here, $T^*$ is the lifting up from the tetrahedral group $T$ as the subgroup of $SO(3)$ to the double cover as the subgroup of $SU(2)$, and consists of 24 elements. To consider the action of the $SU(2)$ rotation to $\zeta$, we define the new tensor $\mathcal{Q}(\zeta) \equiv Q(\zeta)\sigma \otimes \sigma$ and a new operation

$$\mathcal{T}[\phi, U(\mathbf{e},\theta)] \equiv e^{i\phi} U(\mathbf{e},\theta) \otimes U(\mathbf{e},\theta) \mathcal{Q}(\zeta) \left[U(\mathbf{e},\theta) \otimes U(\mathbf{e},\theta)\right]^\dagger, \tag{31}$$

where $\sigma$ is the Pauli matrix and $U(\mathbf{e},\theta)$ is the $SU(2)$ rotation matrix.

Topological invariant defined as $\pi_1[\{U(1)_g \times SU(2)_s\}/T^*_{g+s}]$ can be described by $\mathcal{T}$. In the following, we concentrate on the energetically stable vortices and classify those by the conjugacy class. For one straight vortex along $z$-axis, we consider the order parameter $\Psi$ in the cylindrical coordinate $(r,\varphi,z)$ ($f_{1,2,3}(r)$ satisfy $f_{1,2,3}(r\to 0)=0$ and $f_{1,2,3}(r\to\infty)=1$).

(i) $0-1/2$ vortex
Topological invariants can be written by $\mathcal{T}[0,\{\mathcal{I}_{x,y,z}\}]$, where $\mathcal{I}_{x,y,z}=U(\pi,\{\hat{\mathbf{x}},\hat{\mathbf{y}},\hat{\mathbf{z}}\})$ are

$$\mathcal{I}_x=\begin{pmatrix} 0 & -i \\ -i & 0 \end{pmatrix},\quad \mathcal{I}_y=\begin{pmatrix} 0 & -1 \\ 1 & 0 \end{pmatrix},\quad \mathcal{I}_z=\begin{pmatrix} -i & 0 \\ 0 & i \end{pmatrix}. \tag{32}$$

$\Psi$ is given by

$$\Psi(r,\varphi,z)=\hat{S}\frac{\sqrt{n}}{2}\left(if_1(r)e^{-i\varphi},0,\sqrt{2},0,if_1(r)e^{i\varphi}\right)^T, \tag{33}$$

where $S$ is the arbitrary $U(1)$ gauge transformation and $SO(3)$ spin rotation. 0 and 1/2 of the $0-1/2$ vortex denotes the mass and spin currents $\kappa_{\rm m}$ and $\kappa_{\rm s}^{\mu}$ defined as

$$\begin{aligned}\kappa_{\rm m}&=\kappa\oint d\mathbf{x}\cdot\frac{\sum_m \mathrm{Im}[(\nabla\Psi_m)\Psi_m^*]}{n}\\ \kappa_{\rm s}^{\mu}&=\kappa\oint d\mathbf{x}\cdot\frac{\sum_{m,n}\mathrm{Im}[(\nabla\Psi_m)f^{\mu}_{m,n}\Psi_n^*]}{n},\end{aligned} \tag{34}$$

where $\kappa=h/M$. For $0-1/2$ vortex, $\kappa_{\rm m}=0$ and $\kappa_{\rm s}\equiv\sqrt{(\kappa_{\rm s}^x)^2+(\kappa_{\rm s}^y)^2+(\kappa_{\rm s}^z)^2}=\kappa/2$.

(ii) $0-(-1/2)$ vortex
Topological invariants can be written by $\mathcal{T}[0,-\{\mathcal{I}_{x,y,z}\}]$. $\Psi$ is given by

$$\Psi(r,\varphi,z)=\hat{S}\frac{\sqrt{n}}{2}\left(if_1(r)e^{i\varphi},0,\sqrt{2},0,if_1(r)e^{-i\varphi}\right)^T. \tag{35}$$

Circulations become $\kappa_{\rm m}=0$ and $\kappa_{\rm s}=\kappa/2$.

(iii) $1/3-1/3$ vortex
Topological invariants can be written by $\mathcal{T}[2\pi/3,\{\mathcal{E},-\mathcal{I}_{x,y,z}\}\mathcal{C}]$, where $\mathcal{E}$ and $\mathcal{C}$ are

$$\mathcal{E}=\begin{pmatrix} 1 & 0 \\ 0 & 1 \end{pmatrix},\quad \mathcal{C}=U(\mathbf{e}_1,2\pi/3)=\frac{1}{2}\begin{pmatrix} 1-i & -1-i \\ 1-i & 1+i \end{pmatrix} \tag{36}$$

By using the another form of the cyclic order parameter:

$$e^{i\cos^{-1}(1/\sqrt{3})f^y}e^{i\pi f^z/4}\zeta_{\rm C}=\frac{1}{\sqrt{3}}\left(0,\sqrt{2},0,0,1\right)^T, \tag{37}$$

we obtain

$$\Psi(r,\varphi,z)=\hat{S}\sqrt{\frac{n}{3}}\left(0,\sqrt{2},0,0,f_2(r)e^{i\varphi}\right)^T. \tag{38}$$

Circulations become $\kappa_{\rm m}=\kappa/3$ and $\kappa_{\rm s}=\kappa/3$.

(iv) $(-1/3)-(-1/3)$ vortex
Topological invariants can be written by $\mathcal{T}[-2\pi/3, -\{\mathcal{E}, \mathcal{I}_{x,y,z}\}\mathcal{C}^2]$. $\Psi$ is given by

$$\Psi(r,\varphi,z) = \hat{S}\sqrt{\frac{n}{3}}\left(0, \sqrt{2}, 0, 0, f_2(r)e^{-i\varphi}\right)^T. \tag{39}$$

Circulations become $\kappa_{\rm m} = -\kappa/3$ and $\kappa_{\rm s} = \kappa/3$.

(v) $(-2/3)-1/3$ vortex
Topological invariants can be written by $\mathcal{T}[-4\pi/3, \{\mathcal{E}, -\mathcal{I}_{x,y,z}\}\mathcal{C}]$. $\Psi$ is given by

$$\Psi(r,\varphi,z) = \hat{S}\sqrt{\frac{n}{3}}\left(0, \sqrt{2}f_3(r)e^{-i\varphi}, 0, 0, 1\right)^T. \tag{40}$$

Circulations become $\kappa_{\rm m} = -2\kappa/3$ and $\kappa_{\rm s} = \kappa/3$.

(vi) $2/3-(-1/3)$ vortex
Topological invariants can be written by $\mathcal{T}[4\pi/3, -\{\mathcal{E}, \mathcal{I}_{x,y,z}\}\mathcal{C}^2]$. $\Psi$ is given by

$$\Psi(r,\varphi,z) = \hat{S}\sqrt{\frac{n}{3}}\left(0, \sqrt{2}f_3(r)e^{i\varphi}, 0, 0, 1\right)^T. \tag{41}$$

Circulations become $\kappa_{\rm m} = 2\kappa/3$ and $\kappa_{\rm s} = \kappa/3$.

All other vortices classified by $\pi_1[\{U(1)_g \times SU(2)_s\}/T^*_{g+s}]$ are energetically unstable and split into vortices discussed above.

Order parameters of vortex cores are given by taking the limit of $r \to 0$:

(i) $0-1/2$ and $0-(-1/2)$ vortices:

$$\Psi(r \to 0, \varphi, z) = \hat{S}\sqrt{\frac{n}{2}}\zeta_{\rm UN} \tag{42}$$

(ii) $1/3-1/3$ and $(-1/3)-(-1/3)$ vortices:

$$\Psi(r \to 0, \varphi, z) = \hat{S}\sqrt{\frac{2n}{3}}\zeta_{\rm F2} \tag{43}$$

(iii) $(-2/3)-1/3$ and $2/3-(-1/3)$ vortices:

$$\Psi(r \to 0, \varphi, z) = \hat{S}\sqrt{\frac{n}{3}}\zeta_{\rm F1} \tag{44}$$

Characteristics of all vortices are summarized in TABLE 2. Anti-vortex can be obtained by $\mathcal{T}^{-1}$.

As shown in Fig. 4 (b), the tetrahedral symmetry of the cyclic phase can be visualized by the triad. Around the vortex core, the triad rotates. There are two kinds of the rotation which keeps the triad invariant: $\pi$ rotation along the lobe of the triad and $\pm 2\pi/3$ rotation along the axis crossing the center and the apex of the tetrahedron which correspond to $0-(\pm 1/2)$ vortices, and the other vortices respectively. For example, we plot $\zeta^{\Sigma}(\mathbf{e})$ for $0-1/2$ and $1/3-1/3$ vortices in Figs. 5 (a) and 5 (b) respectively.

| vortex | topological invariant | $\kappa_m/\kappa$ | $\kappa_s/\kappa$ | vortex core | anti-vortex |
|---|---|---|---|---|---|
| $0-1/2$ | $\mathcal{T}[0,\mathcal{I}_{x,y,z}]$ | 0 | 1/2 | $\propto \zeta_{\rm UN}$ | $0-(-1/2)$ |
| $0-(-1/2)$ | $\mathcal{T}[0,-\mathcal{I}_{x,y,z}]$ | 0 | 1/2 | $\propto \zeta_{\rm UN}$ | $0-1/2$ |
| $1/3-1/3$ | $\mathcal{T}[2\pi/3,\{\mathcal{E},-\mathcal{I}_{x,y,z}\}\mathcal{C}]$ | 1/3 | 1/3 | $\propto \zeta_{\rm F2}$ | $(-1/3)-(-1/3)$ |
| $(-1/3)-(-1/3)$ | $\mathcal{T}[-2\pi/3,-\{\mathcal{E},\mathcal{I}_{x,y,z}\}\mathcal{C}^2]$ | -1/3 | 1/3 | $\propto \zeta_{\rm F2}$ | $1/3-1/3$ |
| $(-2/3)-1/3$ | $\mathcal{T}[-4\pi/3,\{\mathcal{E},-\mathcal{I}_{x,y,z}\}\mathcal{C}]$ | -2/3 | 1/3 | $\propto \zeta_{\rm F1}$ | $2/3-(-1/3)$ |
| $2/3-(-1/3)$ | $\mathcal{T}[4\pi/3,-\{\mathcal{E},\mathcal{I}_{x,y,z}\}\mathcal{C}]$ | 2/3 | 1/3 | $\propto \zeta_{\rm F1}$ | $(-2/3)-1/3$ |

**Table 2.** Stable vortices in the cyclic phase. Each vortex is classified by the conjugacy class.

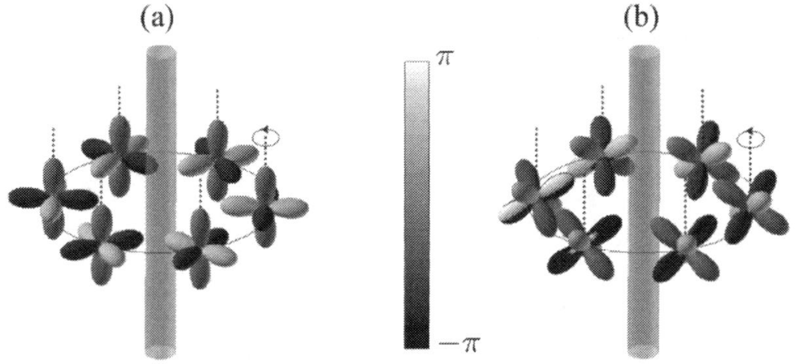


**Figure 5.** Plot of $\zeta^{\Sigma}(\mathbf{e})$ around the vortex. (a) $0-1/2$ vortex. (b) $1/3-1/3$ vortex. Dashed lines in figures show the rotation axes of the triad.

### *3.3. Spin-2 BECs and the cyclic phase in experiments*

In experiments, spinor BECs are realized as optically trapped cold dilute atomic gases. In this system, internal degrees of freedom are given by the hyperfine spin $F = I + S$, where $I$ and $S$ are the nuclear and electron spins. For spin-2 case, $F = 2$ $^{87}$Rb BECs ($I = 3/2$, $S = 1/2$) are usually used because of their long lifetimes [36, 37, 38, 39, 40]. For interaction parameters of $F = 2$ $^{87}$Rb, $c_0/(4\pi\hbar^2 a_{\rm B}/M) = 106\pm4$, $c_1/(4\pi\hbar^2 a_{\rm B}/M) = 0.99\pm0.06$, and $c_2/(4\pi\hbar^2 a_{\rm B}/M) = -0.106\pm0.116$, and the uniaxial nematic phase is suggested for the ground state [39]. However, the large error bar for $c_2$ cannot exclude the possibility of the cyclic ground state because of complications arising from quadratic Zeeman effects and hyperfine-spin-exchanging relaxations [27, 34, 40].

### *3.4. Collision dynamics of non-Abelian vortices*

Here, we perform the numerical simulation of vortices in the cyclic phase and show that the results satisfy the geometrical consideration discussed in the Sec. 2.

To study the dynamics of vortices, we derive the Gross-Pitaevskii equation which describes the time evolution of the order parameter [27, 28]. The Gross-Pitaevskii equation can be obtained by the Hamilton equation $i\hbar\partial\Psi_m/\partial t = \delta H/\delta\Psi_m$ for the mean-field Hamiltonian:

$$
\begin{aligned}
i\hbar\frac{\partial\Psi_{\pm2}}{\partial t} &= \left[-\frac{\hbar^2}{2M}\nabla^2 + c_0 n \pm 2c_1 F^z\right]\Psi_{\pm2} + c_1 F^{\mp}\Psi_{\pm1} + c_2 A_{20}\Psi^*_{\mp2} \\
i\hbar\frac{\partial\Psi_{\pm1}}{\partial t} &= \left[-\frac{\hbar^2}{2M}\nabla^2 + c_0 n \pm c_1 F^z\right]\Psi_{\pm1} + c_1\left[\frac{\sqrt{6}}{2}F^{\mp}\Psi_0 + F^{\pm}\Psi_{\pm2}\right] - c_2 A_{20}\Psi^*_{\mp1} \\
i\hbar\frac{\partial\Psi_0}{\partial t} &= \left[-\frac{\hbar^2}{2M}\nabla^2 + c_0 n\right]\Psi_0 + \frac{\sqrt{6}}{2}c_1\left[F^-\Psi_0 + F^{\pm}\Psi_{-1} + F^+\Psi_1\right] + c_2 A_{20}\Psi^*_0
\end{aligned}
\tag{45}
$$

Simulations are performed by using the initial state with vortices and numerically solving the Gross-Pitaevskii equation in the cubic box with the Neumann boundary [41]. Numerical parameters are:

(i) Interaction coefficients : $c_0 > 0$, $c_1/c_0 = c_2/c_0 = 0.5$.

(ii) The size of the box : $(64\xi_0)^3$ for the healing length $\xi_0 \equiv \sqrt{\hbar^2/Mc_0 n}$.

(iii) Number of numerical grids : $256^3$.

Interaction coefficients which we use make the cyclic state stable. Although these values are not realistic for the actual $^{87}$Rb BECs, we concentrate on the topological features for collisions of vortices, and do not pay attention to the consistency with actual experiments.

As initial states, we use two vortex configurations: (I) two straight vortices with an oblique angle and (II) two twisted vortices. As a pair of topological invariants, we use the following four patterns:

(i) $\mathcal{T}[2\pi/3, \mathcal{C}]$ (1/3 − 1/3 vortex) and $\mathcal{T}[2\pi/3, \mathcal{C}]$ (1/3 − 1/3 vortex) : same topological invariant

(ii) $\mathcal{T}[2\pi/3, \mathcal{C}]$ (1/3 − 1/3 vortex) and $\mathcal{T}[4\pi/3, -\mathcal{C}^2]$ (2/3 − (−1/3) vortex) : different and commutative topological invariants

(iii) $\mathcal{T}[2\pi/3, \mathcal{C}]$ (1/3 − 1/3 vortex) and $\mathcal{T}[2\pi/3, -\mathcal{I}_y\mathcal{C}]$ (1/3 − 1/3 vortex) : non-commutative topological invariants

(iv) $\mathcal{T}[2\pi/3, \mathcal{C}]$ (1/3 − 1/3 vortex) and $\mathcal{T}[4\pi/3, -\mathcal{I}_x\mathcal{C}^2]$ (2/3 − (−1/3) vortex) : non-commutative topological invariants

In the following, we denote the type of collision as (I)-(i) (two straight vortices with the same topological invariants).

Figure 6 shows the collisions of vortices with commutative topological invariants: (I)-(i) (Fig. 6 (a)-(c)) and (I)-(ii) (Fig. 6 (d)-(f)). Figures 6 (a)-(c) and 6 (d)-(f) show the

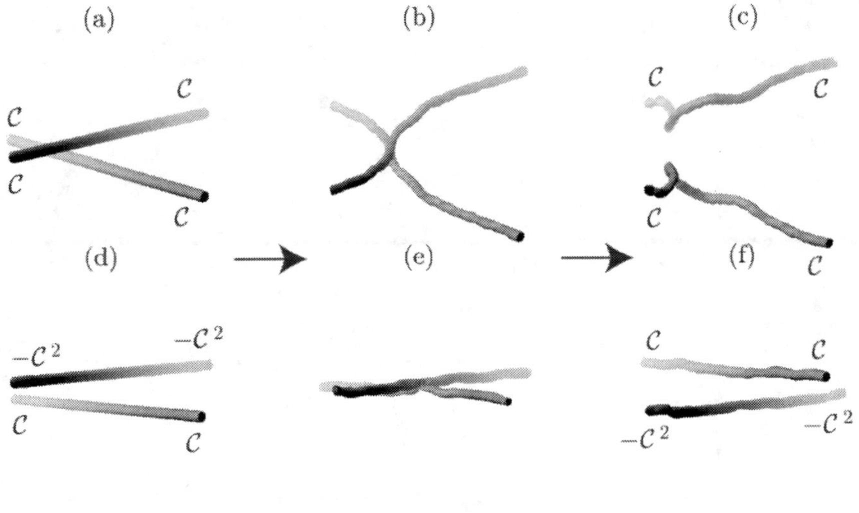


**Figure 6.** Collisions of vortices with commutative topological invariants. (a)-(c) (I)-(i) (reconnection). (d)-(f) (I)-(ii) (passing through). Topological invariant of each vortex is labeled: $\mathcal{C} \equiv \mathcal{T}[2\pi/3, \mathcal{C}]$ and $-\mathcal{C}^2 \equiv \mathcal{T}[4\pi/3, -\mathcal{C}^2]$

reconnection and passing through, which corresponding to Fig. 3 (f) (the rung vortex vanishes: $BA^{-1} = \mathcal{T}[2\pi/3, \mathcal{C}]\,(\mathcal{T}[2\pi/3, \mathcal{C}])^{-1} = \mathcal{T}[0, \mathcal{E}]$) and Fig. 3 (d).

Figure 7 shows the collisions of vortices with non-commutative topological invariants: (I)-(iii) (Fig. 6 (a)-(c)) and (I)-(iv) (Fig. 6 (d)-(f)). Both Figs. 7 (a)-(c) and 7 (d)-(f) show the formations of rung vortices, the topological invariants of which are $BA^{-1}$ =

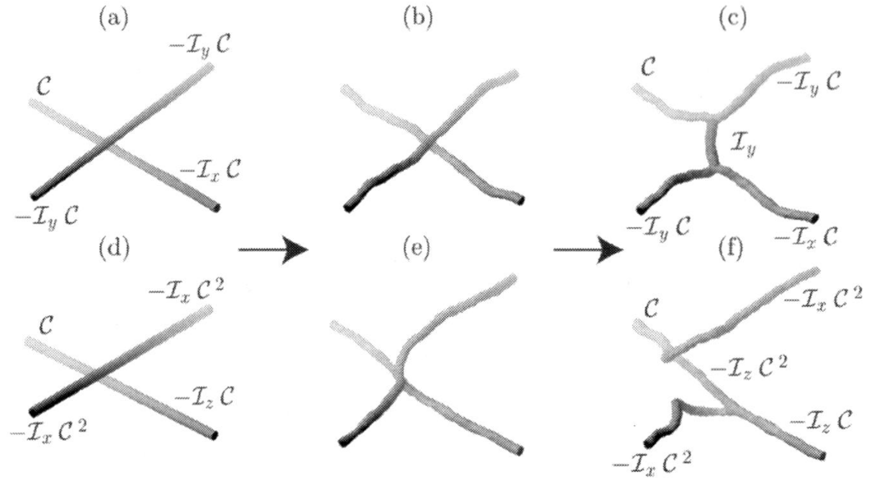


**Figure 7.** Collisions of vortices with non-commutative topological invariants. (a)-(c) (I)-(iii) (rung vortex $\mathcal{T}[0,\mathcal{I}_y]$). (d)-(f) (I)-(iv) (rung vortex $\mathcal{T}[-2\pi/3,-\mathcal{I}_z\mathcal{C}^2]$). Topological invariant of each vortex is labeled: $-\mathcal{I}_{x,y,z}\mathcal{C}\equiv\mathcal{T}[2\pi/3,-\mathcal{I}_{x,y,z}\mathcal{C}]$, $-\mathcal{I}_x\mathcal{C}^2\equiv\mathcal{T}[4\pi/3,-\mathcal{I}_x\mathcal{C}^2]$, $\mathcal{I}_y\equiv\mathcal{T}[0,\mathcal{I}_y]$, and $-\mathcal{I}_z\mathcal{C}^2\equiv\mathcal{T}[-2\pi/3,-\mathcal{I}_z\mathcal{C}^2]$

$\mathcal{T}[2\pi/3,\mathcal{C}]\ (\mathcal{T}[2\pi/3,-\mathcal{I}_y\mathcal{C}])^{-1} = \mathcal{T}[0,\mathcal{I}_y]$ and $BA^{-1} = \mathcal{T}[2\pi/3,\mathcal{C}]\ (\mathcal{T}[4\pi/3,-\mathcal{I}_x\mathcal{C}^2])^{-1} = \mathcal{T}[-2\pi/3,-\mathcal{I}_z\mathcal{C}^2]$ being consistent with Fig. 3 (f).

We have performed numerical simulations with various combinations of topological invariants, relative velocities, and collision angles, and confirmed that passing through and reconnection occur only when the topological invariants of the two vortices are commutative, and that the formation of a rung vortex always occurs when the topological invariants of two vortices are non-commutative.

Figures 8 (a)-(c) and 8 (d)-(f) show the collision of two twisted vortices: (II)-(ii) and (II)-(iii). For commutative topological invariants : (II)-(ii), vortices unravel by the passing through as

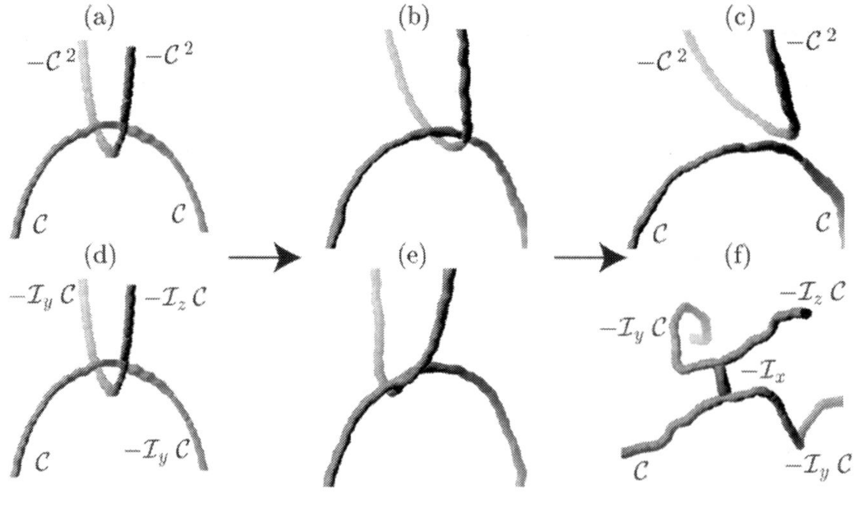


**Figure 8.** Collisions of twisted vortices. (a)-(c) Commutative topological invariants: (II)-(ii) (passing through). (d)-(f) Non-commutative topological invariants: (II)-(iii) (rung vortex $\mathcal{T}[0,-\mathcal{I}_x]$).

shown in Fig. 8 (a)-(c). On the other hand, twisted vortices with non-commutative topological invariants can never unravel because of the formation of the rung vortex as shown in Fig. 8 (d)-(f) which corresponds to Fig. 3 (h), i.e. the topological invariant of the rung vortex is $BAB^{-1}A^{-1} = \mathcal{T}[2\pi/3, -\mathcal{I}_y\mathcal{C}]\,\mathcal{T}[2\pi/3, \mathcal{C}]\,(\mathcal{T}[2\pi/3, -\mathcal{I}_y\mathcal{C}])^{-1}\,(\mathcal{T}[2\pi/3, \mathcal{C}])^{-1} = \mathcal{T}[0, -\mathcal{I}_x]$.

We finally describe a possible experimental manifestation of rung vortices. The phase-contrast imaging experiment [42] enables the measurement of local magnetization, and vortices with ferromagnetic-1 and ferromagnetic-2 cores like Fig. 7 (c)-(f) manifest themselves as bridged structures of localized magnetization.

## 4. Conclusion

In Sec. 2, we explain quantized vortices and their topological invariants based on homotopy theory, and discuss the definition of non-Abelian vortices and their collision dynamics. Non-Abelian vortices are defined as vortices with non-Abelian topological invariants given by the fundamental group of the order-parameter manifold. When two non-Abelian vortices with non-commutative topological invariant collide, they can neither reconnect nor path through each other, and form a rung vortex bridging two colliding vortices. In the case of two twisted vortices with non-commutative topological invariants, they cannot unravel because of the formation of the rung vortex.

In Sec. 3, we discuss the cyclic phase of spin-2 BECs as the system in which non-Abelian vortices can be realized. The fundamental group of the order-parameter manifold in the cyclic phase can be described by the non-Abelian tetrahedral group. We perform numerical simulations of the Gross-Pitaevskii equation for spin-2 BECs and show that collision dynamics of non-Abelian vortices discussed in Sec. 2 can be realized in this system.

In this paper, we concentrate on the cyclic phase of spin-2 BECs as the system for the study of non-Abelian vortices. Non-Abelian properties given by our results are, however, universal for any non-Abelian vortex in other system. We hope that our results lead further understanding of non-Abelian vortices in various kinds of system such as cosmic strings [43, 44, 45, 46, 47, 48, 49, 50, 51, 52, 53, 54, 55, 56], liquid crystals [17, 18, 57, 58, 59, 60, 61, 62], and so on.

## Acknowledgments

This work was supported by KAKENHI (22740219).

## References

[1] Onsager L 1949 *Nuovo Cimento* **6** Suppl. 2 279
[2] Vinen W F 1961 *Proc. Roy. Soc. A* **260** 218
[3] Davis J C, Close J D, Zieve R , and Packard R E 1991 *Phys. Rev. Lett.* **66** 329
[4] Madison K W, Chevy F, Wohlleben W, and Dalibard J 2000 *Phys. Rev. Lett.* **84** 806
[5] Abo-Shaeer J R, Raman C, Vogels J M, and Ketterle W 2001 *Science* **292** 476
[6] Volovik G E 1992, *Exotic Properties of Superfluid Helium 3*, (Singapore: World Scientific)
[7] Leonhardt U and Volovik G E 2000 *JETP Lett.* **72** 46
[8] Zhou F 2001 *Phys. Rev. Lett.* **87** 080401; Zhou F and Snoek M 2003 *Ann. Phys. (N. Y. )* **308** 692
[9] Ji A -C, Liu W M, Song J L, and Zhou F 2008 *Phys. Rev. Lett.* **101** 010402
[10] Koplik J and Levine H 1996 *Phys. Rev. Lett.* **71**, 1375; 1996 **76** 4745
[11] Leadbeater M, Winiecki T, Samuels D C, Barenghi C F and Adams C S 2001 *Phys. Rev. Lett.* **86** 1410
[12] Ogawa S, Tsubota M and Hattori Y 2002 *J. Phys. Soc. Jpn.* **71** 813
[13] Bewley G P, Paoletti M S, Sreenivasan K R, Lathrop D P 2008 *Proc. Natl. Acad. Sci.* **105** 13707
[14] Laguna P and Matzner R A 1989 *Phys. Rev. Lett.* **62** 1948
[15] Bettencourt L M A and Kibble T W B 1994 *Phys. Lett. B* **332** 297; Bettencourt L M A, Laguna P, and Matzner R A 1997 *Phys. Rev. Lett.* **78** 2066
[16] Bevis N and Saffin P M 2008 *Phys. Rev. D* **78** 023503
[17] Poenaru V and Toulouse G 1977 *J. Phys. (Paris)* **38** 887
[18] Mermin N D 1979 *Rev. Mod. Phys.* **51** 591
[19] Kobayashi M, Kawaguchi Y, Nitta M, and Ueda M 2009 *Phys. Rev. Lett.* **103** 115301

[20] Ruostekoski J and Anglin J R 2003 *Phys. Rev. Lett.* **91** 190402
[21] Khawaja U A and Stoof H 2001 *Nature* **411** 918; Khawaja U A and Stoof H 2001 *Phys. Rev. A* **64** 043612
[22] Kawaguchi Y, Nitta M, and Ueda M 2008 *Phys. Rev. Lett.* **100** 180403
[23] Koashi M and Ueda M 2000 *Phys. Rev. Lett.* **84** 1066; Ueda M and Koashi M 2002 *Phys. Rev. A* **65** 063602
[24] Ciobanu C V, Yip S -K, and Ho T -L 2000 *Phys. Rev. A* **61** 033607
[25] Klausen N N, Bohn J L, and Greene C H 2001 *Phys. Rev. A* **64** 053602
[26] Mäkelä H, Zhang Y, and Suominen K -A 2003 *J. Phys. A: Math. Gen.* **36** 8555; Mäkelä H 2006 *J. Phys. A* **39** 7423
[27] Saito H and Ueda M 2005 *Phys. Rev. A* **72** 053628
[28] Pogosov W V, Kawate R, Mizushima T, and Machida K 2005 *Phys. Rev. A* **72** 063605
[29] Zhou F and Semenoff G W 2006 *Phys. Rev. Lett* **97** 180411
[30] Semenoff G W and Zhou F 2007 *Phys. Rev. Lett* **98** 100401
[31] Barnett R, Mukerjee S, and Moore J E 2008 *Phys. Rev. Lett.* **100** 240405
[32] Song J L, Semenoff G W, and Zhou F 2007 *Phys. Rev. Lett.* **98** 160408
[33] Turner A M, Barnett R, Demler E, and Vishwanath A 2007 *Phys. Rev. Lett.* **98** 190404
[34] Uchino S, Kobayashi M, and Ueda M 2010 *Phys. Rev. A* **81** 063632
[35] Uchino S, Kobayashi M, Nitta M, and Ueda M 2010 *Phys. Rev. Lett.* **105** 230406
[36] Schmaljohann H, Erhard M, Kronjäger J, Kottke M, van Staa S, Cacciapuoti L, Arlt J J, Bongs K, and Sengstock K 2004 *Phys. Rev. Lett.* **92** 040402
[37] Chang M -S, Hamley C D, Barrett M D, Sauer J A, Fortier K M, Zhang W, You L, and Chapman M S 2004 *Phys. Rev. Lett.* **92** 140403
[38] Kuwamoto T, Araki K, Eno T, and Hirano T 2004 *Phys. Rev. A* **69** 063604
[39] Widera A, Gerbier F, Fölling S, Gericke T, Mandel O, and Bloch I 2006 *New J. Phys.* **8** 152
[40] Tojo S, Hayashi T, Tanabe T, Hirano T, Kawaguchi Y, Saito H, and Ueda M 2009 *Phys. Rev. A* **80** 042704
[41] Animation can be seen in "http://cat.phys.s.u-tokyo.ac.jp/~michikaz/physics/collision/collision.html"
[42] Higbie J M, Sadler L E, Inouye S, Chikkatur A P, Leslie S R, Moore K L, Savalli V, and Stamper-Kurn D M 2005 *Phys. Rev. Lett.* **95** 050401
[43] Vilenkin A and Shellard E P S 1994, *Cosmic Strings and Other Topological Defects*, (Cambridge, England: Cambridge University Press)
[44] Hindmarsh M B and Kibble T W B 1995 *Rep. Prog. Phys.* **58** 477
[45] Hanany A and Hashimoto K 2005 *J. High Energy Phys.* **06** 021
[46] Achúcarro A and de Putter R 2006 *Phys. Rev. D* **74** 121701(R)
[47] Laguna P, Natchu V, Matzner R A and Vachaspati T 2007 *Phys. Rev. Lett.* **98** 041602
[48] Hashimoto K and Tong D 2005 *J. Cosmol. Astropart. Phys.* **09** 004
[49] Eto M, Hashimoto K, Marmorini G, Nitta M, Ohashi K and Vinci W 2007 *Phys. Rev. Lett.* **98** 091602
[50] Bevis N, Hindmarsh M and Kunz M 2004 *Phys. Rev. D* **70** 043508
[51] Wyman M, Pogosian L and Wasserman I 2005 *Phys. Rev. D* **72** 023513 [2006 Erratum-ibid. D **73** 089905]
[52] Battye R A, Garbrecht B and Moss A 2006 *JCAP* **0609** 007
[53] Bevis N, Hindmarsh M, Kunz M and Urrestilla J 2008 *Phys. Rev. Lett.* **100** 021301
[54] Spergel D and Pen U L 1997 *Astrophys. J.* **491** L67
[55] McGraw P 1998 *Phys. Rev. D* **57** 3317
[56] Bucher M and Spergel D N 1999 *Phys. Rev. D* **60** 043505
[57] de Gennes P G and Prost J 1995, *The Physics of Liquid Crystals*, (New York: Oxford Univ. Press), 2nd ed
[58] Chandrasekhar S, Ratna B R, Sadashiva B K, and Raja V N 1988 *Mol. Cryst. Liq. Cryst.* **165** 123; Chandrasekhar S, Nair G G, Praefcke K, and Singer D 1996 *Mol. Cryst. Liq. Cryst.* **288** 7
[59] Volovik G E and Mineev V P 1977 *JETP* **45** 1186
[60] Thouless D J 1988, *Topological Quantum Numbers in Nonrelativistic Physics*, (Singapore: World Scientific), Chap. 8
[61] Zapotocky M, Goldbart P M, and Goldenfeld N 1995 *Phys. Rev. E* **51** 1216
[62] Priezjev N V and Pelcovits R A 2002 *Phys. Rev. E* **66** 051705

STATPHYS-Kolkata VII
Journal of Physics: Conference Series **297** (2011) 012014 doi:10.1088/1742-6596/297/1/012014

# Polar vs. apolar alignment in systems of polar self-propelled particles

**Fernando Peruani**
Max Planck for the Physics of Complex Systems, Nöthnitzer str. 38, Dresden, Germany

**Francesco Ginelli**
Service de Physique de l'Etat Condensé, CEA-Saclay, 91191 Gif-sur-Yvette, France

**Markus Bär**
Physikalisch-Technische Bundesanstalt, Abbestrasse 2-12, 10587 Berlin, Germany

**Hugues Chaté**
Service de Physique de l'Etat Condensé, CEA-Saclay, 91191 Gif-sur-Yvette, France

**Abstract.** The symmetry of the alignment mechanism in systems of polar self-propelled particles determines the possible macroscopic large-scale patterns that can emerge. Here we compare polar and apolar alignment. These systems share some common features like giant number fluctuations in the ordered phase and self-segregation in the form of bands near the onset of orientational order. Despite these similarities, there are essential differences like the symmetry of the ordered phase and the stability of the bands.

## 1. Introduction

Systems of collectively moving entities are ubiquitous in nature, ranging from flocks of birds [1, 2, 3], insect swarms [5, 6], bacterial collective motion [7, 8], to even driven granular media [9, 10, 11, 14]. Beyond the complexity of each system, it is possible to classify the interaction among the moving entities according to its symmetry, which can be either polar or apolar. Polar (or ferromagnetic) interactions lead to parallel alignment of the velocity of the moving objects, while apolar (or nematic) interactions allow both, parallel as well as antiparallel alignment. Fish, birds, and insects seem to exhibit polar alignment [1, 2, 3, 5, 6]. Bacteria and driven granular media often display apolar alignment [8, 9, 10, 11]. The physical origin of the alignment mechanism varies among these examples and it is difficult to identify a particular physical interaction with a given symmetry in these non-equilibrium systems. For instance, steric interactions among elongated self-propelled objects lead to apolar alignment. This can be easily observed in realistic self-propelled rod models [12], experiments with driven rods [10, 11], and experiments with gliding bacteria [8]. However, if the objects are isotropic, volume exclusion effects can result in a polar alignment, as observed in models [13] and experiments [14] with polarly driven disks.

Here we focus on the collective large-scale patterns emerging in systems of point particles moving in two-dimensions at constant speed. We compare the dynamics and macroscopic properties of self-propelled particles interacting by a polar and an apolar velocity alignment mechanism. We list the common features and highlight the differences. We provide in this way a summary of the two main classes of polar self-propelled particle systems.

## 2. Equation of motion of self-propelled particles

The evolution of the $i$-th particle is given by the following updating rules:

$$\mathbf{x}_i^{t+\Delta t} = \mathbf{x}_i^t + v_0 e^{\mathrm{i}\,\theta_i^t} \Delta t \tag{1}$$

$$\theta_i^{t+\Delta t} = \arg\left( \sum_{|\mathbf{x}_i^t - \mathbf{x}_j^t| \le \epsilon} f\left(\theta_j^t, \theta_i^t\right) e^{\mathrm{i}\,\theta_j^t} \right) + \eta_i^t \tag{2}$$

where $\mathbf{x}_i^t$ is the position of the particle and $\theta_i^t$ its direction of motion at time $t$, $v_0$ the active particle speed, $\arg(\mathbf{b})$ indicates the argument of the imaginary number $\mathbf{b}$ , $\eta_i^t$ is a delta-correlated white noise of strength $\eta$ [15], and $\Delta t$ is the temporal time step. Notice that Eqs. (1) and (2) can be considered the limiting case of very fast angular relaxation of a system of equations of the form:

$$\dot{\mathbf{x}}_i = v_0 e^{\mathrm{i}\,\theta_i}\,, \quad \dot{\theta}_i = -\gamma \frac{\partial U}{\partial \theta_i}(\mathbf{x}_i, \theta_i) + \tilde{\eta}_i(t)\,,$$

as discussed in [16]. The symmetry of the alignment mechanism is contained in the function $f$ which is defined as:

$$f\left(\theta_j^t, \theta_i^t\right) = \begin{cases} 1 & \text{for polar alignment} \\ \operatorname{sign}\left(\cos(\theta_j^t - \theta_i^t)\right) & \text{for apolar alignment} \end{cases}, \tag{3}$$

where $\operatorname{sign}(x)$ return a $+1$ if $x \ge 0$ and $-1$ otherwise. Eqs. (1) and (2) together with Eq. (3) defined the so-called Vicsek model [17, 18] for polar alignment, and the model for (ideal) self-propelled rods introduced in [16, 19] for apolar alignment.

### 2.1. *Order parameters*

The orientational order can be characterized by the following order parameters. The polar (ferromagnetic) order parameter is defined by:

$$\phi = \langle \left| \frac{1}{N} \sum_{k=1}^{N} \exp(\mathrm{i}\,\theta_k^t) \right| \rangle\,, \tag{4}$$

where $\langle \ldots \rangle$ denotes time-average and $N$ stands for the total number of particles in the system. $\phi$ takes the value 1 when all particles move in the same direction, while in the disordered phase, i.e., when particles move in any direction with equal probability, it vanishes. On the other hand, the apolar (nematic) ordered parameter takes the form:

$$S = \langle \left| \frac{1}{N} \sum_{k=1}^{N} \exp(\mathrm{i}\,2\,\theta_k^t) \right| \rangle\,. \tag{5}$$

Formally, $S$ can be derived from the order parameter matrix $Q$ of *liquid crystals* (LC) [20], as the largest eigenvalue (which here we have normalized such that $S^{LC} \in [0,1]$). When the system is perfectly nematically ordered, i.e., when particles move in opposite directions along the same axis, $S$ takes the value 1. In summary, a perfectly polarly ordered phase is characterized by $\phi = S = 1$, while for a genuine apolarly ordered phase, $\phi = 0$ and $S = 1$.

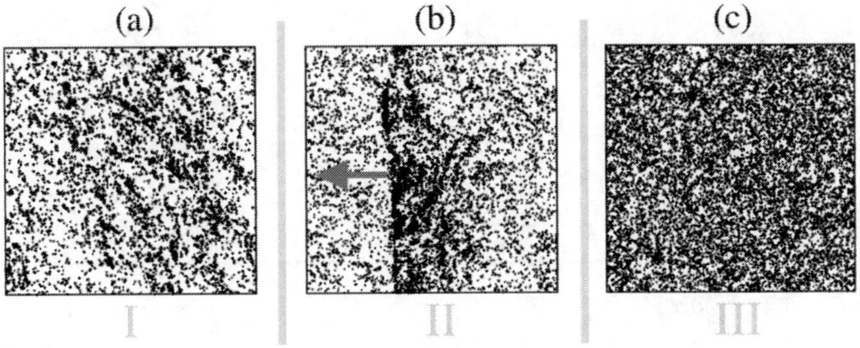


**Figure 1.** Typical steady-state snapshots for SPPs with polar alignment at different noise values (linear size $L = 512$, particle density $\rho = 1/8$, active speed $v_0 = 1/4$). The snapshots correspond to phase I, (a) $\eta = 0.06$, phase II, (b) $\eta = 0.1$, and phase III, (c) $\eta = 0.19$. The arrow in (b) indicates the direction of motion of the high-density band.

## 3. Macroscopic patterns with polar alignment

We start out by reviewing the large-scale properties of SPPs with polar alignment. If we fix the active speed $v_0$ and the density $\rho$, while varying the noise strength $\eta$ we observe the emergence of three statistically distinct stationary states or phases. Two of these phases correspond to orientationally ordered phases, while the third one is a disordered phase. We refer to these phases as phase I, II, and III by increasing value of $\eta$.

Phase I exhibits polar long-range order and is characterized by the absence of regular large-scale, high density structures, see Fig.1(a). Due to this fact, it is often said that this phase is spatially homogeneous albeit with large density fluctuations. The presence of polar long-range order implies that $\phi(\eta) \to K_0(\eta)$ as the system size $L \to \infty$, with $K_0(\eta) > 0$ a constant that depends only on $\eta$. Phase I also exhibits giant number fluctuations (NF). NF are defined as $\Delta n^2(\ell) = \langle (n(\ell) - \langle n \rangle(\ell))^2 \rangle$, where $n(\ell)$ stands for the number of particles in a box of linear size $\ell$. The average number of particles is $\langle n \rangle = \rho \ell^2$ while $\Delta n(\ell)$ is expected to be $\Delta n \propto \langle n \rangle^{\alpha}$. Giant NF correspond to $\alpha > \frac{1}{2}$ [21]. In [18] it was shown that $\alpha \sim 0.8$ for phase I.

By increasing $\eta$ we reach a point at which large-scale, elongated, high-density, high-order solitary structures emerge. These structures, that we refer to as bands, move at roughly constant speed, see Fig.1(b). A system of SPPs with polar alignment can display multiple bands. The low-density region in between bands is disordered and there is no characteristic separation length between bands [18]. In summary, phase II is characterized by polar long-range order and the presence of traveling bands. If $\eta$ is increased further, we reach phase III that exhibits local and global disorder, see Fig.1(c).

## 4. Macroscopic patterns with apolar alignment

Now we focus on a system of SPPs with apolar alignment. While sweeping $\eta$, keeping the active speed and the density fixed, we observe the emergence of four statistically distinct stationary states or phases, see Figs. 3-4. Two of these four regimes correspond to ordered phases while the other two to disordered phases. At the two extremes, very high and very low values of $\eta$, the system is spatially homogeneous, Figs. 3(a,e). At intermediate values of $\eta$, spontaneous density segregation occurs in the form of a high-density ordered region along which the particles move back and forth, Figs. 3(b-d). It is important to stress that macroscopic polar order always remains near zero. Nevertheless, if we look at very short length scales, we observe that the

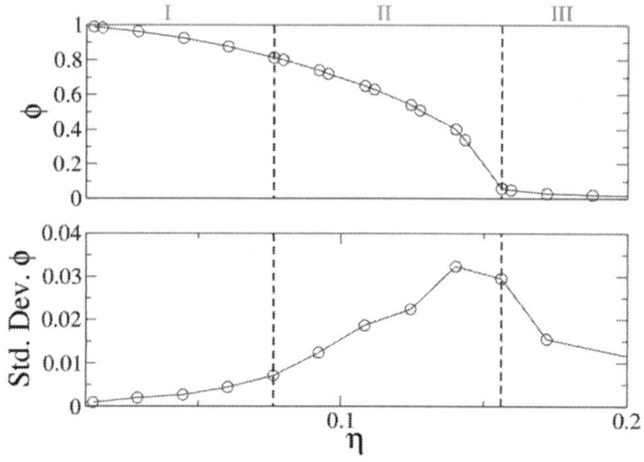


**Figure 2.** Polar order parameter $\phi$ and its standard deviation as function of the noise amplitude $\eta$ (other parameters as in Fig. 1).

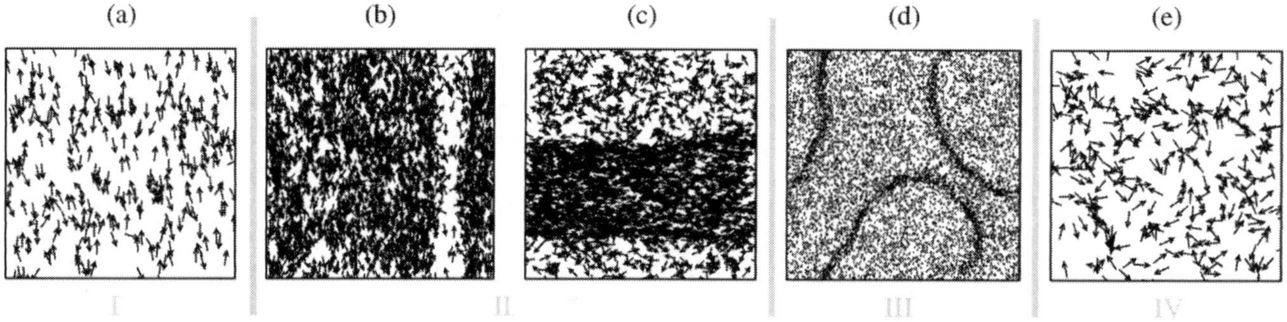


**Figure 3.** Typical steady-state snapshots for SPPs with apolar alignment at different noise values (linear size $L = 2048$, density $\rho = 1/8$, and velocity $v_0 = 1/2$). (a) $\eta = 0.08$, (b) $\eta = 0.10$, (c) $\eta = 0.13$, (d) $\eta = 0.168$, (e) $\eta = 0.20$. Arrows indicate the polar orientation of particles (except in (d)); only a fraction of the particles are shown for clarity reasons.

apolar ordered phase is formed by polarly oriented clusters. In the following we provide a more quantitative description of these four phases, which we name, by increasing value of $\eta$, phase I to IV. Phase I exhibits true long-range apolar (nematic) order and is spatially homogeneous. This apolar order corresponds to two subpopulations of particles of roughly equal size that migrate in opposite directions (Fig. 5a). We conclude that the apolar order is truly long-range by looking at scaling of the apolar order parameter $S$ with the system size. From Fig. 5a, it seems that $S(L)$ approaches a constant asymptotic value $C_0(\eta)$ for $L \to \infty$, with $C_0(\eta) > 0$, indicating the possible existence of true long-range apolar order. Another remarkable feature of phase I is the presence of giant number fluctuations (NF). Fig. 5b shows the scaling $\Delta n \propto \langle n \rangle^{\alpha}$ in phase I, where it is found again that $\alpha \sim 0.8$.

By increasing $\eta$ we move from phase I to phase II. Its onset is characterized by the emergence of a narrow, low-density, disordered channel. This channel becomes wider at larger $\eta$ values, so that one can speak of a high-density ordered region along which particles travel in both directions. The high-density region exhibits apolar order with properties similar to those

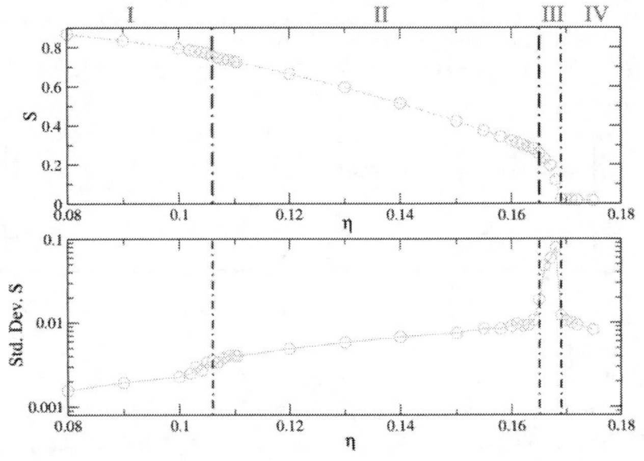


**Figure 4.** Apolar (nematic) order parameter $S$ and its standard deviation as function of the noise amplitude $\eta$ (other parameters as in Fig. 3).

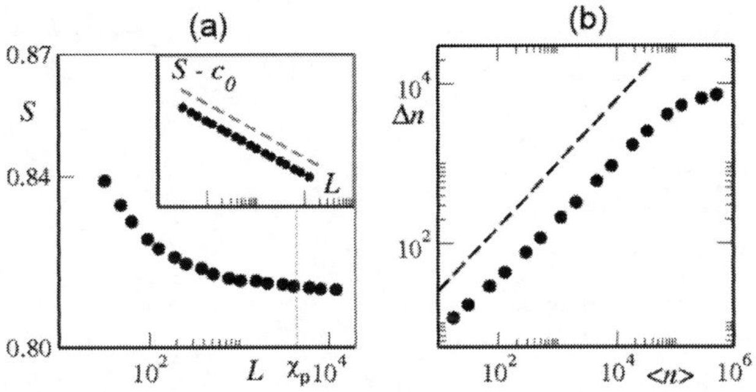


**Figure 5.** Phase I of SPPs with apolar alignment: long-range apolar order and giant number fluctuations ($\rho = 1/8$, $v_0 = 1/2$, and $\eta = 0.095$). (a) Apolar order parameter $S$ vs system size $L$ in square domains. The vertical red dashed line marks the persistence length $\chi \approx 4400$ that particles can travel before performing a U-turn. Inset: $S - C_0 = 0.813063$ vs $L$ (red dashed line: $L^{-2/3}$ decay). (b) Number fluctuations $\Delta n$ as function of $\langle n \rangle$. The dashed line correspond to an algebraic growth with exponent 0.8 (simulations with $L = 4096$). See Ref. [19].

observed in phase I: true long-range order and giant NF. The (rescaled) region possesses a well-defined profile with sharper and sharper edges as $L$ increases (Fig. 6a). The fraction area $\Omega$ occupied by the dense region is thus asymptotically independent of system size, and it decreases continuously as the noise strength $\eta$ increases (Fig. 6b).

By increasing $\eta$ even further, we reach phase III, where the dense region becomes unstable and constantly bend, merge, break and reform. As result of this dynamics, $S(t)$ fluctuates strongly and on very large time scales, but its average decreases as $1/\sqrt{N}$ [19]. Thus, phase III is characterized by the absence of long-range order and the presence of large correlation lengths and times. At even larger $\eta$ values, we find phase IV, which exhibits local and global disorder on small length- and time-scales, and is spatially homogeneous.

It is difficult to accurately locate the three boundaries that separate the four phases. Fig. 7 is the result of an important numerical effort and shows a reliable estimate of $\eta_c^{I-II}(\rho)$, $\eta_c^{II-III}(\rho)$,

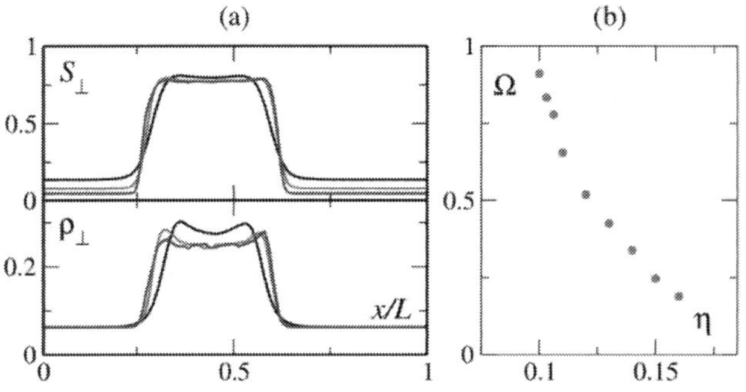


**Figure 6.** Phase II of SPPs with apolar alignment is characterized by the presence of an elongated high-density, high-order region. (a) The rescaled transverse density (top) and nematic order (bottom) profile in square domains of linear size $L = 512$ (black), 1024 (red), and 2048 (blue) at $\eta = 0.14$ (other parameters as in Fig. 5). (b) Surface fraction $\Omega$ occupied by the band as a function of noise amplitude $\eta$. See Ref. [19].

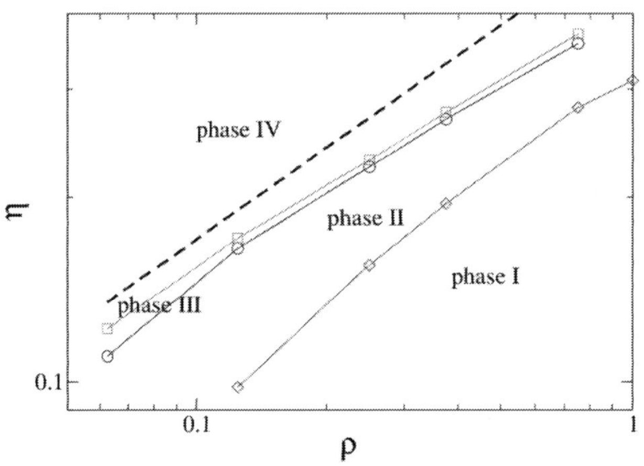


**Figure 7.** Phase diagram - Critical value of $\eta$, separating the four phases, as function of the density. Thresholds obtained for systems of size $L = 1024$ and $v_0 = 1/2$. The dashed line corresponds to the scaling $\eta \propto \rho^{1/2}$.

and $\eta_c^{III-IV}(\rho)$. We are interested in the scaling of $\eta_c^{A-B}(\rho) \propto \rho^{\gamma_{A-B}}$, where $A - B$ refers to some of the three boundaries. Assuming an homogeneous distribution of particles, we can expect $\eta_c^{A-B}(\rho)$ to be proportional to the inverse of the mean free path, i.e., $\propto \rho^{1/2}$. This argument would apply particularly to $\eta_c^{III-IV}(\rho)$. Moreover, for SPPs with polar alignment, this seems to be the case, i.e., $\eta_c^{II-III}(\rho) \propto \rho^{1/2}$, see Fig. 1 and [18]. However, for SPPs with apolar alignment this scaling only seems to hold for $\eta_c^{IV-III}(\rho)$ at very low density. At high density, $\eta_c^{IV-III}(\rho)$ is characterized by $\gamma_{IV-III} \sim 0.43$, suggesting the existence of a more complex mechanism regulating the self-organization of the SPPs particles. Fig. 7 also shows that the parameter range at which phase III is observed shrinks as the density is increased. This finding suggests that at higher densities phase III may not exist. Finally, we notice that $\eta_c^{I-II}(\rho)$ behaves differently than a power-law. Currently, we lack a theoretical understanding of these intriguing numerical results.

Journal of Physics: Conference Series **297** (2011) 012014 doi:10.1088/1742-6596/297/1/012014

## 5. Comparison between polar and apolar alignment

Self-propelled particles moving at constant speed with either polar or apolar alignment exhibit a phase transition from a disordered to orientationally ordered phase. For both alignments the order seems to be long-range, being purely polar for polar alignment and purely apolar (nematic) for apolar alignment. It has been shown theoretically that self-propelled particles with polar alignment can display long-range order [22]. However, the claim about the existence of apolar long-range order is exclusively based on large-scale simulations and a theoretical justification is still missing. Assuming that the findings here reported for large-scale simulations hold in the thermodynamical limit, we conclude that SPPs with polar alignment exhibit three phases, while with apolar alignment, the phases are four. For both alignments, there are two ordered phases, one of them spatially "homogeneous", though with giant number fluctuations with roughly the same critical exponent 0.8, while the other one is characterized by the presence of high-density, high ordered regions. For polar alignment, the system can display multiple solitary bands. These bands are traveling bands whose moving direction is perpendicular to the long axis of the band, i.e., particles are aligned roughly in the same direction exhibited by the band velocity. On the other hand, for apolar alignment, we observe only one dense region. This region does not move and is composed of two populations of particles moving in opposite direction along its axis. For polar alignment there is one disordered phase characterized by local and global disorder. SPPs with apolar alignment also exhibit a similar phase, however these particles also display an intriguing disordered phase where particle self-segregation in the form of highly dynamical bands occur.

Bands in phase II, for both, SPPs with polar as well as with apolar alignment, are strongly influenced by the implemented boundary conditions. More specifically, bands tend to be parallel to the size of the simulation box. It has been suggested in [26] that bands in SPPs with polar alignment may be a finite size effect that becomes particularly evident at high velocity values. Here, we have shown that these bands also emerge at low density and low velocity values. Moreover, transient bands are also observed in simulations using reflecting boundary conditions (data not shown), and even in lattice-gas simulations [28]. In summary, we believe that bands in SPPs with either polar or apolar alignment are a robust property of these systems, provided - and this is the key point - the (linear) system size is larger than a critical $L^*$ value (see discussion in [18]). Nevertheless, we admit that our arguments are just based on numerical evidence and a solid theoretical justification is still lacking. Results by Bertin et al. [25] seem to provide the first theoretical justification of the existence of such bands. To learn about this controversy, see [26, 18, 27, 25]. On the other hand, bands in phase III of SPPs with apolar alignment do not exhibit any influence from the boundary condition, making their existence easier to justify.

Adopting our view, in a coarse-grained description of these systems, the homogeneous disordered phase should get unstable by the emergence of soliton-like structures: traveling bands, for polar alignment, while for apolar alignment, local alignment has to lead to the formation of dynamic bands. In the case of apolar alignment, the onset of global order is not associated to the instability of the disorder homogeneous state, as for polar alignment, but with the stability of the high-density, dynamic band of phase III (Fig. 3d). On the other extreme, well in the "homogeneous" ordered phase, i.e. in phase I of both systems, we can approximate the behavior of the order parameter $\phi$ and $S$ as:

$$\phi = \frac{2}{\eta}\sin(\eta/2) \tag{6}$$

$$S = \frac{1}{\eta^2}\sin^2(\eta), \tag{7}$$

using a simple mean-field argument as discussed in [23]. From this we can presume that for a given set of parameters $v_0$, $\rho$, and $\eta$, $\phi(v_0, \rho, \eta) \geq S(v_0, \rho, \eta)$. Moreover, a simple mean-field analysis of the"homogeneous" disordered phase reveals that, for fixed $v_0$ and $\rho$, this phase looses its stability at lower values of $\eta$ for apolar alignment than for polar alignment [16]. From all this, we learn that is easier to achieve orientational order with polar than with apolar alignment. This comparison also suggests that polar order is more robust than apolar order.

Beyond this rough comparison between both alignment, a theoretical understanding of phase separation that occurs in both systems in the form of high-density, high-order bands is still lacking. For polar alignment, there are some promising recent theoretical results [24, 25] that provide an interesting perspective to this puzzling issue. We hope that in the near future all these open problems will be fully understood.

## References

[1] Parrish J K and Hamner W M 1997, *Three Dimensional Animals Groups*, (Cambridge: Cambridge University Press)
[2] Cavagna A et al. 2010 *Proc. Natl. Acad. Sci.* **107** 11865
[3] Bhattacharya K and Vicsek T 2010 *New J. Phys.* **12** 093019
[4] Helbing D, Farkas I, and Vicsek T 2000 *Nature (London)* **407** 487
[5] Buhl J et al. 2006 *Science* **312** 1402
[6] Romanczkuk P, Couzin I D, and L. Schimansky-Geier L 2009 *Phy. Rev. Lett.* **102** 010602 .
[7] Zhang H P et al. 2010 *Proc. Natl. Acad. Sci.* **107** 13626
[8] Peruani F et al. 2011 *unpublished*
[9] Narayan V, Ramaswamy S, and Menon N 2007 *Science* **317** 105
[10] Kudrolli A, Lumay G, Volfson D, and Tsimring L S 2008 *Phys. Rev. Lett.* **100** 058001
[11] Kudrolli A 2010 *Phys. Rev. Lett.* **104** 088001
[12] Peruani F, Deutsch A, and Bär M 2006 *Phys. Rev. E* **74** 030904
[13] Grossman D, Aranson I S, and Ben Jacob E 2008 *New J. Phys.* **10** 023036
[14] Deseigne J, Dauchot O, and Chaté H 2010 *Phys. Rev. Lett.* **105** 098001
[15] The parameter $\eta$, given in all figures, is defined as half the width of the noise, which is top-hat distributed and centered around 0. The parameter $\eta$ is measured in *units* of $\pi/2$, so that $\eta = \delta$ means that $\eta_i^t \in [-\pi\delta/2, \pi\delta/2]$.
[16] Peruani F, Deutsch A, and Bär M 2008 *Eur. Phys. J. Special Topics* **157** 111
[17] Vicsek T, Czirok A, Ben-Jacob E, Cohen I, and Shochet O 1995 *Phys. Rev. Lett.* **75** 1226
[18] Chaté H, Ginelli F, Grégoire G, and Raynaud F 2008 *Phys. Rev. E* **77** 046113
[19] Ginelli F, Peruani F, Bär M, and Chaté H 2010 *Phys. Rev. Lett.* **104** 184502
[20] Doi M and Edwards S F 1986, *The Theory of Polymer Dynamic* (Oxford: Clarendon Press).
[21] Ramaswamy S, Simha R A and Toner J 2003 *Europhys.Lett.* **62**, 196
[22] Toner J and Tu Y 1995 *Phys. Rev. E* **58** 4828; 1995 *Phys. Rev. Lett.* **75** 4326
[23] Peruani F, Schimansky-Geier L, and Bär M 2010 *Eur. Phys. J. Special Topics* **191** 173
[24] Mishra S, Baskaran A, and Marchetti M C 2010 *Phys. Rev. E* **81** 061916
[25] Bertin E, Droz M, and Grégoire G 2006 *Phys. Rev. E* **74** 022101 (2006); 2009 *J. Phys. A* **42** 445001
[26] Nagy M, Daruka I, and Vicsek T 2007 *Physica A* **373** 445
[27] Baglietto G and Albano E 2009 *Phys. Rev. E* **80** 050103
[28] Schaller V et al. 2010 *Nature* **467** 73

Journal of Physics: Conference Series **297** (2011) 012015 doi:10.1088/1742-6596/297/1/012015

# Hamiltonian Derivations of the Generalized Jarzynski Equalities under Feedback Control

**Takahiro Sagawa**

Department of Physics, University of Tokyo, 7-3-1, Hongo, Bunkyo-ku, Tokyo, 113-8654, Japan

E-mail: sagawa@cat.phys.s.u-tokyo.ac.jp

**Abstract.** In the presence of feedback control by "Maxwell's demon," the second law of thermodynamics and the nonequilibrium equalities such as the Jarzynski equality need to be generalized. In this paper, we derive the generalized Jarzynski equalities for classical Hamiltonian dynamics based on the Liouville's theorem, which is the same approach as the original proof of the Jarzynski equality. The obtained equalities lead to the generalizations of the second law of thermodynamics for the Hamiltonian systems in the presence of feedback control.

## 1. Introduction

Due to the recent advancement of nonequilibrium statistical mechanics, it has been established that the second law of thermodynamics can be expressed not only in terms of inequalities but also in terms of equalities, such as the fluctuation theorem and the Jarzynski equality [1, 2, 3, 4, 5, 6, 7]. These equalities are universally valid even if the state of the system is far from equilibrium. For the case of classical dynamics, the Jarzynski equality [6] is expressed as

$$\langle e^{-\beta(W-\Delta F)}\rangle = 1, \tag{1}$$

where $\langle \dots \rangle$ describes the ensemble average for all microscopic trajectories, $T = (k_{\rm B}\beta)^{-1}$ is the temperature of the initial canonical distribution with $k_{\rm B}$ being the Boltzmann constant, $W$ is the work performed on the system, and $\Delta F$ is the free-energy difference. We note that Eq. (1) holds if the initial distribution of the system is the canonical distribution. The usual expression of the second law of thermodynamics is a straightforward consequence of the Jarzynski equality: by using the concavity of the exponential function, we obtain

$$\Delta F - \langle W \rangle \leq 0. \tag{2}$$

Inequality (2) implies that the amount of the free-energy gain of the system cannot be larger than the amount of the energy-input (work) on the system on average.

However, if a thermodynamic system is subject to feedback control (or equivalently, the system is controlled by "Maxwell's demon" [8, 9, 10]), the second law of thermodynamics (2) needs to be generalized [11, 12, 13, 14, 15, 16, 17, 18, 19, 20, 21]. Here, "feedback" means that the control protocol depends on the outcomes obtained by measurements. For example, for the case of the Szilard engine [9], the demon can extract $k_{\rm B}T\ln 2$ of work from an isothermal cycle by using 1 bit (= $\ln 2$ nat) of information. The essential of the role of the demon is to increase the system's free energy with feedback control by using the information about the system. Recently, the Szilard-type Maxwell's demon has been experimentally realized by using colloidal particles [22].

The Jarzynski equality (1) also needs to be generalized in the presence of feedback control. In a previous work [23], we have derived the generalizations of the Jarzynski equality for classical stochastic dynamics based on the detailed fluctuation theorem [4, 5]. Moreover, several nonequilibrium equalities with feedback control have also been derived for different setups [24, 25, 26, 27, 28], and relevant issues have been studied [29, 30]. We note that one of the generalized Jarzynski equalities has been experimentally verified [22]. However, the derivations of the generalized Jarzynski equalities based on the classical Hamiltonian dynamics have not been studied yet. In this paper, we will derive two generalizations of the Jarzynski equality based on the Hamiltonian dynamics, in particular based on the Liouville's theorem.

## 2. Setup

We consider a $d$-dimensional Hamiltonian system with $N$ classical particles. Let $x := (\boldsymbol{r}, \boldsymbol{p})$ be the phase-space point of the system, where $\boldsymbol{r}$ and $\boldsymbol{p}$ respectively denote the $dN$-dimensional position and momentum. We write the time-reversal of $x$ as $x^* := (\boldsymbol{r}, -\boldsymbol{p})$. The Hamiltonian of the system is given by $H(x, \lambda)$, where $\lambda$ denotes external control parameters such as the frequency of the optical tweezers.

We consider the dynamics of the system from time $t = 0$ to $\tau$. We drive the system from the thermal equilibrium by changing $\lambda$. Let $x_t := x(t)$ and $\lambda_t := \lambda(t)$. The initial distribution of the system is assumed to obey the canonical distribution corresponding to $H(x_0, \lambda_0)$, whose probability density is given by

$$P_{\rm can}[x_0, \lambda_0] := \frac{e^{-\beta H(x_0,\lambda_0)}}{Z_0}, \tag{3}$$

where $Z_0 := \int dx_0 e^{-\beta H(x_0,\lambda_0)}$ is the partition function. The corresponding free energy is $F_0 := -k_{\rm B}T \ln Z_0$. The equation of motion is given by

$$\begin{aligned} \frac{d\boldsymbol{r}(t)}{dt} &= \frac{\partial H}{\partial \boldsymbol{p}}(x(t), \lambda(t)), \\ \frac{d\boldsymbol{p}(t)}{dt} &= -\frac{\partial H}{\partial \boldsymbol{r}}(x(t), \lambda(t)). \end{aligned} \tag{4}$$

We write the formal solution of Eq. (4) as

$$x_t = M_t[x_0], \tag{5}$$

where $M_t$ is a bijective map acting on the phase space. The probability density of $x_t$, denoted as $P[x_t]$, is then given by

$$P[x_t] = P_{\rm can}[M_t^{-1}(x_t), \lambda_0], \tag{6}$$

where $M_t^{-1}$ is the inverse of $M_t$, and we used the Liouville's theorem.

We perform a measurement on the system at time $t_{\rm m}$ ($0 \leq t_{\rm m} < \tau$) and obtain outcome $y$. We assume that the measurement involves a stochastic error characterized by

$$P[y|x_{\rm m}], \tag{7}$$

which describes the probability density of obtaining $y$ under the condition that the true state of the system is given by $x_{\rm m} := x(t_{\rm m})$ at time $t_{\rm m}$. If $P[y|x_{\rm m}] = \delta(y - x_{\rm m})$ holds, the measurement is error-free. On the other hand, if the noise is Gaussian, the conditional probability is given by

Journal of Physics: Conference Series **297** (2011) 012015 doi:10.1088/1742-6596/297/1/012015

$P[y|x_{\rm m}] \propto \exp\left(-(y-x_{\rm m})^2/(2N)\right)$, where $N$ is the intensity of the noise. In general, the joint probability density of $x_{\rm m}$ and $y$ is given by

$$P[x_{\rm m}, y] = P[y|x_{\rm m}]P[x_{\rm m}], \tag{8}$$

and the probability density of obtaining $y$ is given by

$$P[y] = \int dx_{\rm m} P[x_{\rm m}, y]. \tag{9}$$

We note that $y^*$ denotes time-reversal of $y$ corresponding to the time-reversal $x \mapsto x^*$.

We then discuss the concept of the mutual information [31], which is the key quantity to characterize the information content that is obtained by the measurement. The mutual information, denoted as $\langle I \rangle$, is defined as

$$\langle I \rangle := \int dx_{\rm m} dy P[x_{\rm m}, y] \ln \frac{P[x_{\rm m}, y]}{P[x_{\rm m}]P[y]}, \tag{10}$$

where we write

$$I[x_{\rm m}, y] := \ln \frac{P[x_{\rm m}, y]}{P[x_{\rm m}]P[y]}. \tag{11}$$

The mutual information characterizes the correlation between state $x_{\rm m}$ and outcome $y$; the more information we get, the larger $\langle I \rangle$ is. In other words, the larger the error is, the less $\langle I \rangle$ is. In fact, if two probability variables $x_{\rm m}$ and $y$ are independent (i.e., $P[x_{\rm m}, y] = P[x_{\rm m}]P[y]$ holds), the mutual information vanishes so that $I[x_{\rm m}, y] = 0$ holds for all $x_{\rm m}$ and $y$.

After the measurement at time $t_{\rm m}$, we perform feedback control on the system so that the control protocol of $\lambda$ depends on outcome $y$ at time $t$ $(> t_{\rm m})$. To explicitly express the effect of the feedback, we write

$$x_t(y) = M_t(y)[x_0] \ (t > t_{\rm m}), \tag{12}$$

where $x_t(y)$ is the state of the system at time $t$ $(> t_{\rm m})$, under the condition that the initial state is $x_0$ and the measurement outcome is $y$. Equality (12) means that the Hamiltonian evolution of the system is determined by outcome $y$ after time $t_{\rm m}$. We note that map $M_t(y)$ satisfies the Liouville's theorem for each $y$.

The Hamiltonian at time $\tau$ may also depend on outcome $y$ as $H(x_\tau(y), \lambda_\tau(y))$, where $\lambda_\tau(y)$ denotes the value of external parameter $\lambda$ at time $\tau$ under the condition that the outcome is given by $y$. The corresponding canonical distribution is given by

$$P_{\rm can}[x_\tau(y), \lambda_\tau(y)] := \frac{e^{-\beta H(x_\tau(y), \lambda_\tau(y))}}{Z_\tau(y)}, \tag{13}$$

where $Z_\tau(y) := \int dx_\tau(y) e^{-\beta H(x_\tau(y), \lambda_\tau(y))}$ is the partition function which gives the free energy:

$$F_\tau(y) := -k_{\rm B}T \ln Z_\tau(y). \tag{14}$$

We note that the probability distribution of $x_\tau(y)$ is not necessarily given by the canonical distribution; $P[x_\tau(y)] = P_{\rm can}[x_\tau(y), \lambda_\tau(y)]$ does not necessarily hold.

In the total process, the work performed on the system is given by

$$W := H(x_\tau(y), \lambda_\tau(y)) - H(x_0, \lambda_0) = H(M_\tau(y)[x_0], \lambda_\tau(y)) - H(x_0, \lambda_0), \tag{15}$$

which is determined by initial state $x_0$ and outcome $y$. On the other hand, the free-energy difference is given by

$$\Delta F := F_\tau(y) - F_0, \tag{16}$$

which depends on outcome $y$.

## 3. Main Results

We now derive the two generalizations of the Jarzynski equality. The first one is given by

$$\langle e^{-\beta(W-\Delta F)-I} \rangle = 1, \tag{17}$$

where $\langle \cdots \rangle$ describes the ensemble average with respect to $x_0$ and $y$. In fact, $W$ is determined by $x_0$ and $y$, $\Delta F$ by $y$, and $I$ by $y$ and $x_0$ through $x_{\rm m} = M_{t_{\rm m}}[x_0]$. The key feature of Eq. (17) is that the left-hand side involves the term of the mutual information obtained by the measurement.

The proof of Eq. (17) is as follows. We first assume that $P[y|x_{\rm m}] \neq 0$ holds for all $y$ and $x_{\rm m}$. By using the joint distribution $P[x_0, y]$, we obtain

$$\begin{aligned} \langle e^{-\beta(W-\Delta F)-I} \rangle &= \int dx_0 dy \frac{e^{-\beta H(x_0,\lambda_0)}}{Z_0} P[y|x_{\rm m}] e^{-\beta(H(x_\tau(y),\lambda_\tau(y))-H(x_0,\lambda_0))} \frac{Z_0}{Z_\tau(y)} \frac{P[y]}{P[y|x_{\rm m}]} \\ &= \int dx_\tau(y) dy \frac{e^{-\beta H(x_\tau(y),\lambda_\tau(y))}}{Z_\tau(y)} P[y], \end{aligned} \tag{18}$$

where we used the Liouville's theorem $dx_0 = dx_\tau(y)$. By noting that $\int dx_\tau(y) e^{-\beta H(x_\tau(y),\lambda_\tau(y))} = Z_\tau(y)$, we obtain Eq. (17). We note that Eq. (17) has been obtained in Ref. [23] for classical stochastic systems.

By using the concavity of the exponential function, we have

$$\langle e^{-\beta(W-\Delta F)-I} \rangle \geq e^{-\langle \beta(W-\Delta F)-I \rangle}. \tag{19}$$

Therefore, Eq. (17) leads to

$$\langle \Delta F - W \rangle \leq k_{\rm B} T \langle I \rangle, \tag{20}$$

which is the generalized second law of thermodynamics. Inequality (20) implies that, by using feedback control, the free-energy increase can be lager than the performed work by the term proportional to the mutual information obtained by the measurement. The equality in (20) is achieved by the Szilard engine [9], where $\langle \Delta F \rangle = 0$, $\langle W \rangle = -k_{\rm B} T \ln 2$, and $\langle I \rangle = \ln 2$ hold. We note that inequality (20) has been obtained in Ref. [15] for quantum systems and in Ref. [23] for classical stochastic systems.

We next discuss the second generalization of the Jarzynski equality, which is given by

$$\langle e^{-\beta(W-\Delta F)} \rangle = \gamma, \tag{21}$$

where $\gamma$ characterizes the efficacy of feedback control. In fact, $\gamma$ is quantitatively defined as follows. We consider the backward or time-reversed process of the feedback control with outcome $y$. The initial state of the backward process is given by the canonical distribution corresponding to $\lambda_\tau(y)$, and the backward control protocol is given by $\lambda_t^\dagger(y) := \lambda_{\tau-t}(y)$ for each $y$ that is obtained in the forward process. We note that we do not perform any feedback in the backward process. We then perform a measurement on the system at time $\tau - t_{\rm m}$ during the backward process, and obtain outcome $y'$. We note that $y'$ does not necessarily equal to $y^*$. We write as $P_{(y)}^\dagger[y']$ the probability density of obtaining $y'$ by the measurement during the backward process with control protocol $\lambda_t^\dagger(y)$. In particular, the probability density of $y' = y^*$ is written as $P_{(y)}^\dagger[y^*]$. Then, $\gamma$ is defined as

$$\gamma := \int dy P_{(y)}^\dagger[y^*]. \tag{22}$$

Parameter $\gamma$ characterizes how efficiently the feedback is performed, in terms of the sum of the probabilities that the time-reversed outcome $y^*$ is obtained during the backward process with time-reversed protocol $\lambda_t^\dagger(y)$. We note that $\gamma = 1$ holds without feedback control, because $\{P^\dagger[y^*]\}$ becomes a single probability distribution for such cases. On the other hand, $\gamma = 2$ holds for the case the Szilard engine, where the number of the outcomes is two and the feedback control is perfect [23]. We note that Eq. (21) has been experimentally verified for a classical stochastic system [22].

We now prove Eq. (21). We first obtain

$$\begin{aligned}\langle e^{-\beta(W-\Delta F)}\rangle &= \int dx_0 dy \frac{e^{-\beta H(x_0,\lambda_0)}}{Z_0} P[y|x_\mathrm{m}] e^{-\beta(H(x_\tau(y),\lambda_\tau(y))-H(x_0,\lambda_0))} \frac{Z_0}{Z_\tau(y)} \\ &= \int dx_0 dy \frac{e^{-\beta H(x_\tau(y),\lambda_\tau(y))}}{Z_\tau(y)} P[y|x_\mathrm{m}] \\ &= \int dx_0 dy \frac{e^{-\beta H(x_\tau^*(y),\lambda_\tau(y))}}{Z_\tau(y)} P[y^*|x_\mathrm{m}^*],\end{aligned} \tag{23}$$

where we assumed that the error is time-reversal symmetric as

$$P[y|x_\mathrm{m}] = P[y^*|x_\mathrm{m}^*], \tag{24}$$

and that the Hamiltonian is also time-reversed symmetric as $H(x_\tau(y), \lambda_\tau(y)) = H(x_\tau^*(y), \lambda_\tau(y))$. By noting that the Liouville's theorem $dx_0 = dx_\tau(y)$ holds for each $y$, and noting that $dx_\tau(y) = dx_\tau^*(y)$ holds, we have

$$\langle e^{-\beta(W-\Delta F)}\rangle = \int dx_\tau^*(y) dy \frac{e^{-\beta H(x_\tau^*(y))}}{Z_\tau(y)} P[y^*|x_\mathrm{m}^*]. \tag{25}$$

Since the Hamiltonian dynamics is reversible, $x_\mathrm{m}^* = [M_{\tau-t_\mathrm{m}}(y)]^{-1}[x_\tau^*(y)]$ holds. Therefore, we obtain

$$\langle e^{-\beta(W-\Delta F)}\rangle = \int dx_\tau^*(y) dy \frac{e^{-\beta H(x_\tau^*(y))}}{Z_\tau(y)} P[y^*|[M_{\tau-t_\mathrm{m}}(y)]^{-1}(x_\tau^*(y))] = \int dy P_{(y)}^\dagger[y^*] = \gamma, \tag{26}$$

which proves Eq. (21). We note that Eq. (21) has been derived for classical stochastic systems [23] and for quantum systems [28]. We also note that Eq. (21) is a straightforward consequence of a result in Ref. [7] for a special class of measurements. Equality (21) leads to

$$\langle \Delta F - W\rangle \le k_\mathrm{B} T \ln \gamma. \tag{27}$$

The equality in (27) is achieved for the case of the Szilard engine where $\gamma = 2$ holds.

We now discuss the relationship between the two generalizations of the Jarzynski equality. The first one (17) only involves the term of the information $I$ obtained by the measurement, which is independent of the protocol of the feedback control. On the other hand, the second one (21) involves the term $\gamma$ that characterizes the efficacy of feedback control, which describes how efficiently the information is used by the feedback. We then discuss the quantitative relationship between $I$ and $\gamma$. We introduce notation $C[X] := \ln\langle e^{-X}\rangle$ for arbitrary probability variable $X$. We note that $C[I] = 0$ holds. Then, from Eqs. (17) and (21), we obtain

$$C[\sigma + I] - C[\sigma] - C[I] = -\ln \gamma, \tag{28}$$

where $\sigma := \beta(W - \Delta F)$. If the joint probability distribution of $\sigma$ and $I$ is Gaussian, Eq. (28) reduces to

$$\langle \sigma I\rangle - \langle\sigma\rangle\langle I\rangle = -\ln \gamma. \tag{29}$$

Eqs. (28) and (29) imply that $\gamma$ characterizes the correlation between $\sigma$ and $I$; the larger $\gamma$ is, the larger the efficiency of decreasing $\sigma$ by using $I$ is.

In the conventional thermodynamics without feedback control, the free energy is not a probability variable. On the other hand, with feedback control, the final free energy $F(y)$ can be a probability variable, because measurement outcome $y$ is a probability variable. The free-energy difference $\Delta F$ defined in (16) then needs to be inside the statistical average $\langle \cdots \rangle$ in the generalizations of the second law of thermodynamics (20) and (27). Therefore, the generalized second laws work only when we observe the ensemble of thermodynamic systems and take the ensemble average both in terms of phase-space point $x$ and outcome $y$. This is a characteristic of thermodynamics of feedback control. If the control protocol is independent of outcome $y$, the free-energy difference also becomes independent of $y$ and inequalities (20) and (27) reduce to the conventional second law of thermodynamics.

In conclusion, we have derived the two generalizations of the Jarzynski equality, Eqs. (17) and (21), based on the Hamiltonian dynamics. The former involves the term of the obtained mutual information, and the latter involves the term of the feedback efficacy. The key of the present derivations is the initial canonical distribution and the Liouville's theorem. The equalities lead to the two generalizations of the second law of thermodynamics (20) and (27), which give the fundamental bounds of the free-energy gain of thermodynamic systems that are subject to feedback control. We note that our results are consistent with the second law of thermodynamics, if we take into account the energy cost needed for the controller during the measurement and the information erasure [32].

## Acknowledgments

TS acknowledges Prof. Masahito Ueda for a lot of valuable discussions. This work was supported by Grants-in Aid for Scientific Research (KAKENHI 22103005 and 22340114), the Global COE Program "the Physical Sciences Frontier", and the Photon Frontier Network Program of MEXT of Japan. TS also acknowledges support from JSPS (Grant No. 208038).

## References

[1] Evans D J, Cohen E G D, and Morriss G P 1993 *Phys. Rev. Lett.* **71** 2401
[2] Gallavotti G and Cohen E G D 1995 *Phys. Rev. Lett.* **74** 2694
[3] Evans D J and Searles D J 2002 *Adv. Phys.* **51** 1529
[4] Crooks G E 1999 *Phys. Rev. E* **60** 2721
[5] Jarzynski C 2000 *J. Stat. Phys.* **98** 77
[6] Jarzynski C 1997 *Phys. Rev. Lett.* **78** 2690
[7] Kawai R, Parrondo J M R, and Broeck C V D 2007 *Phys. Rev. Lett.* **98** 080602
[8] Maxwell J C 1871 *Theory of Heat*, (Appleton: London)
[9] Szilard L 1929 *Z. Phys.* **53** 840
[10] Leff H S and Rex A F (eds.) 2003 *Maxwell's demon 2: Entropy, Classical and Quantum Information, Computing*, (Princeton University Press: New Jersey)
[11] Lloyd S and Zurek W H 1991 *J. Stat. Phys.* **62** 819
[12] Touchette H and Lloyd S 2000 *Phys. Rev. Lett.* **84**, 1156
[13] Cao F J, Dinis L, Parrondo J M R 2004 *Phys. Rev. Lett.* **93** 040603
[14] Lopez B J *et al.* 2008 *Phys. Rev. Lett.* **101** 220601
[15] Sagawa T and Ueda M 2008 *Phys. Rev. Lett.* **100**, 080403
[16] Cao F J and Feito M 2009 *Phys. Rev. E* **79** 041118
[17] Feito M, Baltanas J P, and Cao F J 2009 *Phys. Rev. E* **80** 031128
[18] Maruyama K, Nori F, and Vedral V 2009 *Rev. Mod. Phys.* **81** 1
[19] Suzuki H and Fujitani Y 2009 *J. Phys. Soc. Jpn.* **78** 074007
[20] Brandes T 2010 *Phys. Rev. Lett.* **105** 060602
[21] Kim S W, Sagawa T, Liberato S D, and Ueda M 2011 *Phys. Rev. Lett.* **106** 070401
[22] Toyabe S, Sagawa T, Ueda M, Muneyuki E, and Sano M 2010 *Nature Physics* **6** 988
[23] Sagawa T and Ueda M 2010 *Phys. Rev. Lett.* **104**, 090602
[24] Kim K H and Qian H, *Phys. Rev. E* **75** 022102

[25] Fujitani Y and Suzuki H *J. Phys. Soc. Jpn.* **79** 104003
[26] Ponmurugan M 2010 *Phys. Rev. E* **82** 031129
[27] Horowitz J M and Vaikuntanathan S 2010 *Phys. Rev. E* **82** 061120
[28] Morikuni Y and Tasaki H 2011 *J. Stat. Phys.* **43** 1
[29] Abreu D and Seifert U 2011 *Europhys. Lett.* **94** 10001
[30] Horowitz J M and Parrondo J M R 2011 arXiv:1104.0332
[31] Cover T M and Thomas J A 1991 *Elements of Information Theory*, (John Wiley and Sons: New York)
[32] Sagawa T and Ueda M 2009 *Phys. Rev. Lett.* **102** 250602

STATPHYS-Kolkata VII
Journal of Physics: Conference Series **297** (2011) 012016 doi:10.1088/1742-6596/297/1/012016

# KPZ, ASEP and Delta-Bose Gas

**Takashi Imamura, Tomohiro Sasamoto, Herbert Spohn**
Research Center for Advanced Science and Technology, The University of Tokyo, 4-6-1 Komaba, Meguro, Tokyo 153-8904, Japan
Chiba University, Yayoi-cho 1-33, Inage, Chiba 263-8522, Japan
Zentrum Mathematik and Physik Department, TU Munchen, D-85747 Garching, Germany

E-mail: imamura@jamology.rcast.u-tokyo.ac.jp, sasamoto@math.s.chiba-u.ac.jp, spohn@ma.tum.de

**Abstract.** We explain the solution of the KPZ equation with narrow wedge initial condition from the point of view of its connections to the ASEP and the $\delta$-Bose gas.

## 1. Introduction

The KPZ equation is a well known equation which describes surface growth phenomena [1]. Its one dimensional version reads

$$\frac{\partial}{\partial t}h = \tfrac{1}{2}\lambda \ \frac{\partial}{\partial x}h \ ^2 + \nu\frac{\partial^2}{\partial x^2}h + \sqrt{D}\eta \,. \tag{1}$$

Here $h = h(x,t)$ is the height pro le at time $t$, $t \geq 0$, and at position $x \in \mathbb{R}$. $\lambda$ is the strength of the nonlinearity, $\nu$ of the di usive relaxation and $D$ of the noise. $\eta$ is a Gaussian white noise with covariance

$$\langle \eta(x,t)\eta(x',t')\rangle = \ (x \quad x') \ (t \quad t') \,. \tag{2}$$

Kardar, Parisi and Zhang applied a dynamical renormalization group analysis and showed that the height uctuations scale as $\mathcal{O}(t^{1/3})$. This exponent supports non-Gaussian nature of the phenomena and is shared by many models. The universality class related to these phenomena (in particular associated with this exponent) is called the KPZ universality class [2].

Compared to its fame, not much information is available about the distribution of the height function. A reason may be that the equation itself is not well de ned due to the strong irregularity of the noise. One has to rst make sense of the equation. Moreover even if one could nd a way to de ne its solutions, it is not clear if and how one can treat the equation to get some explicit information.

From the point of view of statistical mechanics, the most important task is to work out the universal aspects of the model. For this purpose, it is enough to have other models which have no di culty of being well de ned and are accessible by some means. For the KPZ universality class, there are several such models like the asymmetric simple exclusion process (ASEP) and the polynuclear growth (PNG) model. In the last decade, the uctuation properties of the ASEP have been studied extensively and some universal quantities have been obtained [3–6].

Given the big progress for the ASEP and related models, it was natural to ask if one can treat the KPZ equation itself. In 2010 the rst exact solution for the height distribution of the KPZ equation was obtained in the case of sharp wedge initial conditions [7–11]. The method was to use results for the ASEP and to consider the weakly asymmetric limit. On the other hand it is known that the KPZ equation is related to the  -Bose gas which is also exactly solvable [12]. It turned out that one can reproduce the result for the KPZ equation by studying the  -Bose gas [13, 14], for further developments see [15, 16].

In this article, after explaining the connections between KPZ and ASEP on one side and KPZ and  -Bose gas on the other side, we discuss the direct relation between the ASEP and the  -Bose gas based on the duality for the ASEP current  uctuations [17, 18].

## 2. Scaling limit for ASEP

The asymmetric simple exclusion process (ASEP) is a stochastic many particle process. Each particle performs an asymmetric random walk with rate $p$ to the right and $q$ to the left [19–21]. There is an exclusion interaction amongst particles so that a jump is suppressed if the target site is occupied. We assume $q > p \geq 0$, $p + q = 1$. When $p = 0$, all particles hop only to the left, which is then called TASEP (totally ASEP). In this article we only consider the step initial condition, for which all sites on $x < 0$ are empty and $x \geq 0$ are occupied. Let $N(t)$ be the integrated current at the edge between sites 0 and 1, i.e.,

$$N(t) = (\#\text{ hops from 0 to 1 during time [0,t]}) \quad (\#\text{ hops from 1 to 0 during time [0,t]})\,. \tag{3}$$

Let $\eta(x,t)$ denote the particle number (0 or 1) at site $x$ and at time $t$. We de ne the ASEP height function by

$$h_{\rm A}(x,t) = \begin{cases} 2N(t) \quad \sum_{0<y\leq x}(2\eta(y,t) \quad 1)\,, & x > 0\,, \\ 2N(t)\,, & x = 0\,, \\ 2N(t) + \sum_{x<y\leq 0}(2\eta(y,t) \quad 1)\,, & x < 0\,. \end{cases} \tag{4}$$

We are interested in the scaling behavior of this quantity. For the height at the origin, the basic result is the following. As $t \to \infty$,

$$h_{\rm A}(0, t/(q \quad p)) \simeq \quad \frac{1}{2}t + 2^{\quad 4/3}t^{1/3} \quad _{\rm TW}\,. \tag{5}$$

Here the random amplitude $_{\rm TW}$ is GUE Tracy-Widom distributed,

$$F_{\rm TW}(s) = \mathbb{P}[\quad _{\rm TW} \leq s] = \det(1 \quad P_s K_{\rm Ai} P_s)_{L^2(\mathbb{R})}\,, \tag{6}$$

where det on the right denotes the Fredholm determinant, $P_s$ is the projection onto $[s, \infty)$, and $K_{\rm Ai}$ is the Airy kernel

$$K_{\rm Ai}(x,y) = \int_0^\infty {\rm d}\lambda {\rm Ai}(x+\lambda){\rm Ai}(y+\lambda)\,. \tag{7}$$

The function $F_{\rm TW}$ describes the distribution of the largest eigenvalue for $N \quad N$ GUE random matrices in the large $N$ limit [22]. Such a result was obtained  rst for the TASEP by Johansson [23] and then generalized to the ASEP by Tracy and Widom [24, 25]. As will be explained in the following section, the above result holds also for the KPZ equation. It is expected that the  uctuations of the height of a curved growing surface in the KPZ universality class are described by the same distribution function, $F_{\rm TW}$.

## 3. The solution of the KPZ equation for narrow wedge initial conditions

Because of scaling, without loss of generality we can set $\lambda = 1, \nu = 1/2, D = 1$ in the KPZ equation, which then reads

$$\frac{\partial}{\partial t}h = \frac{1}{2}\ \frac{\partial}{\partial x}h\ ^2 + \frac{1}{2}\frac{\partial^2}{\partial x^2}h + \eta\,. \tag{8}$$

We consider this equation with the narrow wedge initial condition,

$$h(x,0) = \ |x|/\ ,\qquad \ll 1\,.$$

The macroscopic shape for  nite $t$ is

$$h(x,t) = \begin{cases} x^2/(2t) & \text{for } |x| \le t/\ , \\ (1/2\ ^2)t\quad |x|/ & \text{for } |x| > t/\ . \end{cases}$$

The exact solution for the distribution of the height at a speci ed reference point is obtained in [7–11] and given by

$$h(x,t) = \ x^2/2t\quad \tfrac{1}{12}\gamma_t^3 + \gamma_t\ _t\,, \tag{9}$$

where

$$\gamma_t = 2\ ^{1/3}t^{1/3}. \tag{10}$$

The distribution of $\ _t$ is de ned by

$$F_t(s) = \mathbb{P}[\ _t \le s] = 1\quad \int_{\ \infty}^{\infty} \exp\quad \mathrm{e}^{\gamma_t(s\ \ u)}\quad \det(1\quad P_u(B_t\quad P_{\mathrm{Ai}})P_u)\quad \det(1\quad P_uB_tP_u)\ du\,.$$

Here $P_{\mathrm{Ai}}(x,y) = \mathrm{Ai}(x)\mathrm{Ai}(y)$, $P_u$ is a projection to $[u,\infty)$, and $B_t$ has the kernel

$$B_t(x,y) = K_{\mathrm{Ai}}(x,y) + \int_0^{\infty} \mathrm{d}\lambda(\mathrm{e}^{\gamma_t\lambda}\quad 1)\ ^1\ \ \mathrm{Ai}(x+\lambda)\mathrm{Ai}(y+\lambda)\quad \mathrm{Ai}(x\quad \lambda)\mathrm{Ai}(y\quad \lambda)\ \ .$$

In the scaling limit $t \to \infty$, we indeed recover $F_{\mathrm{TW}}$. This implies that the KPZ equation is in the KPZ universality class, at the level of the one-point distribution for a curved surface.

## 4. -Bose gas

Using the Cole-Hopf transformation,

$$Z(x,t) = \exp[h(x,t)]\,, \tag{11}$$

the KPZ equation becomes

$$\frac{\partial}{\partial t}Z = \frac{1}{2}\frac{\partial^2}{\partial x^2}Z + \eta Z\,, \tag{12}$$

often called stochastic heat equation [26]. Using the Feynman-Kac formula, one obtains

$$Z(x,t) = \mathbb{E}_x\ \ \exp\ \ \int_0^t \eta(b(s),t\quad s)\mathrm{d}s\ \ Z(b(t),0)\ \ , \tag{13}$$

where $b(t)$ is a standard Brownian motion with $b(0) = x, b(t) = 0$. In this rewriting $b(t)$ can be viewed as a directed polymer in a random potential $\eta$. To make sense of the KPZ equation, one way is to de ne (13) through the expanded exponential with multiple Ito-integrals and then to recover the KPZ height by inverting (11) [11].

Journal of Physics: Conference Series **297** (2011) 012016 doi:10.1088/1742-6596/297/1/012016

By taking the average over randomness, one obtains

$$\Xi_{\text{Bose}}(x_1,\ldots,x_N;t) := \langle Z(x_1,t) \quad Z(x_N,t)\rangle = \langle x_1,\ldots,x_N|\mathrm{e}^{\quad tH_N}|0\rangle\,, \tag{14}$$

where

$$H_N = \quad \sum_{j=1}^{N}\frac{1}{2}\frac{\partial^2}{\partial x_j^2} \quad \frac{1}{2}\sum_{i,j=1,i\neq j}^{N} \quad (x_i \quad x_j)\,. \tag{15}$$

$H_N$ is the Hamiltonian of the -Bose gas with attractive -potential. $\Xi_{\text{Bose}}(x_1,\ldots,x_N;t)$ satis es the Schrodinger equation

$$\frac{\partial}{\partial t}\Xi_{\text{Bose}}(x_1,\ldots,x_N;t) = \sum_{i=1}^{N}\frac{1}{2}\frac{\partial^2}{\partial x_i^2}\Xi_{\text{Bose}}(x_1,\ldots,x_N;t) \tag{16}$$

with boundary conditions

$$\frac{\partial}{\partial x_{i+1}} \quad \frac{\partial}{\partial x_i} + 1 \quad \Xi_{\text{Bose}}(x_1,\ldots,x_N;t)|_{x_{i+1}=x_i+0} = 0\,, \quad i = 1,2,\ldots,N \quad 1\,, \tag{17}$$

and the initial condition

$$\Xi_{\text{Bose}}(x_1,\ldots,x_N;t=0) = \prod_{i=1}^{N} \quad (x_i)\,. \tag{18}$$

The system is known to be exactly solvable. The connection between the KPZ equation and the -Bose gas was noticed long time ago [12]. Recently summation over excited states was carried out for the $Z(x,0) = \quad (x)$ (corresponding to the narrow wedge initial condition for the KPZ equation) by applying the Bethe ansatz. The result (9) in the previous section can be obtained along this route [13, 14], but the original derivation was di erent.

## 5. KPZ as weakly ASEP limit

The derivation of (9) uses the weakly asymmetric limit of the ASEP [7–11] and thus we brie y recall how the KPZ equation can be obtained from the weakly ASEP in a scaling limit. Let us take a small parameter $\epsilon > 0$. We consider the scale of space of $\mathcal{O}(\epsilon^{\quad 1})$, time of $\mathcal{O}(\epsilon^{\quad 2})$, and simultaneously set

$$p = \frac{1}{2} \quad \frac{1}{2}\sqrt{\epsilon}\,, \quad q = \frac{1}{2} + \frac{1}{2}\sqrt{\epsilon}\,. \tag{19}$$

In the end we take the $\epsilon \to 0$ limit. (Taking $\epsilon = 0$ from the outset without scaling space and time, one simply gets the symmetric simple exclusion process (SSEP).) For the ASEP height function $h_{\text{A}}$, let us de ne

$$Z_{\text{A}}(x,t) = \frac{1}{\sqrt{\epsilon}}\exp \quad h_{\text{A}}([\epsilon^{\quad 1}x],\epsilon^{\quad 2}t) + \quad \epsilon^{\quad 2}t \quad , \tag{20}$$

where

$$= p + q \quad 2\sqrt{qp} = \frac{1}{2}\epsilon + \frac{1}{8}\epsilon^2 + \mathcal{O}(\epsilon^3)\,, \tag{21}$$

$$= \frac{1}{2}\log\frac{q}{p} = \epsilon^{1/2} + \frac{1}{3}\epsilon^{3/2} + \mathcal{O}(\epsilon^{5/2})\,. \tag{22}$$

Let us also set

$$D_\epsilon = 2\sqrt{pq} = 1 \quad \frac{\epsilon}{2} + \mathcal{O}(\epsilon^{3/2})\,, \tag{23}$$

$$_\epsilon f(x) = \epsilon^{\quad 2}\,(f(x+\epsilon) \quad 2f(x) + f(x \quad \epsilon))\,. \tag{24}$$

Then it holds [11, 26]

$$dZ_{\rm A} = \frac{1}{2} D_\epsilon \quad {}_\epsilon Z_{\rm A} dt + Z_{\rm A} dM\,, \tag{25}$$

where $M$ is a stochastic process, in fact a martingale, in $t$ with

$$d\langle M(x) M(y)\rangle = \frac{1}{\epsilon} \quad {}_{x,y} b_\epsilon(\eta, x) dt \tag{26}$$

and

$$b_\epsilon(\eta, x) = 1 \quad (2\eta([\epsilon \ {}^1 x], \epsilon \ {}^2 t) \quad 1)(2\eta([\epsilon \ {}^1 x, \epsilon \ {}^2 t] + 1) \quad 1) + \mathcal{O}(\sqrt{\epsilon})\,. \tag{27}$$

By comparing (25) with (12) one notices a similarity. In fact it can be shown that, in the limit $\epsilon \to 0$, (25) converges to (12). This means that in this weakly asymmetric limit the ASEP tends to the KPZ equation. In other words one can obtain results for the KPZ equation by taking the weakly asymmetric limit for the ASEP.

We start from the following formula for ASEP, due to Tracy and Widom [25],

$$\mathbb{P} \quad x_m(t/(q \quad p)) \le x \quad = \int_{\mathcal{C}_0} \prod_{k=0}^{\infty} (1 \qquad \tau^k) \det(1 + J( \ )) \frac{\mathrm{d}}{}\,, \tag{28}$$

where $\tau = p/q$, $x_m(t)$ is the position of the particle starting from site $m$ at time $t = 0$, $\mathcal{C}_0$ is a circle around 0 and with radius in $(\tau, 1)$ and

$$J( \ ; \eta, \eta') = \int_{\mathcal{C}_1} \frac{\varphi_\infty( \ )}{\varphi_\infty(\eta')} \frac{m}{(\eta')^{m+1}} \frac{f( \ , \ /\eta')}{\eta} \mathrm{d} \ , \tag{29}$$

$$\varphi_\infty(\eta) = (1 \quad \eta) \ {}^x \mathrm{e}^{t(\eta/(1 \quad \eta))}, \tag{30}$$

$$f( \ , z) = \sum_{k= \ \infty}^{\infty} \frac{\tau^k}{1 \qquad \tau^k} z^k\,, \tag{31}$$

where $\mathcal{C}_1$ is a circle around 0 and with radius in $(1, r/\tau)$. By carrying out an intricate asymptotic analysis for (28), one obtains

$$\lim_{\varepsilon \to 0} \mathbb{P} \left[ \frac{2\sqrt{\epsilon} N([\epsilon \ {}^1 x], \varepsilon \ {}^2 t) \quad \frac{1}{2} t \varepsilon \ {}^1 \quad \epsilon \ {}^{1/2}|x| + x^2/2t + \log(2\sqrt{\varepsilon})}{\gamma_t} \ge s \right] = F_t(s)\,,$$

where $\gamma_t$ is de ned by (10). This then translates to the result (9) for the KPZ equation.

## 6. ASEP and -Bose

Based on the discussions in the previous two sections, it is natural to ask about the relation between the ASEP and the -Bose gas. In [17] we explain how the duality of the ASEP found by Schutz [18] is useful to compute exponential moments of the ASEP current. Let us de ne

$$N_x = \sum_{j= \ \infty}^{x} n_j\,, \tag{32}$$

where $n_j$ is the number operator at site $j$ and

$$Q_x = \tau^{N_x}\,. \tag{33}$$

Set

$$\Xi_{\rm ASEP}(x_1, \dots, x_N; t) = \langle Q_{x_1} \qquad Q_{x_N}\rangle_t\,. \tag{34}$$

Journal of Physics: Conference Series **297** (2011) 012016 doi:10.1088/1742-6596/297/1/012016

Here $\langle \quad \rangle_t$ means the average of the ASEP dynamics for step initial condition. The duality of the ASEP says that this $N$-point correlation function in $Q_x$ satis es the master equation of the ASEP with $N$ particles,

$$\frac{d}{dt}\Xi_{\mathrm{ASEP}}(x_1,\ldots,x_N;t) = \sum_{i=1}^{N} \quad p\Xi_{\mathrm{ASEP}}(\ldots,x_i \quad 1,\ldots;t) + q\Xi_{\mathrm{ASEP}}(\ldots,x_i+1,\ldots;t)$$
$$(p+q)\Xi_{\mathrm{ASEP}}(\ldots,x_i,\ldots;t) \quad . \tag{35}$$

One has to solve (35) with the boundary condition

$$p\Xi_{\mathrm{ASEP}}(\ldots,x_i,x_i,\ldots;t) + q\Xi_{\mathrm{ASEP}}(\ldots,x_i+1,x_i+1,\ldots;t) = (p+q)\Xi_{\mathrm{ASEP}}(\ldots,x_i,x_{i+1},\ldots;t)\,. \tag{36}$$

For step initial condition $Q_x = \tau^{x\Theta(x)}$, where $\Theta(x) = 0(x \le 0), 1(x \ge 1)$, and hence

$$\Xi_{\mathrm{ASEP}}(x_1, \quad ,x_N;t=0) = \tau^{\sum_{i=1}^{N} x_i\Theta(x_i)}. \tag{37}$$

The ASEP and the XXZ spin chain are related by a simple similarity transformation [18]. Let us set

$$\Xi_{\mathrm{XXZ}}(x_1,\ldots,x_N) = \tau^{\quad \frac{1}{2}\sum_{i=1}^{N} x_i}\Xi_{\mathrm{ASEP}}(x_1,\ldots,x_N)\,. \tag{38}$$

Then (38) satis es

$$\frac{d}{dt}\Xi_{\mathrm{XXZ}}(x_1,\ldots,x_N;t) = \sum_{i=1}^{N} \quad \Xi_{\mathrm{XXZ}}(\ldots,x_i \quad 1,\ldots;t) + \Xi_{\mathrm{XXZ}}(\ldots,x_i+1,\ldots;t)$$
$$2 \quad \Xi_{\mathrm{XXZ}}(\ldots,x_i,\ldots;t) \quad , \tag{39}$$

where $\quad = (\tau^{1/2} + \tau^{\quad 1/2})/2$ and

$$\Xi_{\mathrm{XXZ}}(\ldots,x_i,x_i,\ldots;t) + \Xi_{\mathrm{XXZ}}(\ldots,x_i+1,x_i+1,\ldots;t) = 2 \quad \Xi_{\mathrm{XXZ}}(\ldots,x_i,x_i+1,\ldots;t)\,. \tag{40}$$

The initial condition is

$$\Xi_{\mathrm{XXZ}}(x_1,\ldots,x_N;t=0) = \tau^{\frac{1}{2}\sum_{i=1}^{N}|x_i|}\,. \tag{41}$$

Now let us consider the weakly asymmetric limit

$$x \to \epsilon^{\quad 1}x\,, \quad t \to \epsilon^{\quad 2}t\,, \tag{42}$$

and (19). We have

$$\tau = \frac{1 \quad \sqrt{\epsilon}}{1+\sqrt{\epsilon}}\,, \qquad = 1 \quad \epsilon/2 + O(\epsilon^{3/2})\,. \tag{43}$$

We see that in this limit, the Schrodinger equation (39), (40), (41) tends to the one of the -Bose gas (16), (17), (18). In this way one understands the interrelation between KPZ, ASEP, and -Bose gas.

## 7. Conclusion

In 2010 an important advance was achieved by establishing an analytic solution of the one-dimensional KPZ equation. This will certainly be useful for further studying the equation itself and for comparison with Monte Carlo simulations. Another interesting aspect is the deep connection of the KPZ equation with other models and the resulting similarity in the methods used. In this paper, we explained the connections to the directed polymer problem, -Bose gas, ASEP, and XXZ spin chain. We discussed only the basic relations. It would be of considerable interest to more deeply elucidate these interrelations.

## Acknowledgments
T.I. and T.S. acknowledge the supports from KAKENHI (22740251) and KAKENHI (22740054) respectively.

## References
[1] Kardar M, Parisi G and Zhang Y C 1986 *Phys. Rev. Lett.* **56** 889
[2] Barabasi A-L and Stanley H E 1995 (Cambridge University Press)
[3] Spohn H 2005 *Pranama J. Phys.* **64** 1
[4] Sasamoto T 2007 *J. Stat. Mech.* P07007
[5] Ferrari P L 2010 *J. Stat. Mech.* P10016
[6] Kriecherbauer T and Krug J 2010 *J. Phys. A* **43** 403001
[7] Sasamoto T and Spohn H 2010 *Nucl. Phys. B* **834** 523
[8] Sasamoto T and Spohn H 2010 *J. Stat. Phys.* **140** 209
[9] Sasamoto T and Spohn H 2010 *Phys. Rev. Lett.* **104** 230602
[10] Sasamoto T and Spohn H 2010 *J. Stat. Mech.* P11013
[11] Amir G, Corwin I, and Quastel J 2010 *Comm. Pure Appl. Math.* **64** 466
[12] Kardar M 1987 *Nucl. Phys. B* **290** 582
[13] Calabrese P, Doussal P Le, and Rosso A 2010 *Euro. Phys. Lett.* **90** 20002
[14] Dotsenko V 2010 *Euro. Phys. Lett.* **90** 20003
[15] Prolhac S and Spohn H 2011 *J. Stat. Mech.* P01031
[16] Prolhac S and Spohn H 2011 arXiv:1101.4622
[17] Imamura T and Sasamoto T 2010 *J. Stat. Phys.* **142** 919
[18] Schutz G M 1997 *J. Stat. Phys.* **86** 1265
[19] Liggett T M 1985 (Springer-Verlag)
[20] Liggett T M 1999 (Springer-Verlag)
[21] Schutz G M 2000 (Phase Transitions and Critical Phenomena 19)
[22] Tracy C A and Widom H 1994 *Comm. Math. Phys.* **159** 151
[23] Johansson K 2000 *Comm. Math. Phys.* **209** 437
[24] Tracy C A and Widom H 2008 *Comm. Math. Phys.* **279** 815
[25] Tracy C A and Widom H 2009 *Comm. Math. Phys.* **209** 129
[26] Bertini L and Giacomin G 1997 *Comm. Math. Phys.* **183** 571

STATPHYS-Kolkata VII
Journal of Physics: Conference Series **297** (2011) 012017 doi:10.1088/1742-6596/297/1/012017

# The Fluctuation Theorem and Dissipation Theorem for Poiseuille Flow

**Sarah J. Brookes**
Queensland Micro- and Nanotechnology Centre and School of Biomolecular and Physical Sciences, Griffith University Brisbane, QLD 4111, Australia
E-mail: s.brookes@griffith.edu.au

**James C. Reid**
Queensland Micro- and Nanotechnology Centre and School of Biomolecular and Physical Sciences, Griffith University Brisbane, QLD 4111, Australia
E-mail: j.reid@griffith.edu.au

**Denis J. Evans**
Research School of Chemistry, Australian National University, Canberra, ACT 0200, Australia
E-mail: evans@rsc.anu.edu.au

**Debra J. Searles**
Queensland Micro- and Nanotechnology Centre and School of Biomolecular and Physical Sciences, Griffith University Brisbane, QLD 4111, Australia
E-mail: d.bernhardt@griffith.edu.au

**Abstract.** The fluctuation theorem and the dissipation theorem provide relationships to describe nonequilibrium systems arbitrarily far from, or close to equilibrium. They both rely on definition of a central property, the dissipation function. In this manuscript we apply these theorems to examine a boundary thermostatted system undergoing Poiseuille flow. The relationships are verified computationally and show that the dissipation theorem is potentially useful for study of boundary thermostatted systems consisting of complex molecules undergoing flow in the nonlinear regime.

## 1. Introduction

In the last 20 years, considerable advances have been made in the formal study of nonequilibrium systems. Two key advances have been (i) the development of relationships that are valid arbitrarily far from equilibrium so analyses are no longer restricted to the linear response regime, and (ii) the ability to treat finite sized systems, so the thermodynamic limit is not required. In fact, we are now reaching the stage when thermodynamic behavior can be derived from mechanics rather than being based on thermodynamic laws and axioms, and this provides a route to treat the thermodynamics of nonequilibrium systems.

Two important results that have been obtained are the Fluctuation Theorem (FT) [1, 2, 3, 4] and the Dissipation Theorem (DT) [4, 5]. The FT is results in an expression for the probability ratio for observing positive and negative values of the dissipation and therefore characterizes fluctuations in the dissipation that occurs in an N-particle system, measured over a finite time. Importantly, it can be used to derive the Second Law of Thermodynamics. The DT provides an expression for the response of arbitrary phase variables to a change in system conditions or application of an external field. Both these results can be obtained using the Liouville equation and the microscopic equations of motion of the particles in the system.

A very general form of the FT was derived in 2000 [6] and has been tested on a wide range of systems. In contrast, a general form of the DT was only published in 2008 [5] and it has not been applied as widely. A special case of the dissipation theorem that is applicable to homogeneously thermostatted systems and has been know for some time [7]. It can be expressed in two forms known as the Kawasaki expression for non-linear response and the transient time-correlation function (TTCF) expression for nonlinear response. The TTCF form has been widely used on homogeneously thermostatted systems, and has not only been shown to be valid, but to provide an efficient computational route to determine the response of a system for low fields. However, the DT has not yet been fully tested on other classes of systems. In this manuscript we will consider application of the FT and the DT to systems undergoing Poiseuille flow: gravity driven flow of a fluid parallel to two thermostatted walls. Atomic fluids and chain polymers are considered at fields ranging from the linear to the nonlinear regime.

## 2. The Fluctuation Theorem, the Dissipation Function and the Dissipation Theroem

Consider a system that is initially in some known state, which for the purposes of this manuscript can be considered to be an equilibrium state. The system is then subject to a change, such as the application of a gravity field, and the system is monitored for a period, $t$. If the system is sufficiently small, and the period sufficiently short, the time-averaged properties of the system will take on different values during each time this experiment is repeated. If that is the case, we might be interested in the probability that properties take on particular values during the period $t$, and also the average response of the system to this field. These probabilities and averages are described by the FT and DT, respectively.

The FT is a well known relationship in nonequilibrium statistical mechanics. It states that time integral of the dissipation function, $\Omega_t$, satisfies the symmetry relation:

$$\frac{Pr(\Omega_t = A)}{Pr(\Omega_t = -A)} = e^A. \tag{1}$$

where $Pr(\Omega_t = A)$ is the probability that $\Omega_t$ takes on a value within $A \pm dA$ . The FT, (eq. 1) applies to systems starting with a nonequilibrium distribution and relaxing towards equilibrium as time evolves, or those that start in an equilibrium distribution but for $t > 0$, are driven away from equilibrium by boundary conditions (e.g. moving walls) or a mechanical field (e.g. gravity field) as is the case in this paper. The proof of the FT has been given previously (see, for example [2, 4]) and is exceedingly general because it uses so little. In fact, the only feature of the dynamics that is used in the proof is their time reversal symmetry. All microscopic dynamics (quantum and classical) are time reversal symmetric.

The FT has been widely verified both experimentally and computationally for a range of systems. However, difficulty in verifying a relation of the form (1) has been reported in some studies where the subject of the FT, $\Omega$, has been proposed or guessed for a particular system, rather than carefully derived. This is despite the fact that the approach described in [2, 3, 4] leaves no ambiguity as to the function that should satisfy the FT [8]. The time-integral of the dissipation function, $\Omega_t(\mathbf{\Gamma}(0)) \equiv \int_0^t ds\Omega(\mathbf{\Gamma}(s))$, can be defined as the logarithm of the ratio of the probability of seeing a set of trajectories of duration t, initially centered at a point in N-particle phase space, $(\mathbf{\Gamma}(0))$, to the probability of observing (from the same probability distribution) the conjugate set of anti-trajectories formed from their time-reversal. Here $\mathbf{\Gamma} \equiv (\mathbf{q}, \mathbf{p})$ is a point in phase space where $\mathbf{q}$ is the position and $\mathbf{p}$ is the momentum. Mathematically it is defined as:

$$\Omega_t(\mathbf{\Gamma}(0)) = \ln\left(\frac{f_0(\mathbf{\Gamma}(0))}{f_0(\mathbf{\Gamma}^*(t))}\right) - \Lambda_t(\mathbf{\Gamma}(0)) \tag{2}$$

where the asterisk denotes a time reversal mapping of a point in phase space, $\Lambda = \frac{\partial}{\partial \mathbf{\Gamma}} \cdot \dot{\mathbf{\Gamma}}$ is the rate of phase space compression, $\Lambda_t(\mathbf{\Gamma}(0)) \equiv \int_0^t ds\Lambda(\mathbf{\Gamma}(s))$ and $f_0(\mathbf{\Gamma}(t))$ is the phase space probability density with repect to the distribution at $t = 0$, at a phase $\mathbf{\Gamma}(t)$. Here we will determine the dissipation function for a system undergoing thermostatted Poiseuille flow and find that it is an important physical property of the system.

The DT gives the nonequilibrium average of an arbitrary phase function. Using the FT and the Lagrangian form of the Liouville equation, it only takes a few lines of algebra to derive the DT [4] in its TTCF from as:

$$\langle B(\mathbf{\Gamma}(t))\rangle_{f_0} = \langle B(\mathbf{\Gamma}(0))\rangle_{f_0} + \int_0^t ds \langle \Omega(\mathbf{\Gamma}(0)) B(\mathbf{\Gamma}(s))\rangle_{f_0} \tag{3}$$

where the notion $\langle ...\rangle_{f_0}$ implies that the ensemble average is with respect to the distribution $f(\mathbf{\Gamma}, 0)$. Equation (3) is exact for arbitrary system size and arbitrarily far from equilibrium.

## 3. Simulations

### *3.1. Equations of motion for Poiseuille flow*

The FT for Poiseuille flow has been considered previously [9, 10], however, a detailed derivation of the dissipation function was not given. Furthermore, the DT was not tested for this system.

In Poiseuille flow, the system of interest is contained by a slit pore, and a gravity field is applied parallel to the pore walls. In this work, the fluid particles are purely Newtonian, but the temperature of the wall particles is directly controlled by use of a fictitious thermostat. As argued previously [10], this type of construction can be used to provide a realistic model since we can make the thermostat remote from the system of interest, and the system will become insenstive to the details of the thermostat. We set orient the system so that the walls are normal to the $z$ axis and the field acts in the x direction.The equations of motion for the fluid particles can then be written:

$$\begin{aligned} \dot{\mathbf{q}}_i^F &= \mathbf{p}_i^F/m_i, \\ \dot{\mathbf{p}}_i^F &= \mathbf{F}_i^F + F_e\mathbf{i}, \end{aligned} \tag{4}$$

where $i$ is the particle index, the superscript $F$ means that the particle belongs to the fluid, and $\mathbf{F}_e$ is the external force. The forces $\mathbf{F}_i^F$ are due to i interactions of the fluid particle with other particles in the fluid and the wall particles. Details of these interactions will be given in the next sub-section. The equations of motion for the wall particles are given by:

$$\begin{aligned} \dot{\mathbf{q}}_i^W &= \mathbf{p}_i^W/m_i, \\ \dot{\mathbf{p}}_i^W &= \mathbf{F}_i^W - \alpha\mathbf{p}_i^W, \end{aligned} \tag{5}$$

where the superscript $W$ denotes that the particle belongs to the wall. The temperature is controlled by a Gaussian thermostat which fixes the kinetic temperature of the wall through the Lagrangian multiplier $\alpha = \left(\sum_i^{N_W} \mathbf{F}_i^W \cdot \mathbf{p}_i^W\right) / \left(\sum_i^{N_W} \mathbf{p}_i^W \cdot \mathbf{p}_i^W\right)$. In this case, in addition to the interparticle interactions, $\mathbf{F}_i^W$ contains forces applied to tether the wall particles to lattice positions, to created a solid-like wall.

### *3.2. System parameters*

Two types of system are considered in this work - a purely atomic fluid and a fluid consisting of short chain polymers modelled through the Finite Extendable Nonlinear Extendable Elastic Potential (FENE) [11]. All systems were modelled as a slit pore with four wall layers. Each fluid and wall atom in a system interacts via the modified Lennard-Jones potential, the Weeks, Chandler and Andersen potential (WCA) [12, 13]. The WCA truncates the LJ potential at is the minimum, so the potential is zero at the cuttoff value.

$$\Phi_{WCA}(r_{ij}) = \begin{cases} 4\epsilon\left[\left(\frac{\sigma}{r_{ij}}\right)^{12} - \left(\frac{\sigma}{r_{ij}}\right)^{6}\right] + \epsilon, & r_{ij} \leqslant 2^{\frac{1}{6}}\sigma, \\ 0, & r_{ij} > 2^{\frac{1}{6}}\sigma, \end{cases} \tag{6}$$

where $r_{ij} = |\mathbf{q_i} - \mathbf{q_j}|$ with $\mathbf{q}_i$ being the laboratory particle position, $\sigma$ is the value of $r_{ij}$ for which the LJ interaction potential is zero, and $\epsilon$ is the well-depth of the LJ potential. All the physical units are expressed in reduced units where the unit of mass is the particle mass $m$, the energy unit is the parameter $\epsilon$ and the length unit is $\sigma$. To keep the wall atoms at their equilibrium lattice sites, a retarding force in the form of a harmonic potential constraint $U_{harm}(\mathbf{q}_i)$ was used,

$$U_{harm}(\mathbf{q}_i) = \frac{1}{2}k_s(\mathbf{q}_i - \mathbf{q}_{i,eq})^2, \tag{7}$$

where $\mathbf{q}_{i,eq}$ the lattice position of the $i^{th}$ wall atom. A harmonic force constant of $k_s = 15$ was used in the simulations.

The polymers are modelled as a chain of united-atoms where the united atom model represents each carbon group as a single particle and particle are connected using a FENE potential,

$$\Phi_{FENE}(r_{ij}) = \begin{cases} -\frac{1}{2}kR_0^2 \ln\left[1 - \left(\frac{r_{ij}}{R_0}\right)^2\right], & r_{ij} < R_0, \\ \infty, & r_{ij} \geqslant R_0. \end{cases} \tag{8}$$

This potential allows the united-atoms to remain bound to one another, and unable to extend beyond $r_{ij} = R_0$. A suitable of choice of $k$ and $R_0$ also ensures the polymer chains cannot cross one another. In this work $k = 30$ and $R_0 = 1.5\sigma$ which has shown to be an appropriate choice in several studies [14, 15, 16].

Each system contains a total of 128 atoms (or united-atoms) with 64 in the walls and the remaining 64 in the fluid. Two polymer systems consist of 4 and 8 united atoms per chain length were simulated and have a total of 16 and 8 molecules respectively. All atoms were initially on an body-centred lattice with an atom density of 0.8, and with periodic boundaries in all dimensions. Within the unit cells, the walls had a surface area of $5.4\sigma \times 5.4\sigma$ and a pore of width $1.7\sigma$. The systems were equilibrated for $20^7$ time-steps with the temperature maintained at $T = 1$ using a Gaussian thermostat. After this time, $10^5$ transient nonequilibrium trajectories were spawned from the equilibrium trajectory at regular intervals, and the properties measured as a function of time.

## 4. Results and Discussion

### *4.1. The dissipation function*

In this section we will derive the dissipation function, which is the subject of the FT (eq. 2), for Poiseuille flow. The FT for Poiseuille flow has been considered previously in 1999 [9] and 2001 [10]. The 1999 paper considered an isoenergetic system and the fluctuation relation was determined by considering phase space expansion. This is not possible in thermostatted systems and the more general approach, via the dissipation function needs to be used. The 2001 paper states the isothermal FT, but does not provide a full derivation. Here we give a complete derivation of the dissipation function to illustrate the application of this approach to a realistic system.

In order to apply the FT, the dynamics must be time reversal symmetric, and this is valid for Poiseuille flow. In addition the dynamics and initial ensemble must be ergodically consistent which implies that all points accessed by the nonequilibrium trajectory are part of the initial ensemble. If the initial ensemble is the isokinetic ensemble, and the dynamics is thermostatted using a Gaussian thermostat, this will be true. The reasons why these requirements are necessary have been discussed in detail in [2, 3, 4].

We consider a system that is initially at equilibrium with walls thermostatted using a Gaussian thermostat (i.e. the system would be modelled with the equations of motion above but with $F_e = 0$). The distribution function for this system is,

$$f(\mathbf{\Gamma}, 0) = \frac{e^{-\beta H(\mathbf{\Gamma})}\delta(K^W(\mathbf{\Gamma}) - K_0)}{Z} \tag{9}$$

where $\beta = 1/k_B T$, $k_B$ is the Boltzmann constant, $T$ equilibrium temperature of the system, $H(\mathbf{\Gamma})$ is the internal energy of the system, $K^W(\mathbf{\Gamma})$ is the kinetic energy of the wall particles, $K_0$ is the required kinetic energy of the walls, $\delta$ is a Dirac delta function and $Z$ is the partition function.

Equation (2) then gives,

$$\Omega_t(\mathbf{\Gamma}(0)) = \ln\left(\frac{e^{-\beta H(\mathbf{\Gamma}(0))}\delta(K^W(\mathbf{\Gamma}(0)) - K_0)}{e^{-\beta H(\mathbf{\Gamma}^*(t))}\delta(K^W(\mathbf{\Gamma}^*(t)) - K_0)}\right) - \Lambda_t(\mathbf{\Gamma}(0)). \tag{10}$$

We note that the internal energy and kinetic energy at a point in phase space and its time-reversed image are the same, that is $H(\mathbf{\Gamma}^*) = H(\mathbf{\Gamma})$ and $K(\mathbf{\Gamma}^*) = K(\mathbf{\Gamma})$. It can also be seen that if the kinetic energy of the walls is fixed at $K_0$ at equilibrium and during the nonequilibrium trajectory, then both the delta-functions are non-zero and cancel. Furthermore, the phase space expansion rate of the system can be determined from the equations of motion, $\Lambda = \frac{\partial}{\partial \mathbf{\Gamma}} \cdot \dot{\mathbf{\Gamma}} = -(3N_W - 1)\alpha$, so the dissipation function can be expressed:

$$\begin{aligned} \Omega_t(\mathbf{\Gamma}(0)) &= \ln\left(\frac{e^{-\beta H(\mathbf{\Gamma}(0))}}{e^{-\beta H(\mathbf{\Gamma}(t))}}\right) + (3N_W - 1)\int_0^t ds\alpha(\mathbf{\Gamma}(s)) \\ &= \beta H(\mathbf{\Gamma}(t)) - \beta H(\mathbf{\Gamma}(0)) + (3N_W - 1)\int_0^t ds\alpha(\mathbf{\Gamma}(s)) \\ &= \beta\int_0^t ds\dot{H}(\mathbf{\Gamma}(s)) + (3N_W - 1)\int_0^t ds\alpha(\mathbf{\Gamma}(s)) \end{aligned} \tag{11}$$

From the equations of motion, (4) and (5), we can determine the time-derivative of the internal energy $\dot{H} = \sum_{i=1}^{N_F} p_{xi}F_e - 2K_0\alpha$ and since $2K_0 = (3N_W - 1)k_BT$ we finally obtain,

$$\begin{aligned} \Omega_t(\mathbf{\Gamma}(0)) &= \beta F_e \int_0^t ds \sum_{i=1}^{N_F} p_{xi}(\mathbf{\Gamma}(s)) \\ &\equiv \beta F_e N_F J_{x,t}(\mathbf{\Gamma}(0)) \end{aligned} \tag{12}$$

where we defined the fluid current as $J_x \equiv \sum_{i=1}^{N_F} p_{xi}/N_F$ and therefore $J_{x,t}(\mathbf{\Gamma}(0)) = \int_0^t dsJ_x(\mathbf{\Gamma}(s))$. So,

$$\Omega(\mathbf{\Gamma}) = \beta F_e N_F J_x(\mathbf{\Gamma}). \tag{13}$$

This is the dissipation function for Poiseuille flow under boundary thermostatted flow. Importantly in this case the current is the instantaneous current of the fluid particles only,

Journal of Physics: Conference Series **297** (2011) 012017 doi:10.1088/1742-6596/297/1/012017

and the temperature is the kinetic temperature at which the wall particles are fixed using the Gaussian thermostat. Without such a careful derivation these subtle points might not have been anticipated. The temperature that appears can also be identified as the equilibrium thermodynamic temperature to which the system would relax if the field was removed [17].

### *4.2. Simulation results*

In Figure 1 the FT is tested for an atomic fluid with a field of $F_e = 1$, $T = 1.0$ and $n = 0.8$ and a simulation time of $t = 0.4$ for each of $2 \times 10^5$ trajectories. The FT would predict that a plot of $\ln(Pr(\Omega_t = A)/(Pr(\Omega_t = -A))$ versus $A$ would give a straight line with unit slope. Clearly this is consistent with the results. This provides a useful check that the conditions of ergodic consistency and reversibility are being satisfied, that dissipation function has been correctly derived and the simulation code is working properly. As higher fields and/or longer times are considered, it becomes increasingly difficult to observe currentes that oppose the field, and therefore more difficult to validate the FT.

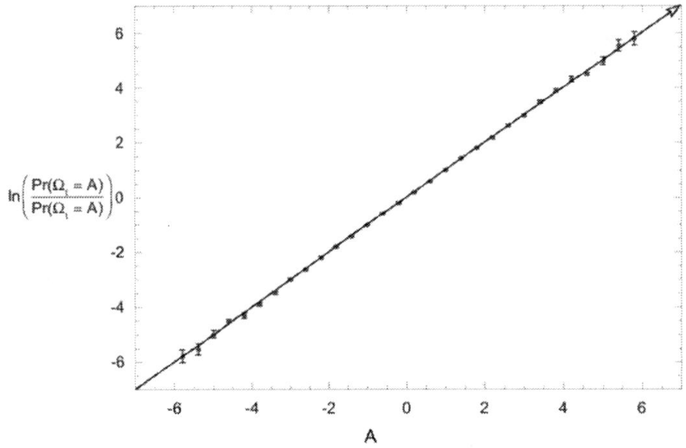


**Figure 1.** A test of the FT for an atomic fluid with $F_e = 1$, $T = 1.0$ and $n = 0.8$. The results were obtained as the mean of 10 sets of $2 \times 10^4$ trajectories of length $t = 0.4$. The error bars are standard error in the mean of these 10 sets of data. The line is the result predicted by the FT.

The fluid current, $J_x$, through the three systems was examined to demonstrate the validity and utility of the DT. Using the DT (eq. 3) and the dissipation function derived in equation (eq. 13) and also noting that $\langle J_x(\mathbf{\Gamma}(0)) = 0\rangle$ is zero since $J_x$ is odd with respect to a time-reversal mapping, the TTCF form of the DT for $J_x$ is;

$$\langle J_x(\mathbf{\Gamma}(t))\rangle_{f_0} = \beta N_F F_e \int_0^t ds \langle J_x(\mathbf{\Gamma}(0)) J_x(\mathbf{\Gamma}(s))\rangle_{f_0}. \tag{14}$$

Figure 2 shows the response of the current, $J_x$, determined using (eq. 14) for all three fluids and for a range of applied external fields between 0.01 to 1. There is a monotonic increase in the current with increase in the applied field at all times. It can be observed that as the length of the polymer increases from a single atom fluid to a 4 atom polymer there is quite a large increase in the current. This is due to the fact that although the atomic number density is held constant, the molecular number density decreases. There is also a smaller increase from the 4 atom polymer to the 8 atom polymer.

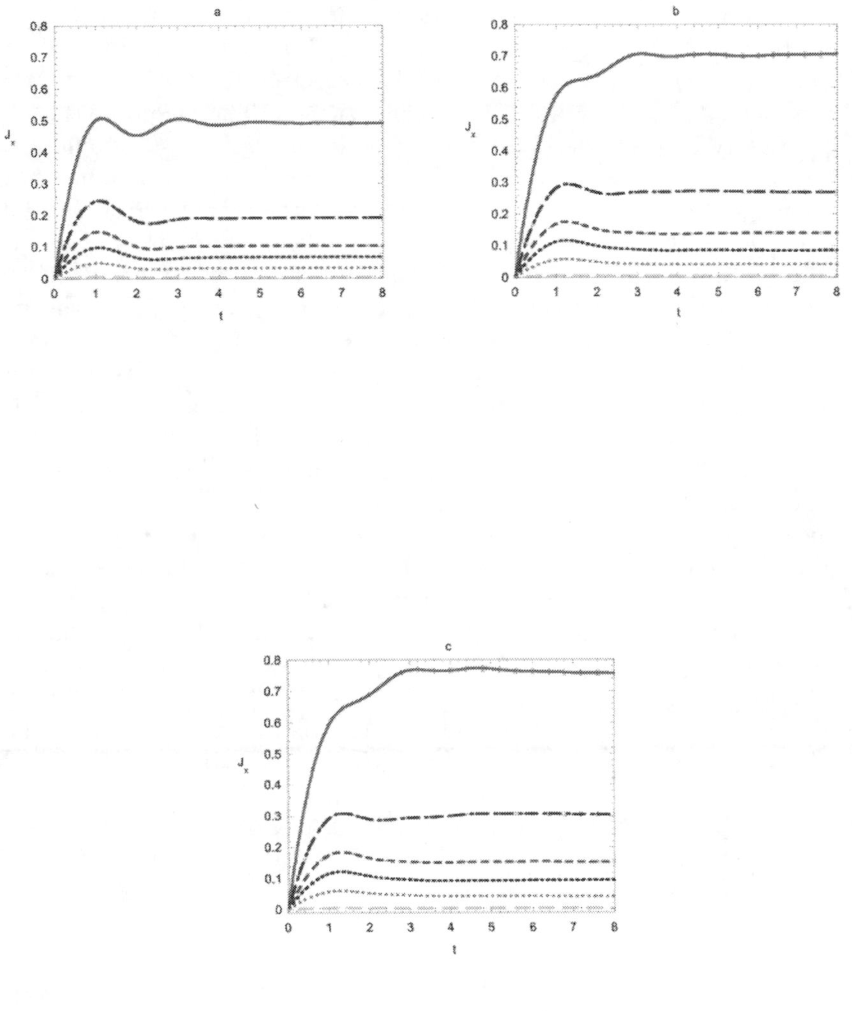


**Figure 2.** The response of $J_x$ to the application of a range of fields: $F_e = 0.01$ (light blue), $F_e = 0.1$ (pink), $F_e = 0.2$ (black), $F_e = 0.3$ (green), $F_e = 0.5$ (blue), $F_e = 1$ (red). The systems studied were at a temperature of $T = 1$ and an atom density $n = 0.8$ in the fluid and walls for a) an atomistic fluid, b) a 4-atom polymer and c) an 8-atom polymer. Error bars are the standard error in the mean from 10 sets of data.

One of the advantages of using (eq. 3) to calculate properties is its ability to treat small fields. In most cases, nonequilibrium molecular dynamics (NEMD) simulations use a direct averaging method to calculate properties of a nonequilibrium state: that is one simply averages the property over the number of trajectories to get $\langle J_x(t) \rangle$. At large fields the signal to noise ratio is large. Consequently, for strong fields evaluation of properties using the TTCF form of the DT (3) is expected to give the same results for the current as obtained from the direct averaging of NEMD simulations. This can be observed in Fig. 3 where the $J_x$ has been plotted over time for $F_e = 1$ for the 4 atom polymer system . The trajectories are very similar both in the transient time period where the field is first turned on and the steady state where the system has relaxed.

In contrast, at small fields the signal to noise ratio can become so small that the signal is hard to distinguish if direct averaging is used. Therefore for weak fields the DT expression allows the average response to be accurately determined, as is illustrated in Fig. 3b) where $J_x$ has been plotted for a 4 atom polymer with an applied external field of $F_e = 0.01$. The directly calculated $J_x$ for the same system and field is also shown. The shape of the graph obtained using the DT is clearly the same as those for the larger fields (Fig. 3a), again with with very little error observed. In contrast the direct calculation give data with a much larger statistical error for the same number of trajectories. The DT can therefore give us reliable data to examine the effects of confinement on the response of a system when a weak external field is applied.

In Figure 4a, $J_x/F_e$ was was determined for all applied external fields ranging from 0.01 to 1 for the atomic and polymer systems. There is a reduction in $J_x/F_e$ as the polymer length decreases. This would indicate that the larger polymer is able to move more freely and is again consistent with the reduced molecular number density. It can also observed that the size of the linear response regime decreases as the polymer length increase. Already for the polymer with 8 united-atoms, the statistical error from direct calculations will be too large to determine the response accurately for low fields using the large number of trajectories averaged in this work. This is not a problem if the system is in the linear regime as the current can simply be extrapolated using the linearity of the response. However, extrapolation is not possible in the nonlinear regime. As much larger polymers are considered, the fields where the system is still in the non-linear regime will become increasingly wide [18]. Then, use of the TTCF form of the dissipation function will be essential. The $J_x/F_e$ response of the 8 atom polymer is best observed in figure 4b where the $J_x/F_e$ is plotted as a function of time. A distinct difference in each field can be seen.

Other properties of the system can be determined in the same manner, using the general equation (3). Here we consider the average Irving-Kirkwood expression for the fluid pressure, $p$ [19]. Figure 5 shows the average pressure of the fluid in the 4 atom polymer with $F_e = 1$ and $F_e = 0.3$. As with the current, when the higher field is applied the direct calculation and the DT expression give similar results. However the DT approach shows a noticeably smoother curve with less error at the lower field.

## 5. Conclusions

The dissipation function for a system undergoing wall-thermostatted Poiseuille Flow has been derived and was shown to satisfy the FT. The form of the dissipation function for this combination of dynamics and ensemble is unique and is found to be $\beta F_e N_F J_x$ where $\beta$ is the kinetic temperature of the walls and corresponds to the temperature that the system would relax to if the field was removed [17]. Clearly this is equal to the average entropy production rate at low fields. At higher fields it becomes impossible to define either the entropy or the temperature. In fact there are large "temperature" gradients in the system at higher field strengths and one does not even know where the "temperature" should be measured! The DT rigorously avoids these pitfalls and gives an exact answer even at arbitrarily high values of the driving field. Had

Journal of Physics: Conference Series **297** (2011) 012017 doi:10.1088/1742-6596/297/1/012017

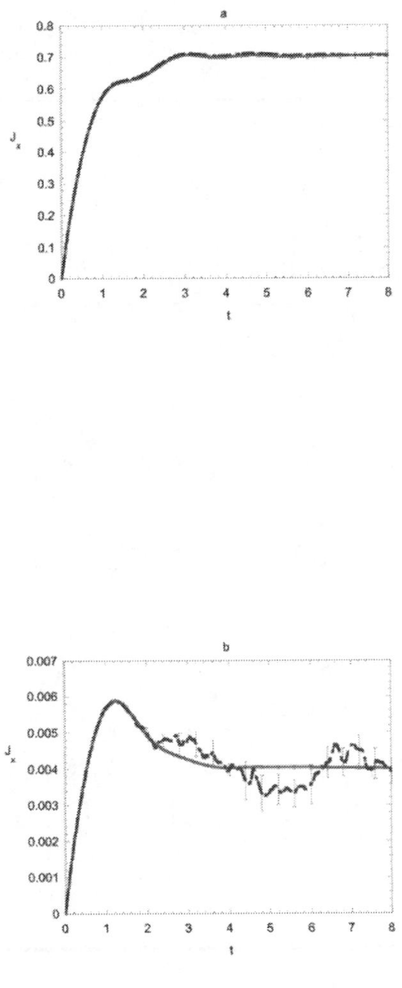


**Figure 3.** The response of $J_x$ to the application of a field determined using direct calculations (blue) and the TTCF form of the DT (red). The systems studied were at a temperature of $T = 1$ and an atom density $n = 0.8$ in the fluid and walls. In a) a field of $F_e = 1$ is used. In this case the agreement between the two approaches is so high that the two lines are superimposed over most of the trajectory. Error bars are shown and are the standard error over 10 sets of data, but are mostly smaller than the line-width. In b) a field of $F_e = 0.01$ is used. Error bars are shown and are the standard error over 10 sets of data. The error bars are mostly smaller than the line-width for the DT results.

we employed a Nosé-Hoover thermostat the kinetic temperature of the thermostatted particles would have been replaced by the target temperature in the Nosé-Hoover feedback mechanism.

Journal of Physics: Conference Series **297** (2011) 012017 doi:10.1088/1742-6596/297/1/012017

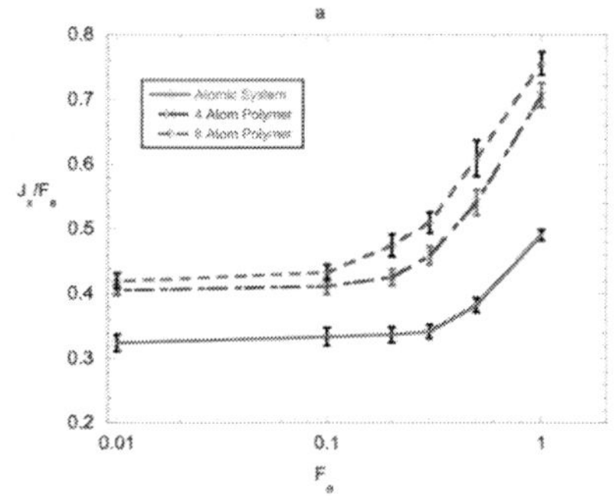


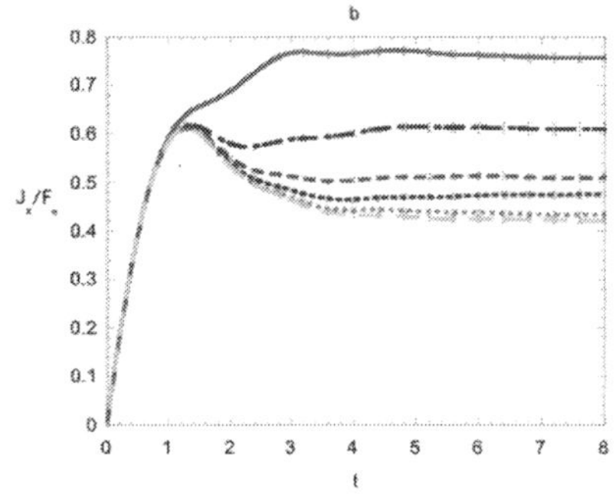


**Figure 4.** a) The value of $J_x/F_e$ as a function of the field for a range of fields between $F_e = 0.01$ and $F_e = 1$ for the atomic fluid, 4-atom polymer fluid and 8-atom polymer fluid. The systems studied were at a temperature of $T = 1$ and an atom density $n = 0.8$ in the fluid and walls. Error bars are shown for all fields and are the standard error over 10 sets of data. The lines provide guides to the eye. b) The full response curve for the 8-atom polymer under the same conditions. Field range from $F_e = 0.01$ (light blue), $F_e = 0.1$ (pink), $F_e = 0.2$ (black), $F_e = 0.3$ (green), $F_e = 0.5$ (blue) to $F_e = 1$ (red).

The kinetic temperature of the thermostatted particles would not necessarily be the same as the corresponding temperature for the isokinetic system. The actual temperature appearing

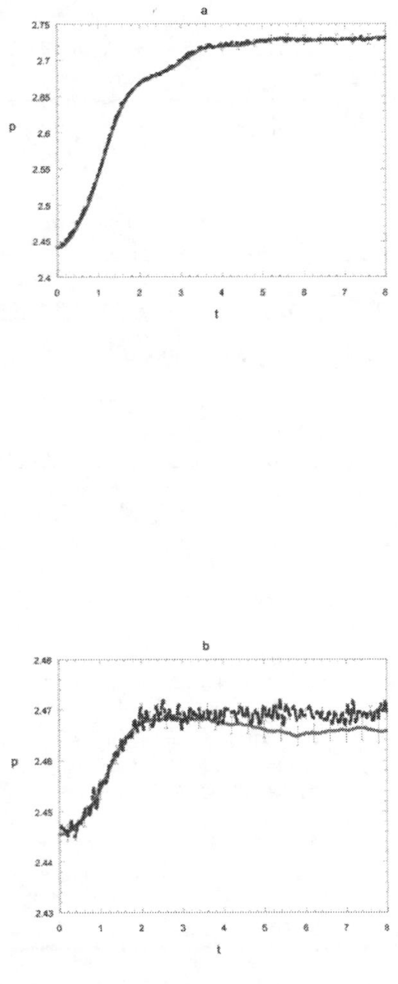


**Figure 5.** The response of the average Irving-Kirkwood pressure to the application fields: a) $F_e = 1$ and b) $F_e = 0.3$ determined using a direct calculation (blue) and the TTCF form of the DT (red). The systems studied were at a temperature of $T = 1$ and an atom density $n = 0.8$ in the fluid and walls. Error bars are shown and are the standard error over 10 sets of data.

in the DT is in fact always equal to the equilibrium thermodynamic temperature the system would relax to if the driving field was set to zero and the entire system was allowed to relax to equilibrium as is proved in the Relaxation Theorem [20]. The dissipation function has also been applied to calculate the current in both atom and polymeric systems using the DT, validating this approach for wall-thermostatted systems for the first time. The results showed that as the size of the molecule increase from a single atom to a 4-atom polymer there is an increase in the current with a smaller increase from the 4-atom to the 8-atom polymer. Furthermore the dependency on molecular size was also noted when the atomic system entered the linear regime at larger applied field. As well as providing accurate results over the range of fields considered, the DT approach for calculating averaged properties of systems was clearly advantageous for small fields. For complex molecules where the nonlinear regime occurs are relatively low fields, it will become essential to use the TTCF form of the DT instead of the direct calculations in order to determine properties accurately.

### Acknowledgments
The authors would like to thank Griffith University for computational time on their V20z Cluster and support through ARC LIEF and Discovery Grants.

### References
[1] Evans D J and Searles D J 1994 *Phys. Rev. E* **50** 1645
[2] Evans D J and Searles D J 2002 *Adv. Phys.* **51** 1529
[3] Sevick E M, Prabhakar R, Williams S R and Searles D J 2008 *An. Rev. Phys. Chem.* **59** 603
[4] Evans D J, Willams S R and Searles D J 2010 *Thermodynamics of Small Systems, in Nonlinear Dynamics of Nanosystems* Editors Radons G, Schuster H G and Rumpf B (Weinheim: Wiley-VCH), 75-109
[5] Evans D J, Searles D J and Willams S R 2008 *J. Chem. Phys.* **128** 014504
[6] Searles D J and Evans D J 2000 *J. Chem. Phys.* **113** 3503.
[7] Evans D J and Morriss G P 2008, *Statistical Mechanics of Nonequilibrium Liquids 2nd Ed.* (Cambridge: Cambridge University Press)
[8] We note that in some cases, although the dissipation function can be formally written, it is difficult to test experimentally. In these cases proposal of a similar, experimentally measurable quantity is justified.
[9] Ayton G S and Evans D J 1999 *J. Stat. Phys.* **97** 811
[10] Ayton G, Evans D J and Searles D J 2001 *J. Chem. Phys.* **115**, 2033
[11] Kröger M 2004 *Phys. Rep.* **390** 453
[12] Weeks J D, Chandler D and Andersen H C 1971 *J. Chem. Phys.* **54** 5237
[13] Heyes D M and Okumura H 2006 *Mol. Sim.* **32** 45
[14] Grest G S and Kremer K 1986 *Phys. Rev. A* **33** 3628
[15] Kremer K and Grest G S 1990 *J. Chem. Phys.* **92** 5057
[16] Bosko J T, Todd B D and Sadus R J 2004 *J. Chem. Phys.* **121** 1091
[17] Evans D J, Williams S R and Searles D J 2011 "A proof of Clausius's Theorem for time reversible deterministic microscopic dynamics" *J. Chem. Phys.* submitted
[18] Cui S T, McCabe C, Cummings P T and Cochran H D *J. Chem. Phys.* **118** 8941
[19] The Irving-Kirkwood expression for the pressure is simply used as a sample phase variable. It should be noted that if an accurate measurement of the pressure is required, the method of planes should be used. See Todd B D and Evans D J 1997 *Mol. Sim.* **17** 317
[20] Evans D J, Searles D J and Williams S R 2009 *J. Stat. Mech.* **2009** P07029

Journal of Physics: Conference Series **297** (2011) 012018 doi:10.1088/1742-6596/297/1/012018

# Simulating Quantum Dynamics with Entanglement Mean Field Theory

**Aditi Sen(De) and Ujjwal Sen**

Harish-Chandra Research Institute, Chhatnag Road, Jhunsi, Allahabad 211 019, India

E-mail: aditi@hri.res.in, ujjwal@hri.res.in

**Abstract.**

Exactly solvable many-body systems are few and far between, and the utility of approximate methods cannot be overestimated. Entanglement mean field theory is an approximate method to handle such systems. While mean field theories reduce the many-body system to an effective single-body one, entanglement mean field theory reduces it to a two-body system. And in contrast to mean field theories where the self-consistency equations are in terms of single-site physical parameters, those in entanglement mean field theory are in terms of both single- and two-site parameters. Hitherto, the theory has been applied to predict properties of the static states, like ground and thermal states, of many-body systems. Here we give a method to employ it to predict properties of time-evolved states. The predictions are then compared with known results of paradigmatic spin Hamiltonians.

## 1. Introduction

Strongly interacting systems, apart from dealing with the properties of magnetic materials, have been enriched by several important discoveries during the last decades including high temperature superconductivity, quantum Hall effects, etc. A major breakthrough in many-body physics came with the experimental realizations of such systems artificially, for example by using ultracold atoms in optical lattices and ion traps [1]. On the other hand, quantum information science have demonstrated an enormous utility in applications in different disciplines, ranging from communication processes to computer science [2]. In the last few years, a strong connection between quantum information and many-body physics has been developed [1, 3]. This includes the study of fundamental properties of many-body systems from the perspective of quantum information, and also the research aimed at an implementation of a quantum computer and other quantum information processing devices in realizable many-body systems, e.g. in ultracold atoms and ions.

One of the main obstacles in such investigations is the lack of analytical methods for handling many-body systems. Indeed only a few interacting many-body systems can be exactly diagonalized [4, 5]. Approximate and numerical methods therefore play a crucial role in studying many-body physics [6]. There are several approximate methods that have been successfully used to evaluate, characterize, and understand the properties of model many-body systems. Among them, the mean field theory (MFT), first introduced by P. Weiss in 1907, and several improvements of that theory, have become an important method for understanding many-body physics, including phase diagrams, critical exponents, etc. Typically, a mean field theory reduces

Journal of Physics: Conference Series **297** (2011) 012018 doi:10.1088/1742-6596/297/1/012018

the parent many-body Hamiltonian into a single-body one, with the latter containing single-site physical parameters as undetermined variables. These undetermined single-site parameters are subsequently obtained by solving certain self-consistency equations, and a distinguishing property of mean field-like theories is a (or a set of) self-consistency equations written in terms of single-site physical parameters, like magnetization [7, 8, 9, 10, 11, 6]. Such mean field theories, although useful in several applications, are not able to properly capture the two-body properties (of the parent many-body Hamiltonian) – essential ingredients in quantum information science [2].

Entanglement mean field theory (EMFT) is an approximate method for studying two-site as well as single-site physical quantities, and critical phenomena of many-body systems [12]. A characteristic trait of EMFT is that it deals with self-consistency equations in terms of two-body physical properties, in contrast to the self-consistency equations in terms of single-site quantities in mean-field theories. The effort required to solve the consistency equation in EMFT is almost or exactly the same as in MFT. Previous works have demonstrated that EMFT can faithfully signal the onset of quantum fluctuations-driven phase transitions as well as temperature-driven phase transitions in a large variety of paradigmatic quantum many-body models. Moreover, the two-site as well as single site physical parameters as predicted by EMFT are qualitatively as well as quantitatively better than the MFT predicted ones. In Ref. [13], an EMFT-inspired theory was applied to classical many-body systems, and it was shown that the critical points predicted by EMFT are significantly better than the MFT-predicted values. EMFT was generalized to a cluster EMFT in Ref. [13], and was shown to provide a significant improvement over the predictions from cluster MFT.

However, both the previous works utilizes EMFT to understand the properties of *static* systems, viz., the ground state or the equilibrium states of time-independent Hamiltonians. Here we apply EMFT to study the the dynamics of a many-body system, as it evolves in time, where the evolution is described by a time-dependent Hamiltonian. In particular, the EMF method is applied to explore the properties of entanglement in the dynamics of the nearest neighbor quantum transverse Ising model. The recipe for the quantum Ising model can also be applied to other interacting spin systems. Exact results are few and far between for many-body systems, even in one-dimension (1D). However, in case of the nearest neighbor one-dimensional quantum transverse Ising model, an exact analytical treatment is possible for the dynamics (and statics), and several important physical parameters, including two-site ones can be exactly obtained in closed form (see Refs. [14, 15] and references therein). It is therefore possible to compare the EMFT predictions in this case with the exact results, and we do so for the nearest neighbor entanglement in this system. Previous work had shown that the EMFT prediction for the nearest neighbor entanglement matches well with the general behavior obtained from exact calculations, in the case of the ground state and thermal states of this model [12]. Here, we make the comparison for the time-evolved state.

## 2. The model

Consider the ferromagnetic quantum transverse Ising model (TIM) with nearest neighbor interactions on a lattice of arbitrary dimension and arbitrary geometry. The model is described by the Hamiltonian

$$H_{TIM} = -J \sum_{\langle \vec{i}\vec{j} \rangle} \sigma_x^{\vec{i}} \sigma_x^{\vec{j}} - h(t) \sum_{\vec{i}} \sigma_z^{\vec{i}}, \tag{1}$$

where $\sigma_i, i = x, y, z$ are the Pauli spin-1/2 matrices, $J > 0$ is the coupling strength, and $h(t)$ is a time-dependent transverse magnetic field. We will be interested in the time-evolution of the system, for which we assume that the initial state is the canonical equilibrium state of the system at time $t = 0$. We will only consider cases where the Hamiltonian has nearest neighbor interactions, as indicated by the notation $\langle \vec{ij} \rangle$ in the first summation in Eq. (1), although the considerations here carry over to more general cases. Since the field considered is time-dependent, the ensuing dynamics will be non-trivial. In this paper, the time-dependence of the magnetic field is assumed to be as follows:

$$\begin{aligned} h(t) &= a, \quad t \leq 0, \\ &= 0, \quad t > 0, \end{aligned} \tag{2}$$

where $a \neq 0$. The time-evolution of the system will therefore be a response to the initial "kick" given to the sytem at zero time.

The one-dimensional version of $H_{TIM}$ can be exactly diagonalized by the Jordan-Wigner transformation [14, 15]. The exact spectrum can be determined. Furthermore, it is possible to obtain several physically relevent quantities of the evolved state, where the time evolution starts off from the canonical equilibrium state, for a wide range of time-dependent transverse fields $h$, including the field that has been considered here.

However, such investigations are not possible in higher dimensional systems. The ground states of even the time-independent versions of the same quantum model in higher dimensions, including square and triangular lattices in two-dimensions, are not known exactly. Therefore, approximate methods or numerical simulations play an important role to study such higher dimensional systems. Numerical simulations [16] include exact finite-size diagonalizations, density matrix renormalizations [17], quantum monte carlo [18], etc., while approximate methods include mean field-like theories [6, 7, 8, 9, 10, 11, 19, 20, 21, 22].

## 3. Entanglement Mean Field Theory

Entanglement mean field theory (EMFT) was introduced in [12] to study the behavior of bipartite entanglement as well as other two-body physical quantities of many-body systems. Hitherto, it has been applied to understand the static properties of such systems [12, 13]. Here we apply it to predict the dynamical properties and in this section, we present the corresponding consistency equation.

For definiteness, suppose that the spin model is decribed on a two-dimensional square lattice. EMFT begins by assuming that a certain pair of two neighboring spins are "special". Suppose now that the pair $\{(i,j),(i,j+1)\}$ of the two-dimensional square lattice is special. We will use the fact that the square of any Pauli matrix is unity. Therefore, the interaction term

$$\sigma_x^{i-1,j}\sigma_x^{i,j}$$

in the Hamiltonian in Eq. (1) can be written as

$$\sigma_x^{i-1,j}\sigma_x^{i,j}\sigma_x^{k,l+1}\sigma_x^{k,l+1},$$

which is equivalent to

$$(\sigma_x^{i-1,j}\sigma_x^{i,j+1})(\sigma_x^{i,j}\sigma_x^{i,j+1}).$$

We now make the entanglement mean field theory approximation by replacing the non-special (operator) pair

$$(\sigma_x^{i-1,j}\sigma_x^{i,j+1})$$

by its mean value $\langle\sigma_x^{i-1,j}\sigma_x^{i,j+1}\rangle/2$. The factor $1/2$ is due to the fact that only one of two spins in the non-special pair touches the special one. Similar approximations can be done for all the other interaction term in the Hamiltonian.

The total number of terms in the Hamiltonian that will have a non-trivial contribution to the self-consistency equation, to be written later, is $2\nu_E$ where the "EMFT coordination number", $\nu_E$, is defined as half of the number of bonds connecting to the special pair, in the Hamiltonian under consideration. As we will see, $\nu_E$ has a similar significance in EMFT, as the coordination number in MFT. Therefore, the EMFT-reduced Hamiltonain in the case of the transverse Ising model in Eq. (1) is

$$H_{TIM}^{EMFT} = -J\nu_E C\sigma_x^{\vec{i}}\sigma_x^{\vec{j}} - \frac{h(t)}{2}(\sigma_z^{\vec{i}} + \sigma_z^{\vec{j}}), \tag{3}$$

where we have assumed that $(\vec{i}, \vec{j})$ is the special pair. There are only two field terms that make non-trivial contributions to the self-consistency equation. The factor 1/2 in the field terms is again due to the fact that each of them only touches one of the special pair. The EMFT-reduced Hamiltonian includes the as-yet-undetermined average correlation

$$C = \langle \sigma_x^{\vec{k}}\sigma_x^{\vec{l}} \rangle,$$

where $\vec{k}$ and $\vec{l}$ are nearby lattice sites. Note that we have not as yet identified the quantum state in which the average is performed.

We will use EMFT to study the properties of the time-evolved state. For definiteness, we assume that the dynamics starts from the EMFT ground state, $|\psi(0)\rangle$, of the system at $t = 0$. Note that the Hamiltonian is time-dependent, without which there will be no non-trivial evolution if the system starts off from its ground state. The EMFT time-evolved state at time $t$ is given by

$$|\psi(t)\rangle = \exp(-iH_{TIM}^{EMFT}(t > 0)t/\hbar)|\psi(0)\rangle. \tag{4}$$

We now assume that the average correlation $C$ was performed for the state $|\psi(t)\rangle$. Therefore, for consistency, we must have

$$C = \langle \psi(t)|\sigma_x^{\vec{i}}\sigma_x^{\vec{j}}|\psi(t)\rangle, \tag{5}$$

and this is the *dynamical* EMFT self-consistency equation. Note that it is written in terms of a two-body physical quantity, in contrast to the consistency equations in any mean-field-like theory. We would like to emphasize here that this consistency equation is one of the main differences of EMF theory with the MF-like theories, where the self-consistency equation is in terms of single-site physical parameters, like magnetization.

## 4. Entanglement Measure: Logarithmic Negativity

As mentioned above, EMFT allows us to predict two-body physical parameters, and of them, we will be primarily interested in nearest-neighbor entanglement of many-body systems. There are several measures of entanglement that are known, and for definiteness, we consider an entanglement measure, called logarithmic negativity [23]. A quantum state, $\rho_{AB}$, of two distinguishable systems $A$ and $B$ is entangled [24] if it cannot be written as

$$\rho_{AB} = \sum_i p_i \rho_A^i \otimes \rho_B^i,$$

where $\rho_A^i$ and $\rho_B^i$ are quantum states of $A$ and $B$ respectively.

*The Peres-Horodecki criterion:* For two spin-1/2 systems, it is known that a quantum state is entangled if and only if there exists a negative eigenvalue of the partially transposed state $\rho_{AB}^{T_A}$ [25, 26], where the partial transposition is defined as follows.

Let us express $\rho_{AB}$ as

$$\rho_{AB} = \sum_{i,j} \sum_{\mu,\nu} a_{ij}^{\mu\nu} (|i\rangle\langle j|)_A \otimes (|\mu\rangle\langle\nu|)_B,$$

where $\{|i\rangle\}$ and $\{|\mu\rangle\}$ are respectively orthonormal bases of the systems at $A$ and $B$ respectively. Then the partial transposition of $\rho_{AB}$ is defined as

$$\rho_{AB}^{T_A} = \sum_{i,j} \sum_{\mu,\nu} a_{ij}^{\mu\nu} (|j\rangle\langle i|)_A \otimes (|\mu\rangle\langle\nu|)_B.$$

One can now define a measure of entanglement, called logarithmic negativity, for quantum states of two spin-1/2 systems, as

$$\mathcal{L}(\rho_{AB}) = \log_2(2N(\rho_{AB}) + 1),$$

where $N(\rho_{AB})$ is the sum of the moduli of the negative eigenvalues of $\rho_{AB}^{T_A}$. Note here that the Peres-Horodecki criterion is still a necessary and sufficient criterion for a state to be entangled in a system composed of one spin-1/2 and one spin-1, although not any longer in higher dimensions [27]. Irrespective of that, one can still define the logarithmic negativity for all such situations. However, we will not have occasion to consider cases other than two spin-1/2 particles.

## 5. EMFT prediction for entanglement of time-evolved state

It is possible to study single-site as well as two-site properties of the many-body system by solving the dynamical EMFT self-consistency equation in (5). One begins by solving Eq. (5) for $C$ for a certain chosen quantum state of the system, and inserting it into the EMFT-reduced Hamiltonian. Thereafter, the properties of that state can be obtained by using the EMFT-reduced Hamiltonian, which, after the insertion of the value of $C$, does not contain any undetermined parameter.

In particular, we can study the behavior of entanglement in the dynamical EMFT state. For the quantum TIM described on a lattice of arbitrary dimension and geometry, characterized by the EMFT coordination number $\nu_E$, we find the bipartite entanglement of the EMFT time-evolved state for which the initial state is the EMFT ground state. This is therefore to be compared with the nearest neighbor entanglement in the time-evolved state for which the initial state is the ground state of the parent Hamiltonian. In Fig. 1, we plot the bipartite entanglement (as quantified by logarithmic negativity) of the EMFT-reduced quantum transverse Ising model against the scaled time $\tau = t\nu_E J/\hbar$ and the scaled initial transverse field $\alpha = a/(\nu_E J)$, when the initial state is the ground state of the EMFT-reduced Hamiltonian.

## 6. Predictions for a chain: EMFT versus exact

The one-dimensional lattice corresponds to $\nu_E = 1$, and the EMFT prediction for bipartite entanglement in that lattice can be read off from the above considerations by substituting $\nu_E = 1$. As mentioned earlier, this 1D model is exactly solvable by the Jordan-Wigner transformation, and it is possible to obtain several single-site as well as two-site physical quantities of the time-dependent Hamiltonian considered here [14]. The single-site properties include magnetization, while two-site quantities include two-body correlations and entanglement. The nearest neighbor (bipartite) entanglement of the time-evolved state in this model, as obtained via Jordan-Wigner transformation, and where the initial state of the evolution is the zero-temperature state (of the parent Hamiltonian), is shown in Fig. 2. Note here that Fig. 2 shows that the bipartite entanglement either vanishes or changes its charateristic (from being convex to concave) at $a/J \approx 1$, irrespective of the time. This is expected as we know that this model undergoes a

Journal of Physics: Conference Series **297** (2011) 012018 doi:10.1088/1742-6596/297/1/012018

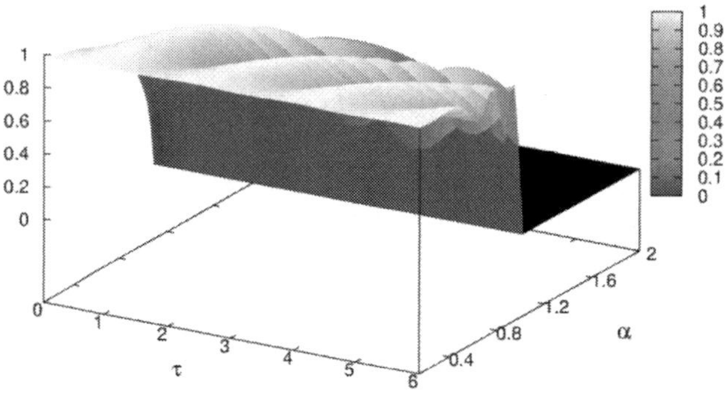


**Figure 1.** (Color online) Behavior of bipartite entanglement of the EMFT time-evolved state, with the initial state of evolution being the ground state of the EMFT-reduced quantum transverse Ising model. The vertical axis represents entanglement, measured in ebits, while the base axes represent the scaled time $\tau = t\nu_E J/\hbar$ and the scaled initial transverse magnetic field $\alpha = a/(\nu_E J)$. The parameters of both the base axes are dimensionless.

quantum phase transition at zero temperature when the (scaled) transverse magnetic field $a/J$ is unity [28]. Compare this with Fig. 1 for $\nu_E = 1$, where we find that the bipartite entanglement as predicted by the EMFT-reduced Hamiltonian also indicates a change of behavior at $a/J = 1$.

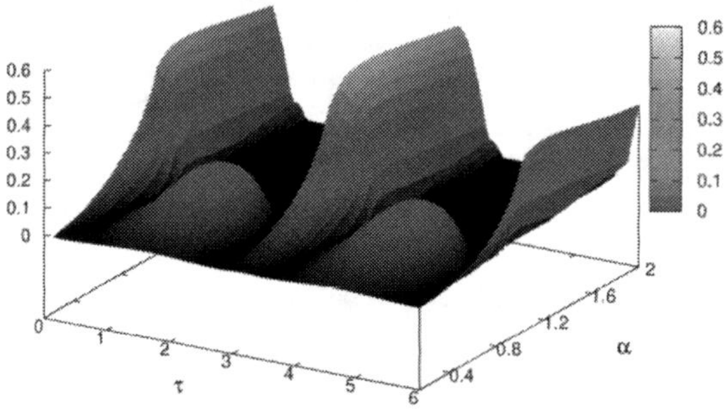


**Figure 2.** (Color online) Nearest-neighbor entanglement of the time-evolved state as obtained in exact calculations (via Jordan-Wigner), with the initial state of the evolution being the zero-temperature state of the transverse Ising model. Entanglement (measured in ebits) is represented along the vertical axis, while the base axes represent the dimensionless variables $\tau = tJ/\hbar$ and $\alpha = a/J$. Compare the entanglement plotted here and that predicted in EMFT in Fig. 1 (for $\nu_E = 1$).

To make the comparison more objective, let us now concentrate on the temporal behavior of entanglement in this model as predicted by EMFT and in the exact case. Specifically, we study the logarithmic negativity of the time-evolved state considered above, for a fixed initial disturbance, with respect to time. In Fig. 2, we see that the nearest neighbor entanglement, obtained via Jordan-Wigner transformation, shows collapses and revivals with respect to time, if we consider a fixed value of the initial transverse field. Interestingly, for $\alpha \leq 1$, the bipartite entanglement of the corresponding EMFT time-evolved state also shows crests and troughs with respect to time for a fixed initial transverse field. In Fig. 3, we plot sections of Fig. 1 and Fig. 2 at an exemplary field parameter $\alpha = 0.594$, and see that the numbers of crests and troughs are equal in the two cases.

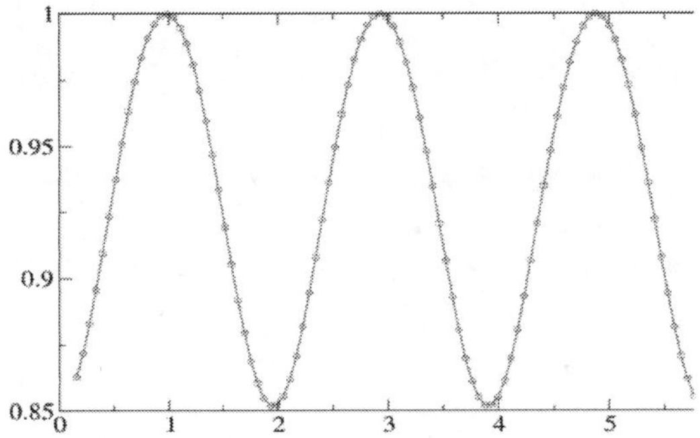


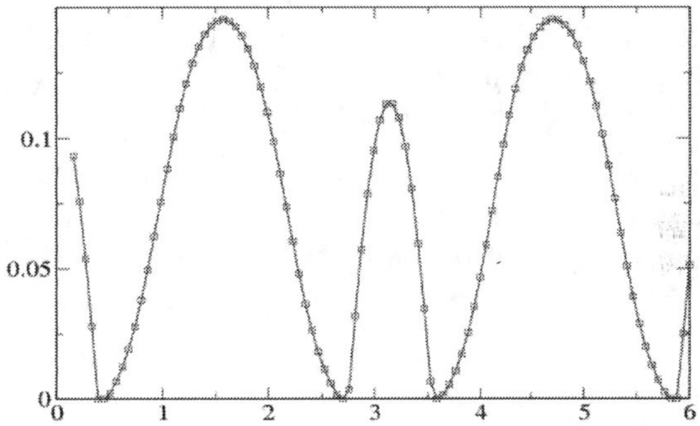


**Figure 3.** One dimensional transverse Ising model: Comparison of temporal behavior of bipartite entanglement of the time-evolved state in EMFT with exact results. The vertical and horizontal axes in both the plots are respectively entanglement (measured in ebits) and the scaled time $\tau = tJ/\hbar$ (dimensionless). Both the plots are for the fixed initial field $\alpha = a/J = 0.594$. The left plot is the EMFT prediction, while the right one is the exact result as obtained via Jordan-Wigner transformation. Both the plots show crests and troughs. More quantitatively, the numbers of crests and troughs are the same in the two cases.

## 7. Conclusion

Entanglement mean field theory is an approximate method to deal with many-body systems. Its main difference with the mean field-like theories is in the self-consistency equations. While the self-consistency equations in mean field-like theories deal with single-site parameters like magnetization, those in entanglement mean field theory deals with single- as well as two-site physical parameters. Using the latter, it is possible to predict two-body correlations as well as entanglement of many-body systems. While previous studies used entanglement mean field theory to predict properties of static states of many-body systems, here we have shown the method to employ it for dynamical states. For definiteness, we have considered the paradigmatic quantum transverse Ising model on a lattice as our many-body system. We have shown that the method can predict the behavior of single- as well as two-site properties of the evolved state. In particular, the behavior of entanglement in the dynamics could be predicted by using the entanglement mean feld theory. In the one-dimensional case, we have compared the behavior of two-site entanglement as predicted by EMFT with exact results obtained via Jordan-Wigner transformation.

## References

[1] Lewenstein M, Sanpera A, Ahufinger V, Damski B, Sen(De) A, and Sen U 2007, *Adv. Phys.* **56** 243

[2] Nielsen M A and Chuang I 2000, *Quantum Computation and Information*, (Cambridge: Cambridge University Press)
[3] Amico L, Fazio R, Osterloh A, and Vedral V 2008, *Rev. Mod. Phys.* **80** 517
[4] Baxter R J 1982, *Exactly Solved Models in Statistical Mechanics*, (New York: Dover)
[5] Sutherland B 2004, *Beautiful Models: 70 Years Of Exactly Solved Quantum Many-Body Problems*, (Singapore: World Scientific)
[6] Mahan G D 2000, *Many-Particle Physics*, (New York: Kluwer/Plenum)
[7] Huang K 1963, *Statistical Mechanics*, (New York: Wiley)
[8] Ziman J M 1972, *Principles of the Theory of Solids*, (Cambridge: Cambridge University Press)
[9] Pathria R K 1972, *Statistical Mechanics*, (New York: Pergamon)
[10] Ashcroft N W and and Mermin N D 1976, *Solid State Physics*, (New York: Holt, Rinehert and Winston)
[11] Chakrabarti B K, Dutta A, and Sen P 1996, *Quantum Ising Phases and Transitions in Transverse Ising Models (Lecture Notes in Physics Monographs)*, (Berlin Heidelberg: Springer)
[12] Sen(De) A and Sen U 2009, arXiv:0911.4856
[13] Sen(De) A and Sen U 2010, arXiv:1008.0965
[14] Barouch E, McCoy B M, and Dresden M 1971, *Phys. Rev. A* **2** 1075
[15] Barouch E and McCoy B M 1971, *Phys. Rev. A* **3** 786
[16] See e.g. Fehske H, Schneider R, Weiße A (Eds.), *Computational Many-Particle Physics*, Lect. Notes. Phys. **739**, (Berlin Heidelberg: Springer)
[17] Schollwöck U 2005, *Rev. Mod. Phys.* **77** 259
[18] Foulkes W M C, Mitas L, Needs R J, and Rajagopal G 2001, *Rev. Mod. Phys.* **73** 33
[19] Stinchcombe R B 1973, *J. Phys. C: Solid State Phys.* **6** 2459
[20] Van Vleck J H 1937, *Phys. Rev.* **52** 1178
[21] Oguchi T 1955, *Prog. Theor. Phys.* **13**, 148
[22] Fisher M A 1998, *Rev. Mod. Phys.* **70** 653
[23] Vidal G and Werner R F 1999, *Phys. Rev. A* **65** 032314
[24] Werner R F 1989, *Phys. Rev. A* **40** 4977
[25] Peres A 1997, *Phys. Rev. Lett.* **77** 1413
[26] Horodecki R, Horodecki P and Horodecki M 1998, *Phys. Lett. A* **223** 1
[27] Horodecki R, Horodecki P, Horodecki M, and Horodecki K 2009, *Rev. Mod. Phys.* **81** 865
[28] Sachdev S 1999, *Quantum Phase Transitions*, (Cambridge: Cambridge University Press)

Journal of Physics: Conference Series **297** (2011) 012019 doi:10.1088/1742-6596/297/1/012019

# First-Principle Derivation of Entropy Production in Transport Phenomena

**Masuo Suzuki**

Tokyo University of Science Kagurazaka 1-3, Shinjuku, Tokyo, 162-8601

E-mail: msuzuki@rs.kagu.tus.ac.jp

**Abstract.**

The linear response framework was established by Kubo a half century ago, but no clear explanation of irreversibility namely entropy production has been given in this scheme. This has been now solved. The serious puzzle up to now is the following. Even using the linear response density matrix $\rho_{\rm lr} = \rho_0 + \rho_1(t)$, it has been difficult to derive the entropy production.

Surprisingly, the correct entropy production is given by the *second-order* term $\rho_2(t)$ as $(dS/dt)_{\rm irr} = (1/T){\rm Tr}\mathcal{H}_0\rho_2'(t) > 0$. It is shown to agree with the ordinary expression $\boldsymbol{J}\cdot\boldsymbol{E}/T = \sigma E^2/T$ in the case of electric conduction for a static electric field $\boldsymbol{E}$, where $\sigma$ denotes the electric conductivity expressed by the famous canonical current-current time correlation functions in equilibrium.

The present article gives a review of the derivation of entropy production (M.S., Physica A 390(2011)1904-1916) based on the first-principle of using the projected density matrix $\rho_2(t)$ or more generally $\rho_{\rm even}(t)$, while the previous standard argument is due to the thermodynamic energy balance. This new derivation clarifies conceptually the physics of irreversibility in transport phenomena. In general, the transport phenomena are described by the odd part $\rho_{\rm odd}(t)$ of the density matrix and the entropy production (namely irreversibility) is described by the even part $\rho_{\rm even}(t)$. These are related to each other through the coupled equations as $i\hbar d\rho_q(t)/dt = [\mathcal{H}_0, \rho_q(t)] + [\mathcal{H}_1(t), (1-P_q)\rho(t)]$, where "$q$" denotes "even" (symmetric) or "odd" (antisymmetric), $P_q$ is the projection operator to the "$q$" part of $\rho(t)$, and $\mathcal{H}_1(t)$ denotes the partial Hamiltonian due to the external force such as $\mathcal{H}_1(t) = -e\sum_j \boldsymbol{r}_j\cdot\boldsymbol{E} \equiv -\boldsymbol{A}\cdot\boldsymbol{E}$ in electric conduction.

The concept of a stationary temperature $T_{\rm st}$ in steady states with current (say electric current) is also proposed by using the projected and symmetry-separated von Newmann equation introduced by the present author. The entropy production of the relevant steady state depends on this stationary temperature. A mechanical formulation of thermal conduction is given by introducing a thermal field $\boldsymbol{E}_{\rm T}$ and its conjugate "heat" operator $\boldsymbol{A}_{\rm H} = \sum_j h_j\boldsymbol{r}_j$ for a local internal energy $h_j$ of the thermal particle $j$.

## 1. Introduction

The present paper gives a review of the theory on irreversibility and entropy production in transport phenomena[1]. The problems of transport phenomena and irreversibility have been studied for long years from many view points[1-25]. Boltzmann's $H$-theorem shows the first attempt to explain the broken symmetry of entropy change in time, namely irreversibility. Green[2] formulated transport coefficients using stochastic equations including the time change of microscopic fluctuations[6-9]. Kubo and Tomita[11] succeeded, for the first time, in formulating magnetic linear responses on the basis of the von Neumann equation. Nakano[12, 13] applied

Journal of Physics: Conference Series **297** (2011) 012019 doi:10.1088/1742-6596/297/1/012019

Kubo-Tomita's method to the problem of electric conduction. Kubo[4] established a general theory of linear responses including the magnetic response and electric conduction.

## 2. Generalized linear response scheme including thermal conductance

To study transport phenomena, we start with the following von Neumann equation for the density matrix $\rho(t)$:

$$i\hbar\frac{\partial}{\partial t}\rho(t) = [\mathcal{H}(t), \rho(t)], \tag{1}$$

where the total Hamiltonian is given by

$$\mathcal{H}(t) = \mathcal{H}_0 + \mathcal{H}_1(t); \quad \mathcal{H}_1(t) = -\boldsymbol{A}\cdot\boldsymbol{F}(t). \tag{2}$$

In Kubo's original formulation [4], the above Hamiltonian $\mathcal{H}(t)$ is purely mechanical. For example, we have

$$\boldsymbol{A} = e\sum_j \boldsymbol{r}_j \quad \text{and} \quad F(t) = \boldsymbol{E} \tag{3}$$

in static electric conduction.

Now we interpret the von Neumann equation (1) so that we may treat thermal conductivity as in Kubo's paper[4]. For this purpose, the Hamiltonian for thermal disturbance should be regarded to be an effective one renormalized in a local but enough large region to catch up the temperature gradient. We assume local equilibrium to define the semi-local temperature $T(\boldsymbol{r})$. Then, as has been introduced in Paper II (Ref.[2]), we define a thermal field $\boldsymbol{E}_{\mathrm{T}} = \nabla\beta(\boldsymbol{r})/\beta_0$ as an external field $\boldsymbol{F}(t)$ in Eq.(2), where $\beta(\boldsymbol{r}) = 1/k_{\mathrm{B}}T(\boldsymbol{r})$ and $\beta_0$ denotes the characteristic $\beta$. Such a renormalized particle as moves receiving a force from the field $\boldsymbol{E}_{\mathrm{T}}$ is called "a thermal particle". If we define the heat operator $\boldsymbol{A}_{\mathrm{H}}$ of the form

$$\boldsymbol{A}_{\mathrm{H}} = \sum_j \{h_j\boldsymbol{r}_j\} \equiv \frac{1}{2}\sum_j (h_j\boldsymbol{r}_j + \boldsymbol{r}_j h_j) \tag{4}$$

when the unperturbed Hamiltonian of the system is given in the form

$$\mathcal{H}_0 = \sum_j h_j; \quad h_j = \epsilon_j + u_j \quad \text{with} \quad \epsilon_j = \frac{1}{2m^*}\boldsymbol{p}_j^2 \quad \text{and} \quad u_j = u(\boldsymbol{r}_j, \{\boldsymbol{r}_k\}), \tag{5}$$

then, from the relation

$$\dot{\boldsymbol{A}}_{\mathrm{H}} \equiv \frac{1}{i\hbar}[\boldsymbol{A}_{\mathrm{H}}, \mathcal{H}_0] \equiv \boldsymbol{j}_{\mathrm{H}}, \tag{6}$$

the heat current is given[2] in the form

$$\boldsymbol{j}_{\mathrm{H}} = \sum_j \{h_j\boldsymbol{v}_j\} + \frac{1}{2i\hbar}\sum_{i,j}\{(\boldsymbol{r}_j - \boldsymbol{r}_k)[u_j(\boldsymbol{r}_j, \{\boldsymbol{r}_k\}), \epsilon_k - \epsilon_j]\}. \tag{7}$$

Now we expand $\rho(t)$ in the von Neumann equation (1) as

$$\rho_{\mathrm{lr}}(t) = \rho_0 + \rho_1(t); \quad \rho_0 = e^{-\beta\mathcal{H}_0}/Z_0(\beta) \tag{8}$$

with $Z_0(\beta) = \mathrm{Tr}\exp(-\beta\mathcal{H}_0)$, up to the first order of an external force $\boldsymbol{F}$. Then, the first-order density matrix $\rho_1(t)$ is expressed as

$$\rho_1(t) = \rho_0\int_{t_0}^{t} ds \int_0^{\beta} d\lambda \boldsymbol{F}(s)\cdot\dot{\boldsymbol{A}}(s - t - i\hbar\lambda), \tag{9}$$

where $\dot{\boldsymbol{A}}$ denotes the current $\boldsymbol{j}$ of $\boldsymbol{A}$, namely $\boldsymbol{j} = \dot{\boldsymbol{A}} = [\boldsymbol{A}, \mathcal{H}_0]/i\hbar$. Thus, we obtain the well-known formula[4, 12]

$$\boldsymbol{J} \equiv \langle \boldsymbol{j} \rangle_t = \mathrm{Tr}\left\{(\rho_0 + \rho_1(t))\,\boldsymbol{j}\right\} = \int_0^\infty ds \int_0^\beta d\lambda e^{-\epsilon s} \langle \boldsymbol{j}\boldsymbol{j}(s + i\hbar\lambda)\rangle_0 \boldsymbol{F}(t-s) \tag{10}$$

by taking the limit $t_0 \to -\infty$ for the initial $t_0$ after the thermodynamic limit. The order of these two limiting procedures and the insertion of the adiabatic factor $e^{-\epsilon s}$ are vital to assure the non-vanishing value of the current, namely to realize the irreversibility in transport phenomena, as was emphasized in [1].

In particular, for the static electric conduction $\boldsymbol{F}(t) = \boldsymbol{E}$(constant), we have

$$\boldsymbol{J} = \sigma \boldsymbol{E}; \quad \sigma = \int_0^\infty dt \int_0^\beta d\lambda e^{-\epsilon t} \langle \boldsymbol{j}\boldsymbol{j}(t + i\hbar\lambda)\rangle_0 \tag{11}$$

with the current $\boldsymbol{j} = \dot{\boldsymbol{A}}$ and $\boldsymbol{A} = e\sum_j \boldsymbol{r}_j$. The finiteness ($\sigma > 0$) of this expression is essential to prove the positivity of the entropy production (namely irreversibility) in electric conduction from the first-principle based on the von Neumann equation.

The present formulation of thermal conductivity is similar to Luttinger's theory[26] based on a gravitational field which causes energy or heat current to flow. However, our formulation of introducing "thermal field" $\boldsymbol{E}_\mathrm{T}$ and its conjugate heat operator $\boldsymbol{A}_\mathrm{H}$ through the idea of renormalization of semi-local thermal effects is more transparent for deriving transport coefficients of the Kubo form. (In fact, some approximations among interactions are included in Luttinger's formulation.)

## 3. Brief comments on the energy dissipation and entropy production in the linear response scheme

The entropy production is a very subtle problem to study even in the linear response scheme. The ordinary formula $S(t) = -k_\mathrm{B}\mathrm{Tr}\rho(t)\log\rho(t)$ is useless for our purpose. There have been proposed many arguments on the entropy production in the linear response scheme.

### *3.1. Energy-balance argument*

The thermodynamic derivation of energy dissipation based on energy balance in a steady state is standard to give the correct energy dissipation $\boldsymbol{J}\cdot\boldsymbol{E} = \sigma E^2$. However, in order to realize this steady-state situation in which the system is stationary, namely the temperature $T$ is constant, we have to modify the von Neumann equation (1) by introducing an appropriate interaction with the heat reservoir and boundary conditions, as will be given later.

### *3.2. Linear-response approximation*

A primitive try to derive entropy production from $S(t) = -k_\mathrm{B}\mathrm{Tr}\rho(t)\log\rho(t)$ approximating $\rho(t)$ by the linear density matrix $\rho_\mathrm{lr}(t)$ is expected to give a correct expression of entropy change from

$$S_\mathrm{lr}(t) = -k_\mathrm{B}\mathrm{Tr}(\rho_0 + \rho_1(t))\log(\rho_0 + \rho_1(t)), \tag{12}$$

but the time derivative of this entropy calculated up to the second order of the electric field $\boldsymbol{E}$ is not equal to $\sigma E^2/T$.

### *3.3. Negative-sign in Zubarev's theory*

Zubarev extended the linear response scheme to a non-linear regime by introducing a non-equilibrium statistical operator of the form[14]

$$\begin{aligned}\rho_{\mathrm{Zub}}(t) &= \exp\left(-\Phi(t) - \beta\mathcal{H}_0 + \beta\int_{t_0}^{t} ds e^{\epsilon(s-t)} \boldsymbol{j}(s-t)\cdot\boldsymbol{E}\right) \\ &\simeq \rho_{\mathrm{lr}}(t) + \cdots = \rho_0 + \rho_1(t) + \cdots \end{aligned} \tag{13}$$

in the case of static electric conduction. Zubarev defined the entropy $S_{\mathrm{Zub}}(t)$ by

$$S_{\mathrm{Zub}}(t) = -k_{\mathrm{B}}\mathrm{Tr}\rho_{\mathrm{Zub}}(t)\log\rho_{\mathrm{loc}}, \tag{14}$$

where the local equilibrium density matrix $\rho_{\mathrm{loc}}$ is given by

$$\rho_{\mathrm{loc}} = e^{-\beta(\mathcal{H}_0+\mathcal{H}_1)}/Z(\beta); \quad Z(\beta) = \mathrm{Tr}e^{-\beta(\mathcal{H}_0+\mathcal{H}_1)} \tag{15}$$

in the present problem with $\mathcal{H}_1 = -\boldsymbol{A}\cdot\boldsymbol{E}$ in static electric conduction. Then, his entropy production is expressed in the form

$$\frac{d}{dt}S_{\mathrm{Zub}}^{(\mathrm{lr})}(t) = \frac{1}{T}\mathrm{Tr}(\mathcal{H}_0+\mathcal{H}_1)\rho'_{\mathrm{Zub}}(t). \tag{16}$$

This is reduced to

$$\frac{d}{dt}S_{\mathrm{Zub}}^{(\mathrm{lr})}(t) = \frac{1}{T}\mathrm{Tr}\mathcal{H}_1\rho'_1(t), \tag{17}$$

in the limit of the linear response scheme, which gives the negative value $-\sigma E^2/T$.

### *3.4. On the relevance of relative entropy*

The relative entropy $S_{\mathrm{rel}}(t)$ defined by

$$S_{\mathrm{rel}} = -k_{\mathrm{B}}\mathrm{Tr}\rho(t)\left(\log\rho(t) - \log\rho_{\mathrm{loc}}\right) \tag{18}$$

has been often used to discuss the entropy change. This time-derivative is given by

$$\frac{d}{dt}S_{\mathrm{rel}}(t) = k_{\mathrm{B}}\frac{d}{dt}\left(\mathrm{Tr}\rho(t)\log\rho_{\mathrm{loc}}\right), \tag{19}$$

and consequently we have

$$\frac{d}{dt}S_{\mathrm{rel}}^{(\mathrm{lr})}(t) = -\frac{1}{T}\mathrm{Tr}\mathcal{H}_1\rho'_1(t) \tag{20}$$

in the linear response scheme. This has a correct sign in contrast to Zubarev's result. However, the usage of the definition (18) is incorrect, because Eq.(19) always vanishes if we include higher-order terms of $\rho(t)$.

### *3.5. Variational arguments*

Onsagar and Machlup[16] discussed phenomenologically linear transport phenomena, using the variational theory concerning the bilinear form of the energy dissipation with respect to the current. Later, Nakano[12] developed a variational theory of energy dissipation based on his microscopic expression of conductivity. These phenomenological theories are, from the beginning, based on the correct expression of the entropy production.

Journal of Physics: Conference Series **297** (2011) 012019 doi:10.1088/1742-6596/297/1/012019

## 4. Derivation of entropy production by the symmetry separation of density matrix

In order to clarify the essence of the entropy production in transport phenomena, we concentrate here our arguments on a non-stationary process in static electric conduction[1] described by the Hamiltonian

$$\mathcal{H} = \mathcal{H}_0 + \mathcal{H}_1; \quad \mathcal{H}_1 = -\boldsymbol{A} \cdot \boldsymbol{E} \text{ and } \boldsymbol{A} = e\sum_j \boldsymbol{r}_j. \tag{21}$$

Here, $\mathcal{H}_0$ denotes the Hamiltonian of electrons together with impurity potentials to scatter electrons or with electron-phonon interactions. Thus, $\mathcal{H}_0$ is assumed here to be *symmetric* with respect to the inversion of space ($\boldsymbol{r} \to -\boldsymbol{r}$), while $\mathcal{H}_1$ is antisymmetric. Therefore, the average $\langle \mathcal{H}_0 \rangle_t \equiv \mathrm{Tr}\mathcal{H}_0\rho(t)$ can be regarded to be such an internal energy $U(t)$ as is identified as the heat energy of the relevant system, for the initial state $\rho_0$ is given by the equilibrium density matrix using $\mathcal{H}_0$ without electric field[1].

Physically, the internal energy $U(t)$ of the system increases owing to the applied electric field, that is, the system heats up in the present non-stationary situation. Thus, the temperature $T$ of the system depends on time $t$ as $T = T(t)$. This is the essence of irreversibility in the electric conduction phenomenon. The entropy change $dS$ is thermodynamically given by $dS = dU/T$ using the internal energy change $dU$. Correspondingly, the entropy production $(dS/dt)_{\mathrm{irr}}$ intrinsically produced inside the system by the generation of heat is defined[1], in our microscopic formulation, by

$$\left(\frac{dS}{dt}\right)_{\mathrm{irr}} = \frac{1}{T}\frac{dU(t)}{dt}, \tag{22}$$

and

$$U(t) = \mathrm{Tr}\mathcal{H}(t)P_{\mathrm{sym}}\rho(t) = \mathrm{Tr}(P_{\mathrm{sym}}\mathcal{H}(t))\rho(t) = \mathrm{Tr}\mathcal{H}_0\rho(t), \tag{23}$$

where $P_{\mathrm{sym}}$ denotes the projection operator to the symmetric (or even) part of the relevant operator with respect to the inversion of space or the external field $E$. The definition of $U(t)$ in terms of Eq.(22) (not defined by the total Hamiltonian) is essential for the following argument. It is also instructive to rewrite Eq.(22) in the form[1]

$$\left(\frac{dS}{dt}\right)_{\mathrm{irr}} = -k_{\mathrm{B}}\frac{d}{dt}\mathrm{Tr}\rho(t)P_{\mathrm{sym}}(\log\rho_{\mathrm{loc}}), \tag{24}$$

using $\rho_{\mathrm{loc}}$ given by Eq.(15).

In order to study the time dependence of $U(t)$ in Eq.(22), we expand the density matrix $\rho(t)$ as

$$\rho(t) = \rho_0 + \rho_1(t) + \rho_2(t) + \cdots + \rho_n(t) + \cdots. \tag{25}$$

From the von Neumann equation (1), we obtain the following hierarchical equations

$$i\hbar\frac{\partial}{\partial t}\rho_n(t) = [\mathcal{H}_0, \rho_n(t)] + [\mathcal{H}_1(t), \rho_{n-1}(t)]. \tag{26}$$

for $n = 1, 2, \ldots$. A solution of Eq.(26) is given in the form[1]

$$\rho_n(t) = \mathcal{F}_t\left(\rho_{n-1}(t)\right) = \mathcal{F}_t^{n-1}\left(\rho_1(t)\right) = \mathcal{F}_t^n\left(\rho_0\right), \tag{27}$$

where

$$\mathcal{F}_t(Q(t)) \equiv \frac{1}{i\hbar}\int_{t_0}^t \mathcal{U}_0(t-s)[\mathcal{H}_1(s), Q(s)]\mathcal{U}_0^\dagger(t-s), \tag{28}$$

with $\mathcal{U}_0(t) = \exp(t\mathcal{H}_0/i\hbar)$. In particular, we have

$$\rho_2(t) = \mathcal{F}_t(\rho_1(t)) = \mathcal{F}_t^2(\rho_0) \\ = \left(\frac{1}{i\hbar}\right)^2 \int_{t_0}^{t} ds \int_{t_0}^{s} ds' \mathcal{U}_0(t-s)[\mathcal{H}_1(s), \mathcal{U}_0(s-s')[\mathcal{H}_1(s'), \rho_0]\mathcal{U}_0^\dagger(s-s')]\mathcal{U}_0^\dagger(t-s). \tag{29}$$

Now, we calculate the change of the internal energy, $U'(t)$, using Eq.(29). Thus, we obtain

$$\left(\frac{dS(t)}{dt}\right)_{\mathrm{irr}} = \frac{1}{T}\frac{dU(t)}{dt} = \frac{1}{T}\frac{d\langle\mathcal{H}_0\rangle_t}{dt} = \frac{1}{T}\mathrm{Tr}\mathcal{H}_0\rho_2'(t) = \frac{\sigma E^2}{T} > 0, \tag{30}$$

using the expression of $\sigma$ in Eq.(11) or using the relation $\boldsymbol{J} = \sigma\boldsymbol{E}$, as far as the second-order of $E$ is concerned.

Now, the time dependence of $T(t)$ is effectively defined[1] by the equation

$$U(t) = \langle\mathcal{H}_0\rangle_0(T(t)) \equiv \mathrm{Tr}\mathcal{H}_0 \exp\left(-\beta(t)\mathcal{H}_0\right)/Z_0(\beta(t)), \tag{31}$$

with $\beta(t) = 1/k_{\mathrm{B}}T(t)$, and $Z_0(\beta) = \mathrm{Tr}\exp(-\beta\mathcal{H}_0)$, in a slowly varying state. This "effective" temperature is a convenient parameter to characterize "thermodynamically" non-equilibrium systems. Then, $T(t)$ is expanded in the form $T(t) = T + a(t)E^2 + \cdots$. Thus, the entropy production (30) is time-independent in the lowest order.

In general, according to the symmetry arguments on $\mathcal{H}_0$ and the current operator $\boldsymbol{j} = \dot{\boldsymbol{A}}$, we find that the average of the current, $\boldsymbol{J}_E(t) = \langle\boldsymbol{j}\rangle_t = \mathrm{Tr}\boldsymbol{j}\rho(t)$, is expressed by

$$\boldsymbol{J}_E(t) = \mathrm{Tr}\boldsymbol{j}\rho_{\mathrm{a}}(t), \tag{32}$$

using the antisymmetric part $\rho_{\mathrm{a}}(t)$ of $\rho(t)$ given by

$$\rho_{\mathrm{a}}(t) \equiv P_{\mathrm{antisym}}\rho(t) = \rho_1(t) + \rho_3(t) + \cdots, \tag{33}$$

and that the entropy production is given by $(dS/dt)_{\mathrm{irr}} = \mathrm{Tr}\mathcal{H}_0\rho_{\mathrm{s}}'(t)/T$ using the symmetric part[1]

$$\rho_{\mathrm{s}}(t) \equiv P_{\mathrm{sym}}\rho(t) = \rho_0 + \rho_2(t) + \cdots. \tag{34}$$

These symmetry-separated density matrices satisfy the following "projected von Neumann equations"

$$i\hbar\frac{\partial}{\partial t}\rho_q(t) = [\mathcal{H}_0, \rho_q(t)] + [\mathcal{H}_1(t), (1-P_q)\rho(t)], \tag{35}$$

where $q$ denotes "sym" or "antisym" and $P_q$ is the projection operator to the "$q$" part of the density matrix $\rho(t)$; $P_q^2 = P_q$ and $P_{\mathrm{sym}} + P_{\mathrm{antisym}} = 1$. This separation of symmetric and antisymmetric parts of $\rho(t)$ plays an essential role in formulating steady states later.

Using these partial (projected) density matrices, the general current $\boldsymbol{J}_E$, conductivity $\sigma_E$ and entropy production $(dS/dt)_{\mathrm{irr}}(E)$ are expressed[1], respectively, as

$$\boldsymbol{J}_E = \mathrm{Tr}\boldsymbol{j}\rho_{\mathrm{a}}(t) = \sigma_E\boldsymbol{E}; \quad \sigma_E = \sigma_0 + \sigma_2E^2 + \cdots + \sigma_{2n}E^{2n} + \cdots, \tag{36}$$

with $\sigma_0 = \sigma$, and

$$\left(\frac{dS}{dt}\right)_{\mathrm{irr}}(E) = \frac{1}{T(t)}\mathrm{Tr}\mathcal{H}_0\rho_{\mathrm{s}}'(t) = \frac{\boldsymbol{J}_E\cdot\boldsymbol{E}}{T(t)} = \frac{\sigma_E E^2}{T(t)} > 0 \tag{37}$$

for an arbitrary strength of the electric field $\boldsymbol{E}$.

Thus, the entropy production associated even with the linear transport phenomena is given[1] by the time derivative of the *second-order* density matrix, $\rho_2'(t)$. That is, the linear response scheme is not closed to the first-order term of the density matrix. As is seen from Eqs.(36) and (37), the current and entropy production have a duality through the conductivity and also through the coupled equations Eq.(35) for $q =$ sym and $q =$ antisym.

Journal of Physics: Conference Series **297** (2011) 012019 doi:10.1088/1742-6596/297/1/012019

## 5. Formulation of steady states using relaxation-type symmetry-separated von Neumann equations

In the present section, we study the entropy production in steady states. For this purpose, we have to introduce the interaction of the system with the heat bath, in order to extract generated heat outside of the system[1]. For this purpose, we extend the von Neumann equation (1) as follows:

$$\frac{\partial \rho(t)}{\partial t} = \frac{1}{i\hbar}[\mathcal{H}(t), \rho(t)] - \Gamma(\rho(t)), \tag{38}$$

where the relaxation term $\Gamma(\rho(t))$ may be given generally in the Lindblad form[18]:

$$\Gamma(\rho) = \sum_{\alpha} \left( V_\alpha \rho V_\alpha^\dagger - \frac{1}{2}\left\{ V_\alpha^\dagger V_\alpha, \rho \right\} \right). \tag{39}$$

As we are here interested only in the conceptual mechanism of the entropy production in steady states, we make use of the simplest form of $\Gamma$. Since the current is related to the conservation of particles or charges, and consequently that it is expressed easily by a kind of non-equilibrium density matrix including current operators[10, 14, 19, 20]. Only the adiabatic factor $e^{-\epsilon t}$ may be necessary for the convergence of the integral of the time-correlation function of the relevant current, which is described by $\rho_{\mathrm{a}}(t)$, as was discussed in [1]. Thus, we assume, for $\Gamma$,

$$\Gamma \rho_{\mathrm{a}}(t) = \epsilon \rho_{\mathrm{a}}(t) \quad (\epsilon \to +0). \tag{40}$$

As the entropy characterizes a qualitative property of the energy in non-equilibrium states, we have to treat the heat change in a way different from treating the current. The generated heat of the system should be extracted outside at a finite speed to construct a steady state and it is expressed by the change of $\rho_{\mathrm{s}}(t)$. Thus, we assume[1]

$$\Gamma \rho_{\mathrm{s}}(t) = \epsilon_{\mathrm{r}}(\rho_{\mathrm{s}}(t) - \rho_0); \quad \epsilon_{\mathrm{r}} = \frac{1}{\tau_{\mathrm{r}}}. \tag{41}$$

The relaxation time $\tau_{\mathrm{r}}$ in Eq.(41) is macroscopic, namely much larger than the collision time $\tau$ of electrons colliding with impurities or phonons (i.e., $\tau \ll \tau_{\mathrm{r}}$).

More explicitly we present Eq.(38) in the following coupled equations

$$\frac{1}{\epsilon_{\mathrm{r}}}\frac{d}{dt}\rho_{\mathrm{s}}(t) = -\mathcal{A}\rho_{\mathrm{s}}(t) + \mathcal{B}\rho_{\mathrm{a}}(t) + \rho_0, \tag{42}$$

and

$$\frac{1}{\epsilon}\frac{d}{dt}\rho_{\mathrm{a}}(t) = -\mathcal{C}\rho_{\mathrm{a}}(t) + \mathcal{D}\rho_{\mathrm{s}}(t). \tag{43}$$

Here, the hyper-operators $\mathcal{A}, \mathcal{B}, \mathcal{C}$ and $\mathcal{D}$ are defined by

$$\mathcal{A} = 1 - \omega_{\mathrm{r}}\delta_{\mathcal{H}_0}, \quad \mathcal{B} = \omega_{\mathrm{r}}\delta_{\mathcal{H}_1}, \quad \mathcal{C} = 1 - \omega_\epsilon \delta_{\mathcal{H}_0}, \quad \text{and} \quad \mathcal{D} = \omega_\epsilon \delta_{\mathcal{H}_1}, \tag{44}$$

where $\omega_\epsilon = 1/(i\hbar\epsilon)$, $\omega_{\mathrm{r}} = 1/(i\hbar\epsilon_{\mathrm{r}}) = \tau_{\mathrm{r}}/(i\hbar)$ and the inner derivation $\delta_Q$ is defined by[17]

$$\delta_Q R = [Q, R] = QR - RQ. \tag{45}$$

In a unified way, we have Eq.(38) with $\Gamma\rho(t) = \Gamma(\rho_{\mathrm{a}}(t)) + \Gamma(\rho_{\mathrm{s}}(t))$. Thus, we obtain the differential equation of the internal energy $U(t)$ for $\mathcal{H}_1 = -\boldsymbol{A} \cdot \boldsymbol{E}$ as

$$\frac{d}{dt}U(t) = \boldsymbol{J}_E(t) \cdot \boldsymbol{E} - \epsilon_{\mathrm{r}}(U(t) - U(t_0)), \tag{46}$$

using the relation[1]

$$\frac{1}{i\hbar}\mathrm{Tr}\mathcal{H}_0[\mathcal{H}_1,\rho_{\mathrm{a}}(t)] = -\mathrm{Tr}\dot{\mathcal{H}}_1\rho_{\mathrm{a}}(t) = \mathrm{Tr}(\boldsymbol{j}\cdot\boldsymbol{E})\rho_{\mathrm{a}}(t) \equiv \boldsymbol{J}_E(t)\cdot\boldsymbol{E}. \tag{47}$$

The solution of Eq.(46) is given by[1]

$$U(t) = \int_{t_0}^{t} e^{-\epsilon_{\mathrm{r}}(t-s)}\boldsymbol{J}_E(s)\cdot\boldsymbol{E}ds + U(t_0). \tag{48}$$

From Eqs.(42) and (43) in the limit $t\to\infty$, we obtain the stationary density matrices $\rho_{\mathrm{s}}^{(\mathrm{st})}$ and $\rho_{\mathrm{a}}^{(\mathrm{st})}$:

$$\rho_{\mathrm{s}}^{(\mathrm{st})} = (\mathbf{1}-\mathcal{A}^{-1}\mathcal{B}\mathcal{C}^{-1}\mathcal{D})^{-1}\rho_0, \text{ and } \rho_{\mathrm{a}}^{(\mathrm{st})} = \mathcal{C}^{-1}\mathcal{D}(\mathbf{1}-\mathcal{A}^{-1}\mathcal{B}\mathcal{C}^{-1}\mathcal{D})^{-1}\rho_0. \tag{49}$$

and thereby we can evaluate the stationary values $\boldsymbol{J}_E^{(\mathrm{st})}$ and $U^{(\mathrm{st})}$. It may be also interesting to note that $\rho_{\mathrm{s}}^{(\mathrm{st})}$ can be expanded in a power series of $E^2$ in the form

$$\rho_{\mathrm{s}}^{(\mathrm{st})} = \left(\mathbf{1}-E^2\mathcal{G}\right)^{-1}\rho_0 = \sum_{n=0}^{\infty}\left(\mathcal{G}^n\rho_0\right)E^{2n}; \quad \mathcal{G} = \mathcal{A}^{-1}\mathcal{B}\mathcal{C}^{-1}\mathcal{D}/E^2 = E\text{-independent}. \tag{50}$$

A similar expansion formula holds also for $\rho_{\mathrm{a}}^{(\mathrm{st})}$:

$$\rho_{\mathrm{a}}^{(\mathrm{st})} = E\mathcal{C}^{-1}\left(\mathcal{D}/E\right)\left(1-E^2\mathcal{G}\right)^{-1}\rho_0 = \sum_{n=0}^{\infty}\left(\mathcal{C}^{-1}(\mathcal{D}/E)\mathcal{G}^n\rho_0\right)E^{2n+1}, \tag{51}$$

where $\mathcal{D}/E$ is independent of $E$. More explicitly, $\rho_1^{(\mathrm{st})}$ and $\rho_2^{(\mathrm{st})}$ are given by

$$\begin{aligned}\rho_1^{(\mathrm{st})} &= \frac{1}{\mathbf{1}-\omega_\epsilon\delta_{\mathcal{H}_0}}\omega_\epsilon\delta_{\mathcal{H}_1}\rho_0 = \int_0^\infty dt\exp\left(\frac{t}{i\hbar}\delta_{\mathcal{H}_0}\right)\left(\frac{1}{i\hbar}[\mathcal{H}_1,\rho_0]\right)e^{-\epsilon t}\\ &= \rho_0\int_0^\infty dt\int_0^\beta d\lambda\boldsymbol{j}(-t-i\hbar\lambda)\cdot\boldsymbol{E}e^{-\epsilon t},\end{aligned} \tag{52}$$

and

$$\rho_2^{(\mathrm{st})} = \int_0^\infty dt\exp\left(-(\epsilon_{\mathrm{r}}-\frac{1}{i\hbar}\delta_{\mathcal{H}_0})t\right)\left(\frac{1}{i\hbar}[\mathcal{H}_1,\rho_1^{(\mathrm{st})}]\right), \tag{53}$$

respectively.

Then, the averages of the current $\boldsymbol{J}_E^{(\mathrm{st})}$ and the internal energy $U^{(\mathrm{st})}$ are given, respectively, by

$$\boldsymbol{J}_E^{(\mathrm{st})} = \mathrm{Tr}\boldsymbol{j}\rho_{\mathrm{a}}^{(\mathrm{st})} \quad \text{and} \quad U^{(\mathrm{st})} = \mathrm{Tr}\mathcal{H}_0\rho_{\mathrm{s}}^{(\mathrm{st})}. \tag{54}$$

Correspondingly, the stationary temperature $T_{\mathrm{st}}$ is expressed using the relation[1]

$$\langle\mathcal{H}_0\rangle_0(T_{\mathrm{st}}) = U^{(\mathrm{st})}, \tag{55}$$

as a function of the relaxation time $\tau_{\mathrm{r}}$. More explicitly, from Eq.(44), we obtain the relation

$$\langle\mathcal{H}_0\rangle_0(T_{\mathrm{st}}) - \langle\mathcal{H}_0\rangle_0(T) = U^{(\mathrm{st})} - U(t_0) = \tau_{\mathrm{r}}\boldsymbol{J}_E^{(\mathrm{st})}\cdot\boldsymbol{E}. \tag{56}$$

This gives the expression of $T_{\mathrm{st}}$ as

$$T_{\mathrm{st}} = T + \tau_{\mathrm{r}}\sigma E^2/C(T) + O(E^4) \tag{57}$$

with the specific heat $C(T)$ of the system described by $\mathcal{H}_0$. The stationary temperature $T_{\rm st}$ is generally expanded in a power series of $E^2$, whose expansion coefficients are expressed in terms of the specific heat $C(T)$ and its temperature derivations $\{C^{(n)}(T)\}$ as well as $\sigma$ and $\tau_{\rm r}$. Note that $T_{\rm st} = T$ only when $\tau_{\rm r} = 0$. This condition ($\tau_{\rm r} = 0$) corresponds to the instantaneous extraction of generated heat outside of the system. Practically we have $T_{\rm st} > T$ (i.e., $\tau_{\rm r} > 0$), as was discussed in [1].

Next, we study the entropy change $(dS/dt)_{\rm irr}$ in the stationary state. We study the right-hand side of Eq.(46) from a physical point of view. The first term $\boldsymbol{J}_E(t)\cdot\boldsymbol{E}$ of it is expressed in the form

$$\boldsymbol{J}_E(t)\cdot\boldsymbol{E} = {\rm Tr}\mathcal{H}_0 D_{\rm d}(\rho(t)) = {\rm Tr}(\tilde{D}_{\rm d}(\mathcal{H}_0))\rho(t), \tag{58}$$

where

$$\tilde{D}_{\rm d}(\mathcal{H}_0) \equiv \frac{1}{i\hbar}[\mathcal{H}_0, \mathcal{H}(t)], \text{ and } D_{\rm d}(\rho(t)) \equiv \frac{1}{i\hbar}[\mathcal{H}(t), \rho(t)] = \frac{\partial\rho(t)}{\partial t} + \Gamma(\rho(t)). \tag{59}$$

That is, the hyper-operators $D_{\rm d}$ and $\tilde{D}_{\rm d}$ denote the time derivatives of the density matrix and an operator, respectively. These are called[1] as "dynamical-derivative representations". Note that the hyper-operator $\tilde{D}_{\rm d}$ defined by $(i/\hbar)\delta_{\mathcal{H}(t)}$ satisfies the Leibniz rule $\tilde{D}_{\rm d}(AB) = (\tilde{D}_{\rm d}(A))B + A(\tilde{D}_{\rm d}(B))$.

The mechanism of heat generation should be the same as in the absence of the relaxation term $\Gamma\rho(t)$. Thus, the entropy production in our new situation described by Eq.(46) is expressed in the form

$$\left(\frac{dS}{dt}\right)_{\rm irr}(t) = {\rm Tr}SD_{\rm d}(\rho(t)) = \frac{1}{T(t)}{\rm Tr}(\tilde{D}_{\rm d}(\mathcal{H}_0))\rho(t) = \left(\boldsymbol{J}_E(t)\cdot\boldsymbol{E}\right)/T(t), \tag{60}$$

with the time-dependent temperature $T(t)$ defined by

$$\langle\mathcal{H}_0\rangle_0(T(t)) = U(t) \equiv \langle\mathcal{H}_0\rangle_t \equiv {\rm Tr}\mathcal{H}_0\rho(t), \tag{61}$$

together with Eq.(48). The stationary entropy production is expressed in the form

$$\left(\frac{dS}{dt}\right)_{\rm irr}^{\rm (st)} = \frac{U^{\rm (st)} - U(t_0)}{T_{\rm st}\tau_{\rm r}}. \tag{62}$$

In order to explain the logic of our theory, we rewrite Eq.(46) in the following form[1]

$$\frac{dU(t)}{dt} = \left({\rm Tr}(\tilde{D}_{\rm d}(\mathcal{H}_0))\rho(t)\right)_{\text{generated heat}} - \left(\frac{U(t) - U(t_0)}{\tau_{\rm r}}\right)_{\text{extracted heat}} \tag{63}$$

which vanishes in the steady state, owing to the balance of the two terms in the right-hand side of Eq.(63). The first term of the right-hand side in Eq.(63) contributes to the entropy production.

## 6. Entropy production for thermal disturbance

According to the general formula (2), the entropy production for the thermal disturbance is given by

$$\left(\frac{dS}{dt}\right)_{\rm irr} = \frac{\sigma_{\rm H}E_{\rm T}^2}{T}. \tag{64}$$

The validity of the above result is easily confirmed[1] thermodynamically by computing the entropy changes at the two sides of the system of the temperatures $T$ and $(T+\Delta T)$ for the heat current $\boldsymbol{J}_{\rm H} = \sigma_{\rm H}\boldsymbol{E}_{\rm T}$.

Thus, the entropy production for electric conduction and thermal conduction is presented in a unified way using the second-order term $\rho_2(t)$ or more generally the symmetric part $\rho_{\rm sym}(t)$ of the density matrix.

## 7. Discussion

In the present review article, two typical examples of static electric conduction and thermal conduction have been discussed in order to elucidate the mechanism of entropy production (and consequently irreversibility) in transport phenomena. However, the present idea and schemes will be easily extended to other more general cases of transport phenomena, which will be reported elsewhere in the near future[29].

It will be instructive for realizing intuitively the role of $\mathcal{H}_0$ in the first-principle approach to apply the present symmetry- separated treatment to relaxation-type Boltzmann equation of the form

$$\frac{\partial f(t)}{\partial t} + e\boldsymbol{E} \cdot \frac{\partial f(t)}{\partial \boldsymbol{E}} = \frac{1}{\tau}(f(t) - f_0) \tag{65}$$

in the uniform case [1]. We separate $f(t)$, by symmetry, as follows:

$$f(t) = f_{\text{sym}}(t) + f_{\text{antisym}}(t), \tag{66}$$

where $f_{\text{sym}}(t) = f_0 + f_2(t) + \cdots$, and $f_{\text{antisym}}(t) = f_1(t) + f_3(t) + \cdots$. As was shown in [1], we can derive easily that

$$\boldsymbol{J} = \int f_{\text{antisym}}(t) e\boldsymbol{v} D(\epsilon) d\epsilon = \sigma_E \boldsymbol{E} = (\sigma + \sigma_2 E^2 + \cdots)\boldsymbol{E} = \frac{ne^2\tau}{m}\boldsymbol{E} + \cdots, \tag{67}$$

and

$$\left(\frac{dS}{dt}\right)_{\text{irr}} = \frac{1}{T}\int \frac{\epsilon}{\tau}\left(f_{\text{sym}}(t) - f_0\right) D(\epsilon) d\epsilon \simeq \frac{\sigma_E E^2}{T}, \tag{68}$$

for large $t$, where $D(\epsilon)$ denotes the density of states and $\epsilon = p^2/2m$ (namely the kinetic energy of an electron with the momentum $\boldsymbol{p}$ and mass $m$). We find again that the second-order term $f_2(t)$ gives the lowest-order contribution to the entropy production.

If we assume that the time correlation of current $\langle \boldsymbol{j}(t)\boldsymbol{j}\rangle_0$ decays in the form $\langle \boldsymbol{j}(t)\boldsymbol{j}\rangle_0 = \langle \boldsymbol{j}^2\rangle_0 \exp(-t/\tau)$ owing to the effect of $\mathcal{H}_0$, then using the formula (11) for $\hbar = 0$ we obtain $\sigma = (ne^2\tau)/m$, which agrees with Eq.(67). This can be also derived phenomenologically by Einstein's theory of Brownian motion of electrons[1].

The symmetry-separated treatment of the von Neumann equation in deriving entropy production as well as electric current through the conductivity may be regarded to be a typical example of the strategy of separation and unification of procedures in theoretical sciences[30].

It should be also remarked that the present treatment of entropy production based on the (symmetry-separated) von Neumann equation yields the unification of Prigogine's non-equilibrium thermodynamical research trend and Kubo's rigorous work based on quantum statistical mechanics.

## Acknowledgments

The author would like to thank Professor B.K.Chakrabarti for valuable comments and to thank Y.Hashizume for discussion and digitization of the present manuscript. This work was supported by the Grant-in-Aid for Science Research on Priority Area "Deepening and Expansion of Statistical Mechanical Informatics".

## References

[1] Suzuki M 2011 *Physica A.* **390** 1904 Paper I.
[2] Suzuki M 2011 *Physica A.* Paper II (submitted).
[3] Green M S 1952 *J.Chem.Phys.* **20** 1281, ibid 1954 **22** 398.
[4] Kubo R 1957 *J.Phys.Soc.Jpn.***12** 570.

[5] Brout R and Prigogine I 1956*Physica***22** 621.
Prigogine I, 1980 *From Being to Becoming* (Freeman), and references cited therein.
Nicolis G and Prigogine I 1997 *Self Organization in Non-Equilibrium Systems.* (Wiley Interscience).
Suzuki M 1981 *Adv.Chem.Phys.***46** 195-278.
Reichl L E 1998*A Modern Course in Statistical Physics* (New York: John Wiley and Sons,INC.).
[6] van Kampen N G 1983 *Stochastic Process in Physics and Chemistry*(Amsterdam:North-Holland),
Kubo R, Matsuo K and Kitahara K 1973 *J.Stat.Phys.***9** 51.
[7] Suzuki M 2011 *Int. J. of Stochastic Analysis* (Hindawi Pub. Corp.).
Suzuki M and Kubo R 1968 *J.Phys.Soc.Jpn.***24** 51.
See Appendix: Fluctuation-Dissipation Theorem.
[8] Suzuki M 1976 *Prog.Theor.Phys.***55** 383.
[9] Chester G V 1963 *Rep.Prog.Phys.***26** 411.
Suzuki M 1980 *Prog.Theor.Phys.Suppl.***69** 160.
[10] Kubo R, Toda M and Hashitsume N, Statistical Physics II, 1991*Non-equilibrium statistical mechanics, Springer Series in Solid State Sciences* (Singapore: Springer).
[11] Kubo R and Tomita K 1954 *J.Phys.Soc.Jpn.***9** 888.
[12] Nakano H 1955 *Prog.Theor.Phys.***15** 77.
[13] Nakano H 1993 *Int.J.Mod.Phys.***B7** 2397.
See also Ichiyanagi M, 1986 *Prog.Theor.Phys.***76** 37.
[14] Zubarev D N, 1971 *Nonequilibrium Statistical Mechanics* (Nauka).
[15] Kuzemsky A L 2005 *Int.J.Mod.Phys.***B19** 1029.
[16] Onsager L 1931 *Phys.Rev.* **37** 405; ibid 1931 **38** 2265.
Onsager L and Machlup 1953 *Phys.Rev.***91** 1505.
[17] Suzuki M 1997 *Comm.Math.Phys.***183** 339.
Suzuki M 1998 *Prog.Theor.Phys.***100** 475.
Suzuki M 1999 *Rev.Math.Phys.***11** 243.
[18] Weiss U 1993 *Quantum Dissipative Systems*(Singapore: World Scientific).
Lindblad G Commun. 1975 *Math.Phys.***40** 147.
Gorini V, Kossakowski A and Sudarshan E C G,1976 *J.Math.Phys.***17** 821.
Shirmer S G and Solomon A I, 2004 *Phys.Rev.***A70** 022107
[19] Mori H 1958 *Phys.Rev.***112** 1829.
[20] Kawasaki K and Gunton J D 1973 *Phys.Rev.***A8** 2048
[21] Kondo J 1964 *Prog.Theor.Phys.***32** 37.
[22] Suzuki M 1977 *Prog.Theor.Phys.***58** 1151.
Note also that there appears a universality class (of critical phenomena) different from that of ferromagnetic phase transitions. This "spin-glass universality class" was also found in metal-insulator transitions [March N, Suzuki M and Parrinello M 1979 *Phys.Rev.***B19** 2027].
Hashizume Y and Suzuki M, 2011*Int.J.Mod.Phys.B* in press.
[23] Suzuki M 1971 *Physica***51** 277.
[24] The present theory was briefly reported in the symposiums on "Duality and Scales in Quantum Sciences" in Kyoto Univ. (Nov.4,2009 and Nov.5,2010).
[25] Miyake S J and Kubo R 1962 *Phys.Rev.Lett.***9** 62.
Tani K 1964 *Prog.Theor.Phys.***32** 167.
Suzuki M 1975 *Prog.Theor.Phys.***53** 1657;ibid 1998 **100** 475; 2000*J.Phys.Soc.Jpn.***69** Suppl.A.
Fujii T , 2007 *J.Phys.Soc.Jpn.***76** 044709.
Komatsu T S, Nakagawa N, Sasa S-I and Tasaki H 2009 *J.Stat.Phys.***134** 401, and references cited therein.
Yuan S, Katsnelson M I and Raedt H.De 2009 *J.Phys.Soc.Jpn.***78** 094003.
Chong S-H, Otsuki M and Hayakawa H 2010 *Phys.Rev.***E81** 041130
Nakamura S et al., 2010 *Phys.Rev.Lett.***104** 080602,and references cited therein.
Saito K and Utsumi Y, 2008 *Phys.Rev.B***78** 115429
[26] Luttinger J M 1964 *Phys.Rev.***135A** 1505.
[27] Einstein A 1912 *Ann.Physik* **38** 443.
[28] Tolman R C and Ehrenfest P 1930 *Phys.Rev.***36** 1791.
[29] Suzuki M in preparation.
[30] Suzuki M 1980 *Prog.Theor.Phys.Suppl.***69** 160.

STATPHYS-Kolkata VII
Journal of Physics: Conference Series **297** (2011) 012020 doi:10.1088/1742-6596/297/1/012020

# Non-adiabatic quench dynamics near anisotropic quantum critical point

**T. Hikichi and *S. Suzuki**

Department of Physics and Mathematics, Aoyama Gakuin University, Fuchinobe, Sagamihara 252-5258, Japan

E-mail: *`sei@phys.aoyama.ac.jp`

**K. Sengupta**

Theoretical Physics Department, Indian Association for the Cultivation of Science, Jadavpur, Kolkata-700032, India

**Abstract.** We study non-equilibrium dynamics induced by a quench of a parameter near a quantum phase transition in the Kitaev model. On the basis of an exact solution for the time-dependent Schrödinger equation, we show that the density of excitations and residual energy after a quench scale with a quench rate $\tau^{-1}$ as $\tau^{-3/4}$ and $\tau^{-5/4}$ respectively, when a parameter is quenched from a gapped phase toward the quantum critical point. These unusual scaling laws originate from an anisotropic nature of the quantum critical point in the Kitaev model. Gerenalizing these results, we provide novel scaling laws for a generic anisotropic quantum critical point in $d$-dimensional systems.

## 1. Introduction

Non-equilibrium dynamics of quantum systems near quantum critical points has been a subject of intense study in recent years [1, 2]. During such dynamics, a quantum system passes from one gapped phase to another via time evolution of a Hamiltonian parameter with a rate $^{-1}$ and an exponent $\alpha$ ( $(t) = {}_0|t/ |^\alpha\mathrm{Sgn}(t)$, where $\mathrm{Sgn}(x) = 1(-1)$ for $x > (<)\,0$) through an intermediate quantum critical point at $= 0$. At the critical point, the energy gap vanishes as $\Delta(\vec{k}) \sim |\vec{k}|^z$ where $z$ is the dynamical critical exponent. Thus the dynamics becomes non-adiabatic around a region near this point and the system fails to remain at the instantaneous ground state leading to formation of defects [3, 4, 5, 6, 7, 8, 9]. The density of these defects ($n$) and the residual energy produced in the process ($Q$) scale with universal exponents: $n \sim {}^{-\nu d\alpha/(z\nu\alpha+1)}$ and $Q \sim {}^{-(d+z)\nu\alpha/(z\nu\alpha+1)}$, where is the correlation length exponent and $d$ is the system dimension [5, 7]. It is well-known that scaling laws do not change if the dynamics terminate at the critical point [8]. All of the above-mentioned studies apply to isotropic critical points where the scaling of the energy gap with the momentum is described by a single exponent $z$. Recently, the anisotropic Dirac model with an anisotropic critical point is studied and it was shown that one needs multiple exponents to describe the scaling of the energy gap [10]. However such studies have not been carried out in the context of the Kitaev model and generic expressions for the scaling laws for $n$ and $Q$ for such critical points in arbitrary dimensions have not been provided.

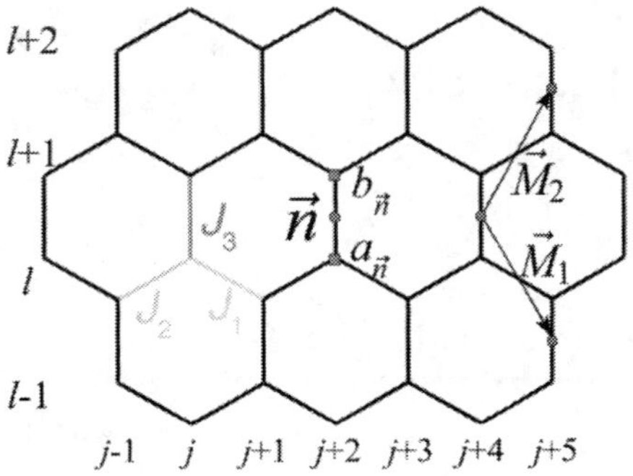


**Figure 1.** Schematic representation of the Kitaev model on a honeycomb lattice. The bonds $J_1$, $J_2$ and $J_3$ shows nearest neighbor couplings between $x$, $y$ and $z$ components of the spins respectively. $\vec{n}$ represents the position vector of the midpoint of each vertical bond (unit cell). The vectors $\vec{M}_1$ and $\vec{M}_2$ are spanning vectors of the lattice. In the Fermionic representation of the model, the Majorana Fermions $a_{\vec{n}}$ and $b_{\vec{n}}$ sit at the bottom and top sites respectively of the vertical bond with center coordinate $\vec{n}$ as shown.

In this work, we study the aspect of non-equilibrium slow dynamics in the vicinity of anisotropic critical points with specific focus on the 2D Kitaev model. We derive a generic model-independent expression for the scaling of $n$ and $Q$ for such dynamics which takes a $d$-dimensional system from a gapped phase to the vicinity of an anisotropic critical point. We consider a scenario where the energy gap $\Delta_{\vec{k}}$ vanishes as $k_i^z$ for $m$ momentum components $(i = 1, \cdots, m)$ and as $k_i^{z'}$ for the rest $d - m$ components $(i = m+1, \cdots, d)$ with $z'$ $z$ at the critical point and show that the time-evolution of the Hamiltonian parameter $(t)$, which brings the system at the critical point at $t = 0$, leads to novel scaling laws for $n$ and $Q$:

$$n \sim \quad ^{-[m+(d-m)z/z']\nu\alpha/(z\nu\alpha+1)},$$
$$Q \sim \quad ^{-[(m+z)+(d-m)z/z']\nu\alpha/(z\nu\alpha+1)}.$$

Our results reproduce their well-known counterparts for the isotropic case $(z = z')$ as special cases. We also show, by exact analytical solution for linear time evolution $(\alpha = 1)$, that the two-dimensional (2D) Kitaev model provides an explicit realization of the scaling laws mentioned above with $d = z' = 2$ and $m = \quad = z = 1$ leading to $n \sim \quad ^{-3/4}$ and $Q \sim \quad ^{-5/4}$. We also corroborate the scaling laws mentioned above by numerical studies of the Kitaev model for arbitrary power-law time evolution.

## 2. Anisotropic critical points

We begin with the study of slow dynamics in the Kitaev model [12, 13, 14, 15]. The Hamiltonian for this model, schematically represented in Fig. 1, is given by

$$H_K = \sum_{j+l=\mathrm{even}} (J_1 \, {}^x_{j,l} \, {}^x_{j+1,l} + J_2 \, {}^y_{j-1,l} \, {}^y_{j,l} + J_3 \, {}^z_{j,l} \, {}^z_{j,l+1}), \tag{1}$$

where $\vec{\ }_{jl} = ( {}^x_{jl}, {}^y_{jl}, {}^z_{jl})$ denote Pauli matrices at the site $(j, l)$ of the honeycomb lattice, $J_1$, $J_2$, and $J_3$ represent nearest-neighbor couplings between $x$, $y$ and $z$ components of the spins respectively. It is well-known that $H_K$ can be represented in terms of Fermionic fields by a straightforward Majorana transformation: $a_{jl} = (\prod_{k=-\infty}^{l-1} \prod_i \, {}^z_{ik} \prod_{i=-\infty}^{j-1} \, {}^z_{il}) \, {}^y_{jl}$ for even $j + l$

and $b_{jl} = (\prod_{k=-\infty}^{l-1} \prod_i \quad {}^{z}_{ik} \prod_{i=-\infty}^{j-1} \quad {}^{z}_{il}) \quad {}^{x}_{jl}$ for odd $j+l$ [13, 14, 15]. This leads to the Fermionic Hamiltonian

$$H_F = i \sum_{\vec{n}} \left( J_1 \, b_{\vec{n}} a_{\vec{n}-\vec{M}_1} \; + \; J_2 \, b_{\vec{n}} a_{\vec{n}+\vec{M}_2} + J_3 D_{\vec{n}} \, b_{\vec{n}} a_{\vec{n}} \right), \tag{2}$$

where $\vec{n} = \sqrt{3}\hat{i}\; n_1 + (\frac{\sqrt{3}}{2}\hat{i} + \frac{3}{2}\hat{j})\; n_2$ denote the midpoints of the vertical bonds. Here $n_1, n_2$ run over all integers so that the vectors $\vec{n}$ form a triangular lattice whose vertices lie at the centers of the vertical bonds of the underlying honeycomb lattice. The Majorana Fermions $a_{\vec{n}}$ and $b_{\vec{n}}$ sit at the bottom and top sites respectively of the bond labeled $\vec{n}$. The vectors $\vec{M}_1 = \frac{\sqrt{3}}{2}\hat{i} - \frac{3}{2}\hat{j}$ and $\vec{M}_2 = \frac{\sqrt{3}}{2}\hat{i} + \frac{3}{2}\hat{j}$ are spanning vectors for the lattice, and $D_{\vec{n}}$ can take the values $\pm 1$ independently for each $\vec{n}$. The crucial point that makes the solution of Kitaev model feasible is that $D_{\vec{n}}$ commutes with $H_F$, so that all the eigenstates of $H_F$ can be labeled by specific values of $D_{\vec{n}}$. It is well-known that the ground state of the model corresponds to $D_{\vec{n}} = 1$ on all links [12, 16].

For $D_{\vec{n}} = 1$, Eq. (2) can be diagonalized as

$$H_F = \sum_{\vec{k}} \psi^{\dagger}_{\vec{k}} H_{\vec{k}} \psi_{\vec{k}}, \tag{3}$$

where $\psi^{\dagger}_{\vec{k}} = (a^{\dagger}_{\vec{k}},\; b^{\dagger}_{\vec{k}})$ are Fourier transforms of $a^{\dagger}_{\vec{n}}$ and $b^{\dagger}_{\vec{n}}$, the sum over $\vec{k}$ extends over half the Brillouin zone (BZ) of the triangular lattice formed by the vectors $\vec{n}$, and $H_{\vec{k}}$ can be expressed in terms of the Pauli matrices $\sigma^i$ in particle-hole space as

$$H_{\vec{k}} = 2\left\{ J_1 \sin(\vec{k}\cdot\vec{M}_1) - J_2 \sin(\vec{k}\cdot\vec{M}_2) \right\} \sigma^1 + 2\left\{ J_3 + J_1 \cos(\vec{k}\cdot\vec{M}_1) + J_2 \cos(\vec{k}\cdot\vec{M}_2) \right\} \sigma^2. \tag{4}$$

The spectrum consists of two bands with energies $E^{\pm}_{\vec{k}} = \pm E_{\vec{k}}$ [15], where

$$E_{\vec{k}} = 2\left[ \{J_1 \sin(\vec{k}\cdot\vec{M}_1) - J_2 \sin(\vec{k}\cdot\vec{M}_2)\}^2 + \{J_3 + J_1 \cos(\vec{k}\cdot\vec{M}_1) + J_2 \cos(\vec{k}\cdot\vec{M}_2)\}^2 \right]^{1/2}. \tag{5}$$

For $|J_1 - J_2| \quad J_3 \quad J_1 + J_2$, the bands touch each other, and the energy gap $\Delta_{\vec{k}} = E^{+}_{\vec{k}} - E^{-}_{\vec{k}}$ vanishes for special values of $\vec{k}$ leading to a gapless phase. In particular we note that for $J_1 = J_2 = 1$ and $J_3 = 2$, the gap vanishes at $\vec{k}_c = (2\pi/\sqrt{3}, 0)$ and around this point $\Delta_{\vec{k}} \sim k_y$ and $\Delta_{\vec{k}} \sim (k_x - 2\pi/\sqrt{3})^2$. Thus this critical point constitutes an example of an anisotropic critical point with $z = m = 1$ and $d = z' = 2$. We note that such an anisotropic scaling occurs for any non-zero value of $J_1$ and $J_2$ at $J_3 = (J_1 + J_2)$.

We now consider a dynamics in this model $J_3(t) = (J_1 + J_2 - Jt/\quad)$ from $t = -\infty$ to $t = 0$ at a fixed rate $1/\quad$ which brings the system from a gapped phase to the anisotropic critical point at $\vec{k}_c$. Although this quench problem can be solved for any $J_1$ and $J_2$, we shall fix $J_1 = J_2 = J$ for simplicity and scale all energies (times) by $J$ ($\hbar/J$) in the subsequent analysis. This choice does not change the scaling properties which we seek. Also, to study the time evolution of the system, we note that after an unitary transformation $U = \exp(-i\sigma^1\pi/4)$, we obtain $H_F = \sum_{\vec{k}} \psi'^{\dagger}_{\vec{k}} H'_{\vec{k}} \psi'_{\vec{k}}$, where $H'_{\vec{k}} = U H_{\vec{k}} U^{\dagger}$ is given by

$$H'_{\vec{k}} = 2\left\{ (g_{\vec{k}} - t/\quad)\sigma^3 + \alpha_{\vec{k}}\sigma^1 \right\}, \tag{6}$$

where $\alpha_{\vec{k}} = \sin(\vec{k}\cdot\vec{M}_1) - \sin(\vec{k}\cdot\vec{M}_2)$ and $g_{\vec{k}} = 2 + \cos(\vec{k}\cdot\vec{M}_1) + \cos(\vec{k}\cdot\vec{M}_2)$. Hence the off-diagonal elements of $H'_{\vec{k}}$ remain time independent, and the quench problem reduces to a Landau-Zener problem for each $\vec{k}$.

The state of the system after the quench at $t = 0$ can be found by solving the Landau-Zener problem at each $\vec{k}$ with the initial condition $\psi^G_{\vec{k}}(t = -\infty) = |1\rangle = (0, 1)^T$ for all $\vec{k}$. After some algebra, one obtains for a given $\vec{k}$ and at $t = 0$ [17]

$$|\psi_{\vec{k}}\rangle^d = e^{-\pi\alpha^2_{\vec{k}}\tau/4}\Big(e^{3i\pi/4}D_{\mu_{\vec{k}}}(\ _{\vec{k}})|1\rangle + \alpha_{\vec{k}}\sqrt{\ }\,D_{\mu_{\vec{k}}-1}(\ _{\vec{k}})|0\rangle\Big), \tag{7}$$

where $_{\vec{k}} = 2ig_{\vec{k}}\sqrt{\ }\exp(-i\pi/4)$, $\mu_{\vec{k}} = -i\alpha^2_{\vec{k}}$ and $D_\mu$ are parabolic cylinder functions. The excited state at $t = 0$, solved by diagonalizing $H'_{\vec{k}}(t = 0)$, yields, for a given $\vec{k}$, $|\psi^+_{\vec{k}}\rangle = [(E^+_{\vec{k}} - 2g_{\vec{k}})|1\rangle + 2\alpha_{\vec{k}}|0\rangle]/\mathcal{D}_{\vec{k}}$, where $\mathcal{D}_{\vec{k}} = [(E^+_{\vec{k}} - 2g_{\vec{k}})^2 + 4\alpha^2_{\vec{k}}]^{1/2}$. Thus the probability of defect formation, given by $p_{\vec{k}} = |\langle\psi^+_{\vec{k}}|\psi_{\vec{k}}\rangle^d|^2$, can be obtained as

$$p_{\vec{k}} = \frac{4\alpha^2_{\vec{k}}e^{-\pi\alpha^2_{\vec{k}}\tau/2}}{\mathcal{D}^2_{\vec{k}}}\left|\alpha_{\vec{k}}\sqrt{\ }\,D_{\mu_{\vec{k}}-1}(\ _{\vec{k}}) + \frac{E^+_{\vec{k}} - 2g_{\vec{k}}}{2\alpha_{\vec{k}}} \times e^{-3i\pi/4}D_{\mu_{\vec{k}}}(\ _{\vec{k}})\right|^2. \tag{8}$$

Since  is large for slow dynamics, the contribution to the defect formation comes from a small region near the critical point where $\Delta_{\vec{k}}$ is sufficiently small for $\vec{k} \simeq \vec{k}_c$. The density of defects can be thus estimated by expanding $p_{\vec{k}}$ about $\vec{k} = \vec{k}_c$: $n \simeq \int d\delta k_x\, d\delta k_y\, p_{\vec{k}=\vec{k}_c+\delta\vec{k}}$, where the limits of integration can now be safely extended to infinity. To compute this integral, we note that around $\vec{k} = \vec{k}_c$, $\alpha_{\delta\vec{k}} \simeq 3\delta k_y$ and $g_{\delta\vec{k}} \sim 3(\delta k_x^2 + 3\delta k_y^2)/4$. Thus a redefinition of variables $\delta k_x \to \delta k'_x = \delta k_x\ ^{1/4}$ and $\delta k_y \to \delta k'_y = \delta k_y\ ^{1/2}$ allows us to extract the  dependence of the defect density

$$n \simeq \int d\delta k_x\, d\delta k_y\, p_{\delta\vec{k}}, \sim\ ^{-3/4}\int d\delta k'_x\, d\delta k'_y\, p_{\delta\vec{k}'}. \tag{9}$$

A similar analysis can be carried out for computation of residual energy $Q = (2\pi)^{-2}\int d^2k p_{\vec{k}}\Delta_{\vec{k}}$. Here we note that near the critical point $\vec{k} = \vec{k}_c$, $\Delta_{\vec{k}} \simeq 4\sqrt{9\delta k_y^2 + 9(\delta k_x^2 + 3\delta k_y^2)^2/16}$ and thus scale as $\ ^{-1/2}$. Thus one obtains

$$Q \simeq \int d\delta k_x\, d\delta k_y\, \Delta_{\delta\vec{k}}p_{\delta\vec{k}}, \sim\ ^{-5/4}\int d\delta k'_x\, d\delta k'_y\, \Delta_{\delta\vec{k}'}p_{\delta\vec{k}'}. \tag{10}$$

Equations (9) and (10) show that $n \sim\ ^{-3/4}$ and $Q \sim\ ^{-5/4}$ at the critical point. These scaling laws do not conform to the predictions of earlier works on defect production during passage through isotropic quantum critical points [5] or critical surfaces [15]; their origin lies in the anisotropic scaling of $\delta k_x$ and $\delta k_y$ with the quench time  .

To generalize these results for arbitrary $d$-dimensional anisotropic critical points, where the energy gap $\Delta_{\vec{k}} \sim k_i^z$ for $m$ directions and $\sim k_i^{z'}$ for $d - m$ directions, we provide a simple phase space argument as first proposed in Ref. [4]. We consider a general power-law quench with $(t) =\ _0|t/\ |^\alpha \mathrm{Sgn}(t)$ which starts at $t = -\infty$ and reaches the critical point at $t = 0$. We first note that the adiabaticity condition breaks down when the rate of change of the energy gap become equivalent to the square of the gap: $d\Delta_{\vec{k}}/dt\ \ \Delta^2_{\vec{k}}$. Since $\Delta_{\vec{k}} \sim\ ^{z\nu}|t/\ |^{z\nu\alpha}$, we find that the time spent by the system in the non-adiabatic regime is given by $\hat{t} \sim\ ^{z\nu\alpha/(z\nu\alpha+1)}$. The scaling of the energy gap in this regime can thus be written as $\Delta_{\vec{k}} \sim\ ^{-z\nu\alpha/(z\nu\alpha+1)}$. The phase space for defect production is given by $\Omega_n \sim k_1 \cdots k_d$. Since $\Delta_{\vec{k}} \sim k_i^z$ for $i = 1, \cdots, m$ and $k_i^{z'}$ for $i = m + 1, \cdots, d$, we finally obtain

$$n \sim\ ^{-[m+(d-m)z/z']\nu\alpha/(z\nu\alpha+1)}. \tag{11}$$

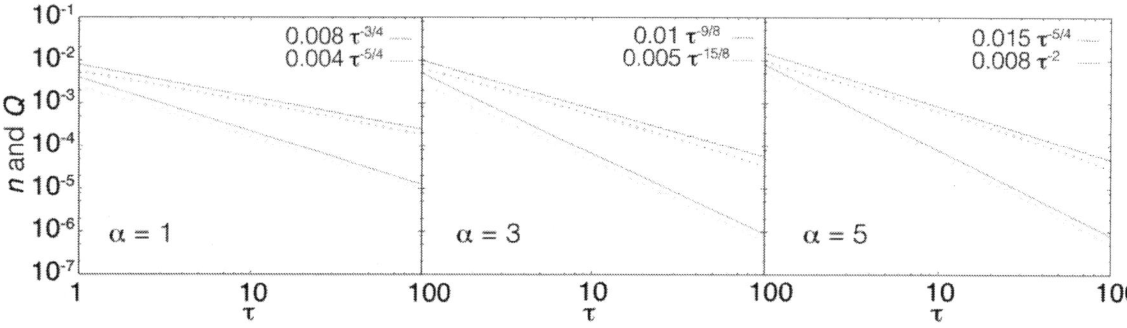


**Figure 2.** Numerical results on the defect density $n$ and residual energy $Q$. The time-dependent Schrödinger equation is solved in the momentum space for systems with size up to $512 \times 512$ unit cells. The parameter $\alpha$ specifying the evolution of $J_3 = 2 - |t/\ \ |^{\alpha}\mathrm{Sgn}(t)$ is chosen as $\alpha = 1$, 3 and 5. The lines indicate the power laws expected from Eqs. (11) and (12) for $d = z' = 2$ and $z = \ \ = m = 1$, $n \sim \ \ ^{-3\alpha/[2(\alpha+1)]}$ and $Q \sim \ \ ^{-5\alpha/[2(\alpha+1)]}$. The agreement between curves obtained numerically and the corresponding power laws is remarkable. In all plots, $t$ varies from an initial value $t_{\mathrm{in}} = -3$ to a final value $t_{\mathrm{f}} = 0$.

A similar argument can also be presented for the residual energy. We note that for $z \quad z'$, the leading behavior of the energy gap near the quantum critical point, where the defects are produced, is $\Delta_{\vec{k}} \sim k_i^z$ for $1 \quad i \quad m$. Thus the phase space for the residual energy production is $\Omega_Q \sim \Delta_{\vec{k}} k_1 \cdots k_d$ leading to a scaling of $Q$ as

$$Q \sim \ \ ^{-[(m+z)+(d-m)z/z']\nu\alpha/(z\nu\alpha+1)}. \tag{12}$$

We note that the scaling laws, Eqs. (11) and (12), reproduce their isotropic counterparts for $z = z'$ leading to $n \sim \ \ ^{-d\nu\alpha/(z\nu\alpha+1)}$ and $Q \sim \ \ ^{-(d+z)\nu\alpha/(z\nu\alpha+1)}$ [7, 5, 8]. Also, the scaling of the Kitaev model for linear time evolution elaborated in this work is reproduced for $d = z' = 2$, and $z = \ \ = \alpha = 1$ leading $n \sim \ \ ^{-3/4}$ and $Q \sim \ \ ^{-5/4}$. Moreover, we note that the scaling of defect density for a linear quench through a gapless surface can also be obtained from Eq. (11) by noting that for such quenches the energy gap depends only on the $m$ momenta components orthogonal to the $d-m$ dimensional gapless surface. This can be represented by putting $z' \to \infty$ (since $k_{\|} \sim \Delta_{\vec{k}}^{1/z'}$) leading to the scaling law $n \sim \ \ ^{-m\nu/(z\nu+1)}$ [15]. Thus Eqs. (11) and (12) reproduce all earlier results on defect production for slow dynamics across quantum critical lines and surfaces as special cases. Finally, we would like to point out that the maximum values of these exponents is 2 which can be obtained by similar considerations as in the cases of isotropic critical points [8].

To verify these scaling laws, we now study non-linear power-law dynamics in the Kitaev model numerically. To this end, we again restrict ourselves to $J_1 = J_2 = 1$ and evolve $J_3(t) = (2 - |t/\ \ |^{\alpha}\mathrm{Sgn}(t))$ for $-\infty \quad t \quad 0$ so that the anisotropic critical point is reached at $t = 0$. The corresponding time-dependent Hamiltonian is given by $H(\vec{k};t) = \sum_{\vec{k}} \psi_{\vec{k}}^{\dagger} \left[ (g_{\vec{k}} - |t/\ \ |^{\alpha}\mathrm{Sgn}(t))\sigma^3 + \alpha_{\vec{k}}\sigma^1 \right] \psi_{\vec{k}}$. We solve the time-dependent Schrödinger equation $i\partial_t \psi_{\vec{k}} = H(\vec{k};t)\psi(\vec{k})$ numerically for each $k$, compute $p_{\vec{k}}$, and use it to obtain the defect density $n = \int d^2k p_{\vec{k}}$ and $Q = \int d^2k \Delta_{\vec{k}} p_{\vec{k}}$ numerically as a function of and $\alpha$. The plots of $n$ and $Q$ vs are shown in Fig. 2 for several representative values of $\alpha$. The lines in the figure indicate the power laws expected from Eqs. (11) and (12) ($n \sim \ \ ^{-3\alpha/[2(\alpha+1)]}$ and $Q \sim \ \ ^{-5\alpha/[2(\alpha+1)]}$) for $d = z' = 2$ and $z = \ \ = m = 1$. The agreement between the numerical and theoretical results corroborates the scaling theory proposed in this work.

## 3. Conclusion

In conclusion, we have shown that the Kitaev model constitutes an example of a two-dimensional model with an anisotropic critical point. We have also demonstrated that the presence of such an anisotropic critical point leads to novel scaling laws defect density and residual energy during slow power-law dynamics which takes the system from a gapped phase to the vicinity of such a critical point. We have generalized our results for such scaling laws for $d$-dimensional systems with such anisotropic critical point. We note that there has been suggestions of experimental realization of the Kitaev model using ultracold atomic system [19]. In the event of such a realization, the simplest experimental test of our theory would involve measurement of defect density $n$ following a slow ramp. Such experiments has recently been performed for standard ultracold boson systems [20].

The authors thank A. Dutta, M. Koller, K. Kubo, A. Polkovnikov, G. Santoro and D. Sen for discussions. KS thanks DST, India for support through grant SR/S2/CMP-001/2009. SS acknowledges support from Grant-in-Aid for Scientific Research (20740225) from MEXT, Japan.

## References

[1] Polkovnikov A, Sengupta K, Silva A and Vengallatore M 2010 *Preprint* 1007.5331
[2] Dziarmaga J 2010 *Adv. Phys.* **59** 1063
[3] Kibble T W B 1976 *J. Phys. A* **9** 1387
[4] Zurek W H 1985 *Nature* **317** 505; Zurek W H 2006 *Rev. Mod. Phys.* **75** 515; Zurek W H, Dorner U and Zoller P 2005 *Phys. Rev. Lett.* **95** 105701
[5] Polkovnikov A 2005 *Phys. Rev.* B **72** 161201(R); Polkovnikov A and Gritsev V 2006 *Nat. Phys.* **4** 477
[6] Damski B 2005 *Phys. Rev. Lett.* **95** 035701; Dziarmaga J 2005 *Phys. Rev. Lett.* **95** 245701; Dziarmaga J, Meisner J and Zurek W H 2008 *Phys. Rev. Lett.* **101** 115701; Cherng R W and Levitov L S 2006 *Phys. Rev.* A **73** 043614
[7] Sen D, Sengupta K and Mondal S 2008 *Phys. Rev. Lett.* **101** 016806; Mondal S, Sengupta K and Sen D 2009 *Phys. Rev.* B **79** 045128
[8] De Grandi C, Gritsev V and Polkovnikov A 2010 *Phys. Rev.* B **81** 012303; De Grandi C and Polkovnikov A 2010 in *Quantum Quenching, Annealing and Computation* (Lect. Notes in Phys. **802**), ed Das A, Chandra A and Chakrabarti B K (Heidelberg: Springer)
[9] Divakaran U, Dutta A and Sen D 2008 *Phys. Rev.* B **78** 144301; Mukherjee V, Dutta A and Sen D 2008 *Phys. Rev.* B **77** 214427; Divakaran U, Mukherjee V, Dutta A and Sen D 2009 *J. Stat. Mech.* P02007; Divakaran U, Mukherjee V, Dutta A and Sen D 2010 in *Quantum Quenching, Annealing and Computation* (Lect. Notes in Phys. **802**), ed Das A, Chandra A and Chakrabarti B K (Heidelberg: Springer); Mukherjee V and Dutta A 2010 *EPL* **92** 37004
[10] Dutta A, Singh R R P and Divakaran U 2010 *EPL* **89**, 67001
[11] Sachdev S 1999 *Quantum Phase Transitions* (Cambridge: Cambridge University Press)
[12] Kitaev A 2003 *Ann. Phys. (N.Y.)* **303** 2
[13] Feng X Y, Zhang G M and Xiang T 2007 *Phys. Rev. Lett.* **98** 087204
[14] Chen H D and Nussinov Z 2008 *J. Phys. A: Math. Theor.* **41** 075001
[15] Sengupta K, Sen D and Mondal S 2008 *Phys. Rev. Lett.* **100** 077204; Mondal S, Sen D and Sengupta K 2008 *Phys. Rev. B* **78** 045101; Mondal S, Sen D and Sengupta K 2010 in *Quantum Quenching, Annealing and Computation* (Lect. Notes in Phys. **802**), ed Das A, Chandra A and Chakrabarti B K (Heidelberg: Springer)
[16] Lieb E H 1994 *Phys. Rev. Lett.* **73** 2158
[17] Vitanov N V 1999 *Phys. Rev.* A **59** 988; Suzuki S and Okada M 2005 in *Quantum Annealing and Related Optimization Methods*, ed Das A and Chakrabarti B K (Berlin: Springer-Verlag)
[18] Barouch E, McCoy B M and Dresden M 1970 *Phys. Rev.* A **2**, 1075 (1970)
[19] Bloch I, Dalibard J and Zwerger W 2008 *Rev. Mod. Phys.* **80** 885
[20] Bakr W S, Peng A, Tai M E, Ma R, Simon J, Gillen J I, Fölling S, Pollet L and Greiner M 2010 *Science* **329** 547

Journal of Physics: Conference Series **297** (2011) 012021 doi:10.1088/1742-6596/297/1/012021

# Quantum phase transitions and dynamical correlations in spin-glass systems

**Kazutaka Takahashi and Yoshiki Matsuda**

Department of Physics, Tokyo Institute of Technology, Tokyo 152-8551, Japan

**Abstract.** We study effects of random fluctuations in quantum spin-glass systems. We show that the expression of the magnetic susceptibility is divided into the classical and quantum parts. The former can be analyzed by the standard replica method with the static approximation. The latter incorporates dynamical correlations and is determined by the energy gap distribution. By using the many-body interacting Ising spin-glass model with transverse field, we discuss possible mechanisms of the quantum phase transitions.

## 1. Introduction

Spin glasses have attracted interest for many years for their fundamental importance to the statistical mechanics and for a wide range of applications [1, 2, 3, 4]. Now the mean-field theory of the spin glasses has been well understood and the Sherrington-Kirkpatrick model [5], which is a typical mean-field spin-glass model, was solved exactly by using the replica method [6, 7, 8, 9].

However, once if we apply the transverse magnetic field, the quantum fluctuation effect becomes important and the model cannot be solved exactly [10]. Generally speaking, the quantum nature becomes important at low temperatures and the picture of the classical spin-glass transition can be changed significantly [11].

For classical thermodynamic systems, the phase transition is determined by the free energy $F = E - TS$. For a given temperature $T$, the state of the system is given by the configuration that gives the smallest free energy, which implies the state with low energy $E$ and large entropy $S$. However, they are not compatible with each other and the competition between them determines the state. When the temperature goes to zero, the entropy becomes irrelevant and the system takes its ground state. Therefore, the quantum phase transition at zero temperature can be found by looking at the energy gap between the ground and first excited states. In spite of this simple picture, the problem is not so simple to solve since the quantum fluctuation makes the ground state nontrivial. Concerning the Sherrington-Kirkpatrick model in a transverse field, there are many studies on the quantum phase transition [12, 13, 14, 15, 16, 17, 18, 19, 20, 21, 22, 23, 24, 25]. However, due to the lack of the effective method, we do not even know the exact value of the transition point.

In this paper, we study the quantum phase transition of the mean-field spin-glass systems. Comparing several models, we discuss possible pictures of the transition. We pay special attention to the role of the quantum fluctuations and see how the effect affects the phase transition at zero temperature [26, 27].

In section 2, we discuss the role of the quantum effect from the expression of the susceptibility. Based on this observation, in section 3, we analyze the many-body interacting Ising spin-glass model in a transverse field by using the replica method. The method used there is a classical one and we study the effect of the quantum fluctuations by the energy gap distribution in section 4. We discuss possible pictures of the quantum phase transition in section 5. Finally, section 6 is devoted to conclusions.

## 2. Susceptibility and quantum fluctuations

Magnetic properties of spin systems can be studied by the response of the system to the applied magnetic field. The magnetic field is included to the Hamiltonian $\hat{H}$ as $\hat{H}(h) = \hat{H} - h\sum_{i=1}^{N}\sigma_i^z$ where $\sigma_i^z$ is the Pauli matrix on site $i$. The magnetization is given by the thermal average of the spin as $m = \sum_{i=1}^{N}\langle\sigma_i^z\rangle/N$. It is expanded in $h$ as $m = \chi_l h - \chi_{nl}h^3 + \cdots$ and the magnetic susceptibility is defined as the coefficients of each term. To study spin-glass systems, it is useful to treat not only the linear susceptibility $\chi_l$ but also the nonlinear one $\chi_{nl}$ [28, 29]. $\chi_l$ is calculated from the partition function $Z$ as

$$\chi_l = \frac{1}{N\beta}\frac{\partial^2}{\partial h^2}[\ln Z]\bigg|_{h=0} = \frac{2N}{\beta}\int_0^\beta d\tau \int_0^\tau d\tau' \, [\langle\sigma(\tau)\sigma(\tau')\rangle] - \beta q, \tag{1}$$

where $\beta$ is the inverse temperature and the square bracket represents the ensemble average. In the last expression, we introduced the imaginary time representation

$$\sigma(\tau) = \frac{1}{N}\sum_{i=1}^{N} e^{\hat{H}\tau}\sigma_i^z e^{-\hat{H}\tau}, \tag{2}$$

and the spin-glass order parameter

$$q = \frac{1}{N}\left[\left\langle\sum_{i=1}^{N}\sigma_i^z\right\rangle^2\right]. \tag{3}$$

The nonlinear susceptibility can also be expressed in a similar way [26, 27].

The imaginary-time formulation clarifies the role of quantum fluctuations since $\tau$ does not appear in classical systems. To see the quantum effects explicitly, we utilize the spectral representation. We insert the completeness relation of eigenstates of the Hamiltonian to the imaginary-time representation. We have the expression

$$\chi_l = \beta(\chi - q) + \frac{2}{N}\left[\frac{1}{Z}\sum_{m,n\,(m\neq n)} e^{-\beta E_n}\frac{\left|\langle n|\sum_{i=1}^{N}\sigma_i^z|m\rangle\right|^2}{E_m - E_n}\right], \tag{4}$$

where $|n\rangle$ is the eigenstate with the eigenvalue $E_n$.

The expression is divided into the classical and quantum parts. The first term represents the classical part and $\chi$ is equal to the spectral function

$$\chi(\omega) = \frac{1}{N}\left[\frac{1}{Z}\sum_{nm} e^{-\beta E_n}|\langle n|\sum_i \sigma_i^z|m\rangle|^2\delta(\omega - (E_m - E_n))\right], \tag{5}$$

at $\omega = 0$. At zero temperature, the state $|n\rangle$ is restricted to the ground state, which means that the excited states do not contribute to the classical part. We note that in the classical system with $\Gamma = 0$, $\chi = 1$ and the linear susceptibility is given by $\chi_l = \beta(1-q)$. It implies that the first term clarifies properties of the classical spin-glass state.

Journal of Physics: Conference Series **297** (2011) 012021 doi:10.1088/1742-6596/297/1/012021

On the other hand, the second term in (4) represents the purely quantum effect since it includes states with different energies. At zero temperature, it reduces to the expression

$$\chi_1^{(\mathrm{q})} = \frac{2}{N}\left[\sum_{n\neq 0}\frac{\left|\langle 0|\sum_{i=1}^N \sigma_i^z|n\rangle\right|^2}{E_n - E_0}\right], \tag{6}$$

where $|0\rangle$ denotes the ground state and $|n\rangle$ the excited states. Transitions between different energies are prohibited in the classical system due to the energy conservation. The above expression shows that, in the quantum system, the excited states affect physical quantities.

Thus, to study the quantum spin-glass systems, we need to take into account both the contributions. If we can neglect the quantum part, the classical analysis is possible and we can use various methods developed in the previous studies. However, when the second term becomes important, the quantum fluctuation effects dominate the susceptibility and we must take a different way. In the following sections, we discuss the role of each part.

## 3. Dynamical mean field theory and the static approximation

### *3.1. Replica method for quantum spin-glasses*

In the path integral formalism, quantum operators can be treated as classical variables. Instead of that, an additional dimension is introduced to represent quantum fluctuations. The partition function of quantum spin Hamiltonian $\hat{H} = H[\hat{\boldsymbol{S}}]$ can be written generally as

$$Z = \int \mathcal{D}\boldsymbol{S}\exp\left\{\int_0^\beta d\tau\,(i\Phi(\tau) - H[\boldsymbol{S}(\tau)])\right\}, \tag{7}$$

where $\boldsymbol{S}(\tau)$ denotes the classical spin variables on the Bloch sphere. The phase factor $\Phi(\tau) = i\langle\boldsymbol{S}(\tau)|(d/d\tau)|\boldsymbol{S}(\tau)\rangle$ includes the derivative with respect to the imaginary time $\tau$, which assures the dynamical nature of the system. Thus, fluctuations in $\tau$ represent quantum effects.

In order to proceed further, we define the model. Our motivation is to discuss quantum effects for the spin-glass transition. As the simplest possible model, we define the $p$-body interacting Ising spin-glass model with transverse field

$$\hat{H} = -\sum_{i_1<i_2<\cdots<i_p}^{N} J_{i_1 i_2\cdots i_p}\sigma_{i_1}^z\sigma_{i_2}^z\cdots\sigma_{i_p}^z - \Gamma\sum_{i=1}^N \sigma_i^x, \tag{8}$$

where $\sigma_i^{z,x}$ are Pauli matrices on site $i$, $N$ is the number of the site and $\Gamma$ is the transverse field. We take the ensemble average over random interaction $J_{i_1 i_2\cdots i_p}$ with the Gaussian distribution

$$\mathrm{P}(J_{i_1\cdots i_p}) = \sqrt{\frac{N^{p-1}}{\pi p! J^2}}\exp\left\{-\frac{N^{p-1}}{p! J^2}(J_{i_1 i_2\cdots i_p})^2\right\}. \tag{9}$$

The model at $p = 2$ is known as the Sherrington-Kirkpatrick model [5] and $p = \infty$ as the Derrida's random energy model [30, 31]. The reason why we study this mean-field model is that the model is solvable at the classical limit $\Gamma \to 0$ and the essential features of the spin-glass state can be found by analyzing this model. With the transverse field, the model cannot be solved exactly except the case of the limit $p \to \infty$ [32].

To handle the ensemble average, we use the replica method which is the standard prescription for the mean-field spin-glasses. We calculate the average of the $n$th power of the partition function and the average free energy per spin is obtained from the relation

$$f = -\frac{1}{N\beta}[\ln Z] = -\frac{1}{N\beta}\lim_{n\to 0}\frac{[Z^n]-1}{n}. \tag{10}$$

Journal of Physics: Conference Series **297** (2011) 012021 doi:10.1088/1742-6596/297/1/012021

The calculation of $[Z^n]$ goes along the same line as the classical case. The main difference is that the variables have the imaginary-time dependence. We obtain the expression

$$[Z^n] = \exp\left\{-\frac{NJ^2}{4}(p-1)\sum_{a,b=1}^{n}\int_0^\beta d\tau d\tau'\left(q_{ab}(\tau,\tau')\right)^p + N\ln \mathrm{Tr}\, e^L\right\}, \tag{11}$$

$$\mathrm{Tr}\, e^L = \int \mathcal{D}\boldsymbol{S}\exp\left\{\sum_{a=1}^{n}\int_0^\beta d\tau\left(i\Phi_a(\tau)+\Gamma S_x^a(\tau)\right)\right.$$
$$\left.+\frac{J^2}{4}p\sum_{a,b=1}^{n}\int_0^\beta d\tau d\tau' S_z^a(\tau)\left(q_{ab}(\tau,\tau')\right)^{p-1}S_z^b(\tau')\right\}, \tag{12}$$

where $a$ and $b$ denote the replica indices and the order parameter $q$ is obtained from the saddle-point equation

$$q_{ab}(\tau,\tau') = \frac{\mathrm{Tr}\, S_z^a(\tau)S_z^b(\tau')e^L}{\mathrm{Tr}\, e^L}. \tag{13}$$

*3.2. Static approximation*

To solve the saddle-point equation without imposing any ansatz is a formidable task and we need a reduction of the equation. If the transverse field is sufficiently large, the perturbative expansion is useful to find the imaginary-time dependence of the order parameter [33]. However, the perturbation is not an appropriate way to treat the spin-glass transition. The possible simplest approximation is to neglect the time dependence of the order parameter [34] as

$$q_{ab}(\tau,\tau') \to \left\{\begin{array}{ll}\chi_a & a=b\\ q_{ab} & a\neq b\end{array}\right. . \tag{14}$$

It was discussed in [25] that the static approximation is justified for $q_{ab}$ and not for $\chi_a$. On the other hand, we may drop the replica dependence of $\chi$ and not of $q$. The replica dependence of $q_{ab}$ is important when we find the replica symmetry breaking solution. However, it is known for the classical limit $\Gamma=0$ at $p=2$ that we can find the transition point from the replica symmetric solution replacing $q_{ab}$ by $q$. For $p\geq 3$, the one-step replica symmetry breaking solution is relevant and we use this ansatz in the following calculation. Using the static approximation and the one-step replica symmetry breaking ansatz, we can write the free energy

$$f = \frac{\beta J^2}{4}(p-1)\left(\chi^p - q^p + mq^p\right) - \frac{1}{\beta m}\ln\int Dz_1\left\{\int Dz_2\left(e^{\beta h}+e^{-\beta h}\right)\right\}^m, \tag{15}$$

and the saddle-point equations

$$\chi = \frac{\int Dz_1\left(\int Dz_2\cosh\beta h\right)^{m-1}\int Dz_2\left(\frac{M^2}{h^2}+\frac{\tanh\beta h}{\beta h}\frac{\Gamma^2}{h^2}\right)\cosh\beta h}{\int Dz_1\left(\int Dz_2\cosh\beta h\right)^m}, \tag{16}$$

$$q = \frac{\int Dz_1\left(\int Dz_2\cosh\beta h\right)^{m-2}\left(\int Dz_2\frac{M}{h}\sinh\beta h\right)^2}{\int Dz_1\left(\int Dz_2\cosh\beta h\right)^m}, \tag{17}$$

where

$$h = \sqrt{\Gamma^2+M^2}, \tag{18}$$

$$M = J\sqrt{\frac{p}{2}}\left(\sqrt{q^{p-1}}z_1 + \sqrt{\chi^{p-1}-q^{p-1}}z_2\right), \tag{19}$$

$$\int Dz_{1,2}(\cdots) = \int_{-\infty}^{\infty}\frac{dz_{1,2}}{\sqrt{2\pi}}e^{-z_{1,2}^2/2}(\cdots). \tag{20}$$

The Parisi parameter $m$ is also found from the saddle-point equation. The replica symmetric solution is given by setting $m = 0$.

Using this expression of the free energy, we can derive the linear susceptibility as $\chi_1 = \beta(\chi - q)$. Therefore, the static approximation corresponds to neglecting the quantum part. If we can neglect the quantum fluctuation effects, the present approximation is justified and we can study the spin-glass state in the same way as the classical case. In fact, at the limit of $p \to \infty$, this approximation is justified. Using the static and one-step replica symmetry breaking ansatz, we can solve the model exactly [32].

At $p = 2$, the static approximation and the replica symmetric ansatz gives the transition point at zero temperature as $\Gamma/J = 2$. This value is larger than that estimated in the following studies. In figure 1, we plot the phase diagram in $\Gamma - T$ plane. The phase boundary separates the paramagnetic phase with $q = 0$ and the spin-glass one with $q \neq 0$.

We also plot the case of $p = 3$ in the figure. In this case, the static approximation and the one-step replica symmetry breaking ansatz is used. At temperatures lower than $T/J \sim 0.35$, the phase transition is of the first order.

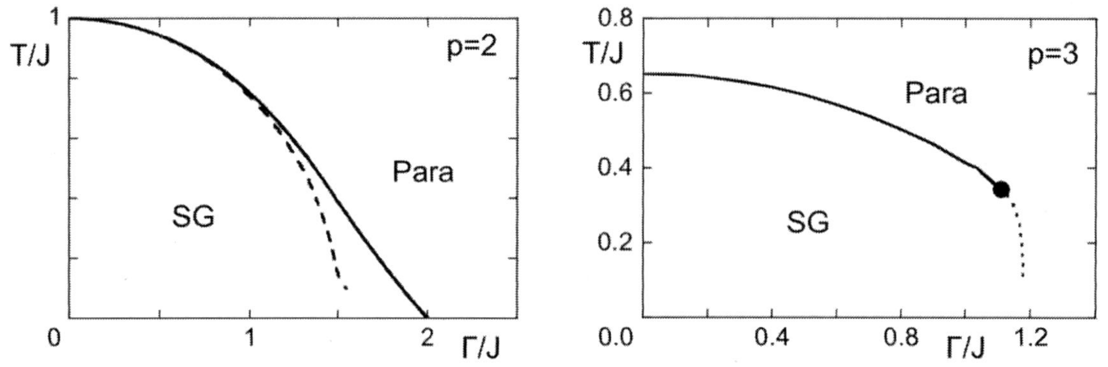


**Figure 1.** Phase diagram at $p = 2$ (left) and $p = 3$ (right) obtained by the static approximation. The solid line denotes the second order phase transition between the paramagnetic (Para) and spin-glass (SG) phases. The dashed line in the left figure is the improved result mentioned in section 3.3. The dotted line in the right figure represents the first order transition. For the improved result at $p = 2$ and the static result at $p = 3$, the phase boundaries are not shown at low temperatures due to numerical problems.

### *3.3. Renormalization of dynamical fluctuations*

The static approximation seems to give a reasonable result for the phase structure, at least, qualitatively. In principle, we can examine the validity of the approximation by using the stability analysis. We write the order parameter as

$$q_{ab}(\tau, \tau') = \begin{cases} \chi + \tilde{\chi}_a(\tau - \tau') & a = b \\ q + \tilde{q}_{ab}(\tau, \tau') & a \neq b \end{cases}, \tag{21}$$

and expand the free energy with respect to the dynamical parts $\tilde{\chi}$ and $\tilde{q}$.

Here, for practical purpose, we consider the renormalization of the static free energy [25]. We write the order parameter as (21) and integrate over the dynamical part as

$$[Z^n] = \int \mathcal{D}\tilde{\chi}\mathcal{D}\tilde{q} \exp\{-Nn\beta f[\chi, q, \tilde{\chi}, \tilde{q}]\} = \exp\{-Nn\beta f_{\text{eff}}[\chi, q]\}. \tag{22}$$

The resultant effective static free energy $f_{\text{eff}}$ includes the dynamical fluctuation effect and can be analyzed by the classical picture. The static order parameters $\chi$ and $q$ are determined from

Journal of Physics: Conference Series **297** (2011) 012021 doi:10.1088/1742-6596/297/1/012021

the saddle-point equation. Of course, it is difficult to perform the integrations exactly and we expand the dynamical variables up to second order. Then, using the Gaussian integration, we obtain $f_{\rm eff}$ [25]. Since the expansion incorporates the factor $\beta J$, it is not justified at low temperatures.

We plot the improved result in figure 1. We see that the result is significantly modified at low temperatures. Since we are using the perturbative approximation, the saddle-point equation has no solution at low-$T$. We need a more reliable method. However, the extrapolation to the zero temperature limit implies the transition point $\Gamma/J \sim 1.5$, which is consistent with the results in [14, 21, 23, 24, 26, 27].

## 4. Energy gap and quantum effects

Next, we study the quantum part of the susceptibility (6). We easily see that this term diverges when the energy gap $E_n - E_0$ goes to zero. This is consistent with the picture that the quantum phase transition occurs when the gap goes to zero. However, in random systems, the location of energy levels depends on the sample and it is not clear how the spin-glass transition point is specified. It should be checked whether the self-averaging property holds for the energy gap distribution.

We calculate the average gap as a function of the transverse field at $p = 2, 3, 4$ and $\infty$ which is obtained by diagonalizing the quantum Hamiltonian numerically. We also consider the magnetization in $x$-direction at $T = 0$ defined as

$$m_x = -\frac{1}{N}\frac{\partial F}{\partial \Gamma}. \tag{23}$$

This is not an order parameter since there is not spontaneous symmetry breaking in the $x$-direction. Rather, this can be understood a kind of entropy at zero temperature. The thermodynamic entropy is defined as $S = -\partial F/\partial T$. Replacing $T$ by $\Gamma$, we obtain $m_x$.

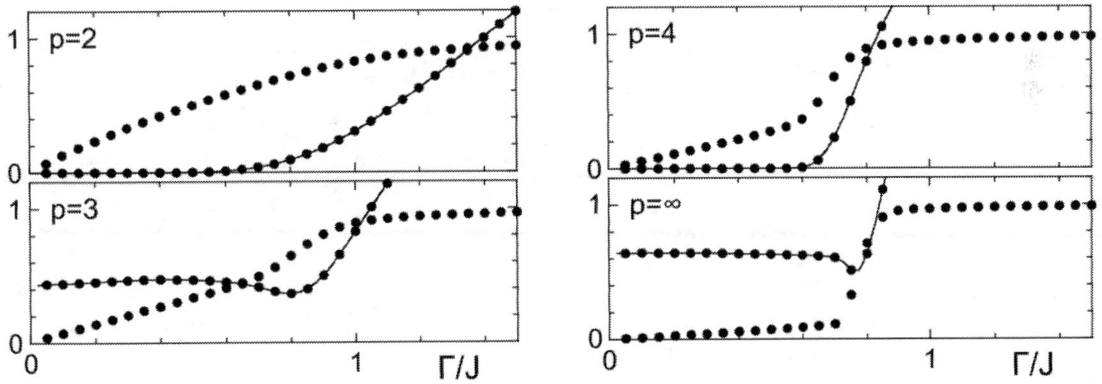


**Figure 2.** Average energy gap (solid line with dots) and magnetization in the transverse direction $m_x$ (dotted line) for $p = 2, 3, 4$ and $\infty$. The size of the system is $N = 20$ and the average is taken over 20000 samples.

We plot the results in figure 2. At $p = 2$, the average gap is an increasing function with respect to the transverse field. The gap vanishing point is around $\Gamma/J \sim 0.7$ which is considerably smaller than the result obtained in the previous section. $m_x$ is also a smooth function and does not show any singular behavior. This result implies that the quantum phase transition cannot be obtained from the average quantity. On the other hand, the results at $p = 3, 4$ and $\infty$ show notable features. At some point, $m_x$ shows a discontinuous change and the energy gap shows a significant change ($p = 4$) or becomes minimum ($p = 3$ and $\infty$). This discontinuous change implies the first order phase transition.

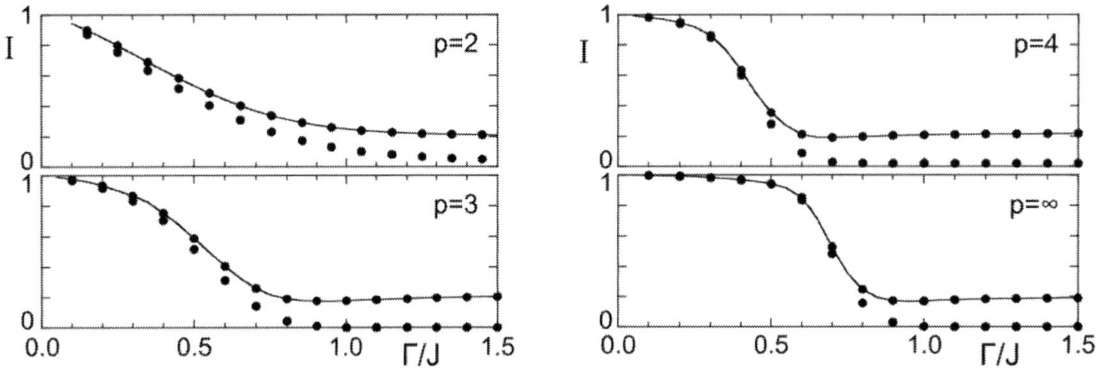


**Figure 3.** Inverse participation ratio for $N = 12$ ($p = 2$ and $\infty$) and $N = 10$ ($p = 3$ and 4). The solid line with dots denotes $I$ in (24) and the dotted line is the contribution from the first excited state ($p = 2$ and 4) or the ground state ($p = 3$ and $\infty$). The average is taken over 20000 samples.

The form of the average energy gap shows a difference between even and odd $p$. In fact, the Hamiltonian commutes with $\prod_{i=1}^{N} \sigma_i^x$ for even $p$ and not for odd $p$ and $p = \infty$. This spin flip symmetry is important to see whether the quantum effect is important or not. Equation (6) includes the matrix element $\langle n|\sigma_i^z|0\rangle$. This quantity measures how much excited states are contained in the ground state multiplied by $\sigma_i^z$. To study this, we calculate the inverse participation ratio

$$I = \sum_n |\langle n|\sigma_i^z|0\rangle|^4. \tag{24}$$

This quantity measures the inverse number of states included in $\sigma_i^z|0\rangle$ [35]. We plot the result in figure 3, which shows that when $\Gamma$ is small, the state $\sigma_i^z|0\rangle$ is almost equal to the first excited state (even $p$) or the ground state (odd $p$). Therefore, for odd $p$, there are not excitations to the higher states and the quantum part gives a small contribution.

## 5. Quantum phase transitions for random systems

Combining the results in the previous sections, we can discuss possible pictures of the quantum spin-glass phase transitions in each case.

- $p = 2$
  In this case, the quantum effect is important to find the phase transition at zero temperature. The second term of the susceptibility (4) dominates the critical behavior and the average gap is not useful to find the transition point.
  The distribution function of the energy gap has a power-law tail $\mathrm{P}(\Delta) \sim \Delta^k$ at small $\Delta$ and the power index $k$ determines the nature of the phase transition since the quantum part of the linear susceptibility is written as $\chi_1^{(q)} \sim \int d\Delta \mathrm{P}(\Delta)/\Delta$. The spin-glass transition can be found by the divergence of the spin-glass susceptibility $\chi_{\rm sg}$. In classical systems, it is directly related to the nonlinear susceptibility $\chi_{\rm nl}$. However, this is not the case for quantum systems. The spectral representation shows that their quantum parts are written as $\chi_{\rm sg} \sim \int d\Delta \mathrm{P}(\Delta)/\Delta^2$ and $\chi_{\rm nl} \sim \int d\Delta \mathrm{P}(\Delta)/\Delta^3$. Thus, these three susceptibilities diverge at different points [26, 27]. In fact, it was also found in finite dimensional systems that the nonlinear susceptibility is divergent within the paramagnetic phase [36, 37, 38, 39, 40, 41]. Thus, the critical behavior of the model is determined by the rare event of samples. In classical systems, such rare configurations of disorder give a weak singularity which is known as the Griffiths-McCoy singularity [42, 43]. The main difference in the present case is that

Journal of Physics: Conference Series **297** (2011) 012021 doi:10.1088/1742-6596/297/1/012021

it is crucial to find the phase transition. In addition to that, such effect is considered not to appear in mean-field systems since such a weak singularity comes from the ordered clusters in a disordered state. In our mean-field system, cluster can be formed in the imaginary time direction. This quantum Griffiths singularity is a purely quantum effect and is considered the main factor of the quantum phase transition of the present model.

- $p = 3$
  The analysis of the inverse participation ratio shows that the quantum part of the susceptibility is small compared to the classical part. Therefore, the classical static approximation is expected to be justified for the replica analysis. At low temperatures, the analysis gives the first order transition at $\Gamma/J \sim 1.18$ while the gap analysis in figure 2 implies a smaller value $\Gamma/J \sim 0.8$. The numerical analysis of the gap was done for a finite size system and we need a more detailed analysis to clarify this small discrepancy.
- $p = 4$
  In this case, the quantum part gives a relevant contribution and the gap fluctuation can be important. However, we see from figure 2 that the behavior of the average gap is related to a discontinuous change of $m_x$ implying a classical first order transition. Although the classical static approximation is not justified, the nature of the phase transition can be classical. We consider that the quantum fluctuation effect is hidden by the abrupt change of the ground state.
- $p = \infty$
  This corresponds to the random energy model and can be solved exactly. The transition point at zero temperature is given by $\Gamma/J = \sqrt{\ln 2} \sim 0.83$. This value can also be found from figure 2. The property of the system can be determined by the ground state and the static approximation is justified in the replica analysis.

## 6. Conclusions

We have discussed quantum fluctuation effects on spin-glass transitions by using the $p$-body interacting Ising spin-glass model in a transverse field. The spectral representation of the susceptibility clarifies the role of quantum fluctuations to determine the state of the system.

The quantum effect can be best seen in the case of $p = 2$ where the gap distribution plays a crucial role for the phase transition at zero temperature. The average gap cannot be any help to characterize the state of the system, which means that the quantum system must be analyzed by a totally different way as the classical one.

For $p \geq 3$, the order parameter shows a discontinuous change. The existence of the quasi-stationary state implies a different type of quantum effects since the tunneling between different states is possible due to the quantum fluctuation.

Thus, compared to the classical spin-glass transitions, we see that the quantum spin-glass systems display different pictures for the phase transition. Our analysis is only for the Ising system with the transverse field and it is interesting to study other quantum systems. Such an analysis is left for future studies.

## Acknowledgments

YM acknowledges support from the Japan Society for the Promotion of Science.

## References

[1] Mézard M, Parisi G and Virasoro M A 1987 *Spin Glass Theory and Beyond* (Singapore: World Scientific)
[2] Fischer K H and Hertz J A 1991 *Spin Glasses* (Cambridge: Cambridge University Press)
[3] Nishimori H 2001 *Statistical Physics of Spin Glasses and Information Processing: An Introduction* (Oxford: Oxford University Press)
[4] Mézard M and Montanari A 2009 *Information, Physics, and Computation* (Oxford: Oxford University Press)

[5] Sherrington D and Kirkpatrick S 1975 *Phys. Rev. Lett.* **35** 1792
[6] Parisi G 1980 *J. Phys. A: Math. Gen.* **13** L115
[7] Parisi G 1980 *J. Phys. A: Math. Gen.* **13** 1101
[8] Parisi G 1980 *J. Phys. A: Math. Gen.* **13** 1887
[9] Talagrand M 2006 *Ann. Math.* **163** 221
[10] Chakrabarti B K, Dutta A and Sen P 1996 *Quantum Ising Phases and Transitions in Transverse Ising Models* (Berlin: Springer)
[11] Sachdev S 1999 *Quantum Phase Transitions* (Cambridge: Cambridge University Press)
[12] Ishii H and Yamamoto T 1986 *J. Phys. C: Solid State Phys.* **18** 6225
[13] Usadel K D 1986 Solid State Commun. **58** 629
[14] Yamamoto T and Ishii H 1987 *J. Phys. C: Solid State Phys.* **20** 6053
[15] Walasek K and Lukierska-Walasek K 1988 *Phys. Rev.* B **38** 725
[16] Ray P, Chakrabarti B K and Chakrabarti A 1989 *Phys. Rev.* B **39** 11828
[17] Yokota T 1989 *Phys. Rev.* B **40** 9321
[18] Thirumalai D, Li Q and Kirkpatrick T R 1989 *J. Phys. A: Math. Gen.* **22** 3339
[19] Goldschmidt Y Y and Lai P Y 1990 *Phys. Rev. Lett.* **64** 2467
[20] Usadel K D, Büttner G and Kopeć T K 1991 *Phys. Rev.* B **44** 12583
[21] Miller J and Huse D A 1993 *Phys. Rev. Lett.* **70** 3147
[22] Rozenberg M J and Grempel D R 1998 *Phys. Rev. Lett.* **81** 2550
[23] Arrachea L and Rozenberg M J 2001 *Phys. Rev. Lett.* **86** 5172
[24] Arrachea L, Dalidovich D, Dobrosavljević V and Rozenberg M J 2004 *Phys. Rev.* B **69** 064419
[25] Takahashi K 2007 *Phys. Rev.* B **76** 184422
[26] Takahashi K and Matsuda Y 2010 *J. Phys. Soc. Japan* **79** 043712
[27] Takahashi K and Matsuda Y 2010 *J. Phys.: Conf. Ser.* **233** 012008
[28] Chalupa J 1977 *Solid State Commun.* **22** 315
[29] Suzuki M 1977 *Prog. Theor. Phys.* **58** 1151
[30] Derrida B 1980 *Phys. Rev. Lett.* **45** 79
[31] Derrida B 1981 *Phys. Rev.* B **24** 2613
[32] Goldschmidt Y Y 1990 *Phys. Rev.* B **41** 4858
[33] Takahashi K and Takeda K 2008 *Phys. Rev.* B **78** 174415
[34] Bray A J and Moore M A 1980 *J. Phys. C: Solid State Phys.* **13** L655
[35] Wegner F 1980 *Z. Phys.* B **36** 209
[36] Fisher D S 1992 *Phys. Rev. Lett.* **69** 534
[37] Guo M, Bhatt R N and Huse D A 1994 *Phys. Rev. Lett.* **72** 4137
[38] Rieger H and Young A P 1994 *Phys. Rev. Lett.* **72** 4141
[39] Fisher D S 1995 *Phys. Rev.* B **51** 6411
[40] Rieger H and Young A P 1996 *Phys. Rev.* B **54** 3328
[41] Guo M, Bhatt R N and Huse D A 1996 *Phys. Rev.* B **54** 3336
[42] Griffiths R B 1969 *Phys. Rev. Lett.* **23** 17
[43] McCoy B M 1969 *Phys. Rev. Lett.* **23** 383

Journal of Physics: Conference Series **297** (2011) 012022 doi:10.1088/1742-6596/297/1/012022

# Phase Transition of Generalized Ferromagnetic Potts Model – Effect of Invisible States –

**Shu Tanaka**

Research Center for Quantum Computing, Interdisciplinary Graduate School of Science and Engineering, Kinki University. 3-4-1, Kowakae, Higashi-Osaka, Osaka 577-8502, Japan

E-mail: `shu-t@alice.math.kindai.ac.jp`

**Ryo Tamura**

Institute for Solid State Physics, University of Tokyo. 5-1-5, Kashiwanoha Kashiwa, Chiba 277-8581, Japan

E-mail: `r.tamura@issp.u-tokyo.ac.jp`

**Naoki Kawashima**

Institute for Solid State Physics, University of Tokyo. 5-1-5, Kashiwanoha Kashiwa, Chiba 277-8581, Japan

E-mail: `kawashima@issp.u-tokyo.ac.jp`

**Abstract.** We investigate the nature of the phase transition of the ferromagnetic Potts model with invisible states. The ferromagnetic Potts model with invisible states can be regarded as a straightforward extension of the standard ferromagnetic Potts model. The invisible states contribute the entropy, however they do not affect the internal energy. They also do not change the symmetry which breaks at the transition temperature. The invisible states stimulate a first-order phase transition. We confirm that the first-order phase transition with spontaneous $q$-fold symmetry breaking for $q = 2, 3$, and 4 takes place even on two-dimensional lattice by Monte Carlo simulation. We also find that the transition temperature decreases and the latent heat increases as the number of invisible states increases.

## 1. Introduction

In statistical physics, it has been a central issue to clarify a relationship between the symmetry which breaks at the transition point and the nature of the phase transition such as the order of the phase transition and the universality class. In order to consider such a relationship, the standard Potts model is often adopted[1, 2]. In the standard Potts model, spins have $q$ values and its interaction is represented by Kronecker's delta. Suppose we consider only two-body interaction, the Hamiltonian of the standard Potts model is given as

$$\mathcal{H}_{\rm s} = -\sum_{\langle i,j \rangle} J_{ij}\delta_{s_i,s_j}, \quad s_i = 1, \cdots, q, \tag{1}$$

where $\langle i, j\rangle$ denotes the pair of the interacted sites. Note that the two-state Potts model is equivalent to the Ising model. Then the $q$-state Potts model can be regarded as the standard generalization of the Ising model. In the standard $q$-state ferromagnetic ($J_{ij}$ 0) Potts model on the two dimensional lattice, second-order and first-order phase transition with spontaneous breaking of $q$-fold symmetry occur when $q$ 4 and otherwise, respectively. The standard ferromagnetic Potts model has sometimes succeeded to explain the nature of phase transition in complicated theoretical models and real materials[2]. For example, the phase transition of the absorption of $^3$He on graphite and the orbital order of transition metal oxides such as $LaMnO_3$ can be explained by three-state Potts model[3, 4, 5].

Recently, there have been found novel examples that the nature of the phase transition is not consistent with that of the standard $q$-state Potts model[6, 7, 8]. For instance, a first-order phase transition with spontaneous threefold symmetry breaking takes place in some two-dimensional frustrated systems with a number of competing short-range interactions. This phase transition seems to be a counterexample of the concept of the standard ferromagnetic $q$-state Potts model, since the standard ferromagnetic three-state Potts model on two-dimensional lattice has a second-order phase transition as mentioned above. Since mechanism of such a phase transition is not clear, from a viewpoint of statistical physics, it is important to understand why a first-order phase transition with threefold symmetry breaking occurs even on two-dimensional lattice.

Purpose of our study is to construct a simple model which can explain such a nature of phase transition by modifying the standard Potts model and to investigate a microscopic mechanism of changing the order of phase transition. We found that the order of phase transition of three-state Potts model changes by adding invisible states[9]. In this paper, we study the detail of the phase transition of Potts model with invisible states and consider the effect of invisible states. The organization of this paper is as follows. In Section 2, we introduce the Potts model with invisible states and review some properties of this model. In Section 3, we show the mean-field analysis and result obtained by Monte Carlo simulation. In Section 4, we summarize our study and show future perspective.

## 2. Model

We consider the following Hamiltonian which is called the ($q$,$r$)-state Potts model[9, 10]:

$$\mathcal{H} = -J \sum_{\langle i,j\rangle} \delta_{s_i,s_j} \sum_{\alpha=1}^{q} \delta_{s_i,\alpha}\delta_{s_j,\alpha}, \qquad s_i = 1, \cdots, q, q+1, \cdots, q+r, \tag{2}$$

where $\langle i, j\rangle$ denotes the pair of the nearest neighbor sites on square lattice. Note that spins can have discrete values from 1 to $q + r$ in this model, while the spins in the standard $q$-state ferromagnetic Potts model can have a value from 1 to $q$. We call the states where 1 $s_i$ $q$ "colored states" and the states where $q + 1$ $s_i$ $q + r$ "invisible states"[9]. In this paper, we only consider the case for ferromagnetic coupling $J > 0$. If and only if 1 $s_i = s_j$ $q$, the interaction $J$ works and it is an energetically favorable state.

First we consider two spin system. Figure 1 (a) shows the ground states for the (3,2)-state Potts model. Figure 1 (b) and (c) denote excited states for the (3,2)-state Potts model. States depicted in Fig. 1 (a) and (b) appear in the standard ferromagnetic $q$-state Potts model, whereas states depicted in Fig. 1 (c) appear only when the invisible states are introduced. Figure 1 (d) shows the energy level structure of two spin system for the ($q$,$r$)-state Potts model. The number of ground states is $q$ and that of the excited states is $q^2 - q + 2qr + r^2$. Contribution of the invisible states in excited states is $2qr + r^2$. This energy level structure indicates that $q$-fold symmetry breaks at the transition point.

 doi:10.1088/1742-6596/297/1/012022

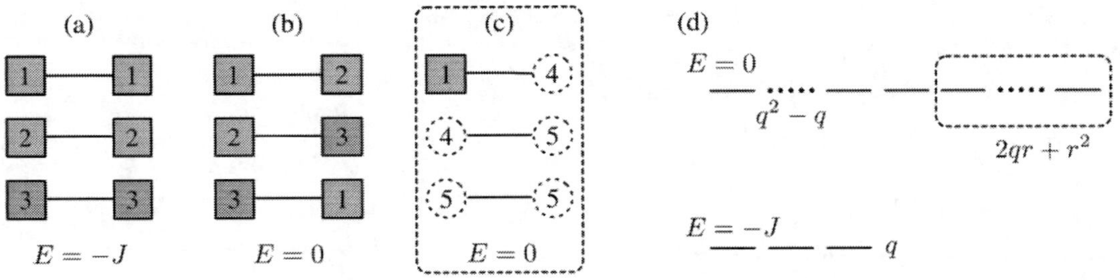


**Figure 1.** (a)-(c): The (3,2)-state Potts model. The squares and dotted circles indicate colored state and invisible states, respectively. (a) Ground states of two spin system. (b) Excited states of two spin system. Both (a) and (b) also appear in the case of the standard ferromagnetic $q$-state Potts model. (c) Excited states of two spin system due to invisible states. (d) Energy level structure of two spin system for the ($q$,$r$)-state Potts model. The number of ground states and excited states are $q$ and $q^2 - q + 2qr + r^2$, respectively. The dotted boxes in (c) and (d) denote the contribution of invisible states.

By comparing the partition function, we can obtain equivalent Hamiltonian:

$$\mathcal{H}' = -J \sum_{\langle i,j \rangle} \delta_{\sigma_i,\sigma_j} \sum_{\alpha=1}^{q} \delta_{\sigma_i,\alpha}\delta_{\sigma_j,\alpha} - T \log r \sum_i \delta_{\sigma_i,0}, \quad \sigma_i = 0, 1, \cdots, q, \tag{3}$$

where we rename the label of invisible state from $q+1$  $s_i$  $q+r$ to $\sigma_i = 0$. The second term of Eq. (3) represents the chemical potential of invisible state. The temperature-dependent chemical potential comes from entropy effect of invisible states. Thus, the invisible state does not contribute the internal energy but affect the entropy.

## 3. Result

In the first half of this section, we consider the Bragg-Williams approximation. In the last half of this section, we study a nature of phase transition by Monte Carlo simulation.

### *3.1. Mean-field analysis*

First we consider the order of the phase transition of the Potts model with invisible states by the Bragg-Williams approximation. Let $x_\alpha$ be the fraction of $\alpha$-th state ($0$  $\alpha$  $q$) with condition $\sum_{\alpha=0}^{q} x_\alpha = 1$. We first define the fraction of invisible states ($x_0 = t$). If the $q$-fold symmetry breaking occurs, one of $q$-states is particularly selected. Here we define the label of the selected state is $\alpha = 1$. Then, it should be noted that the fractions $x_\alpha$ for $2$  $\alpha$  $q$ is independent of $\alpha$ but different from that for $\alpha = 1$. This situation is explicitly summarized as follows:

$$x_0 = t, \quad x_1 = \frac{1}{q}(1-t)\left[1+(q-1)s\right], \quad x_\alpha = \frac{1}{q}(1-t)(1-s) \quad (2 \quad \alpha \quad q), \tag{4}$$

where $0$  $s, t$  $1$.

Let $z$ be the number of the nearest neighbor sites. The free energy is given by

$$\begin{aligned} F^{\mathrm{BW}}(s,t) &= -\frac{zJ(1-t)^2}{2q}\left[(q-1)s^2+1\right] + tT\log\frac{t}{r} \\ &+ (1-t)T\left[\frac{1+(q-1)s}{q}\log\frac{1+(q-1)s}{1-s} + \log\frac{(1-t)(1-s)}{q}\right]. \end{aligned} \tag{5}$$

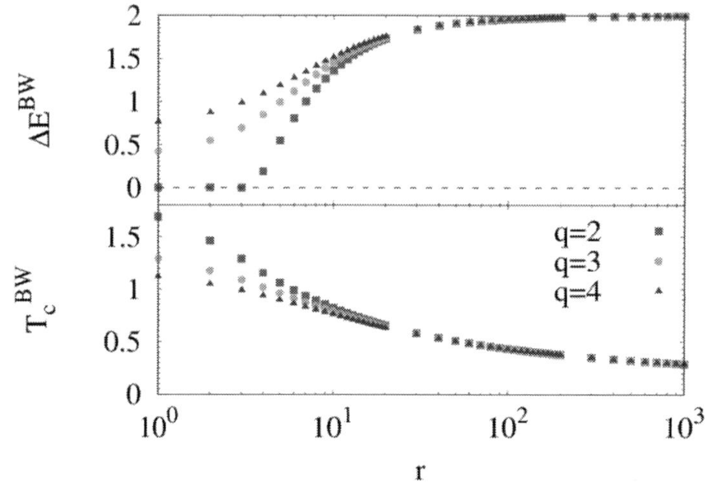


**Figure 2.** The transition temperature $T_{\mathrm{c}}^{\mathrm{BW}}$ and the latent heat $\Delta E^{\mathrm{BW}}$ as functions of the number of invisible states $r$ for $z = 4$ obtained by the Bragg-Williams approximation.

From this free energy, we can obtain the latent heat and transition temperature in the cases of $q = 2, 3$, and 4. Figure 2 shows $r$-dependency of the latent heat and transition temperature for $z = 4$. From Fig. 2, we find a second-order phase transition takes place only for (2,1), (2,2), and (2,3)-state Potts model. Note that the transition temperature and the latent heat for $r = 0$ (*i.e.* the standard Potts model) are given as

$$T_{\mathrm{c}}^{\mathrm{BW}}\left(q=2, r=0\right) = \frac{zJ}{2}, \qquad T_{\mathrm{c}}^{\mathrm{BW}}\left(q, r=0\right) = \frac{zJ}{2\log\left(q-1\right)}\left(\frac{q-2}{q-1}\right) \text{ (for } q \quad 3), \tag{6}$$

$$\Delta E^{\mathrm{BW}}\left(q, r=0\right) = \frac{zJ\left(q-2\right)^2}{2q\left(q-1\right)} \text{ (for } q \quad 3), \tag{7}$$

respectively[11]. As the number of invisible states $r$ increases, the latent heat $\Delta E^{\mathrm{BW}}$ increases and the transition point $T_{\mathrm{c}}^{\mathrm{BW}}$ decreases. From this fact, it can be considered that the invisible state drives a first-order phase transition.

### *3.2. Monte Carlo simulation*

To investigate the effect of invisible states on two dimensional lattice, we study equilibrium properties of the ($q$,$r$)-state Potts model on square lattice by Monte Carlo simulation. Figure 3 shows typical snapshots of the (3,27)-state Potts model for several temperatures. The color points and white points indicate the colored states and the invisible states, respectively. The red, green, and blue points depict $\sigma_i = 1, 2$, and 3, respectively. At zero temperature, all spins take the same colored state. As the temperature increases, size of the ordered domains become small. Above the transition temperature, small colored islands are floating in an invisible sea.

Figure 4 (a) shows the specific heat as a function of temperature. In order to obtain the data with high accuracy, we adopt reweighting method. The maximum value of specific heat increases almost linear dependence with system size. To clarify whether a first-order phase transition occurs or not, next we calculate the energy distribution and implement finite size scaling. Figure 4 (b) shows the energy histogram at the temperature where the specific heat has the maximum value. Energy histogram in Fig. 4 (b) is a bimodal distribution which is

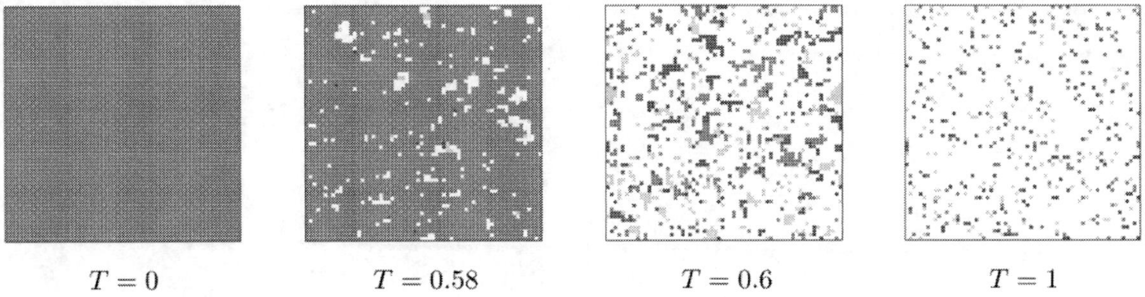


**Figure 3.** Typical snapshots for the (3,27)-state ferromagnetic Potts model. As the temperature decreases, the density of invisible states decreases and the ordering domain grows.

a characteristic behavior of the first-order phase transition. As the system size increases, the peaks become high.

Next we implement the finite size scaling which is established by Challa *et al.*[12]. The finite size scaling for a first-order phase transition in $d$-dimensional system is treated as:

$$T_c(L) = aL^{-d} + T_c(\infty), \quad C_{\max}(L) \propto \frac{(\Delta E)^2 L^d}{4T_c^2(\infty)}, \tag{8}$$

where $T_c(L)$ denotes the temperature where the specific heat takes the maximum value $C_{\max}(L)$ in finite size $L^d$ system. The values $T_c(\infty)$ and $\Delta E$ denote the transition temperature and the latent heat for the thermodynamic limit, respectively. We fit $T_c(L)$ as a function of $L^{-2}$ as shown in Fig. 5(a). From the intercept of fitting curve, we obtain $T_c(\infty) = 0.58513(1)$. Next we fit $C_{\max}(L)$ as a function of $L^2$. We estimate the latent heat $\Delta E = 1.05(2)$ from the gradient of the fitting curve and the obtained transition temperature $T_c(\infty)$ as shown in Fig. 5(b).

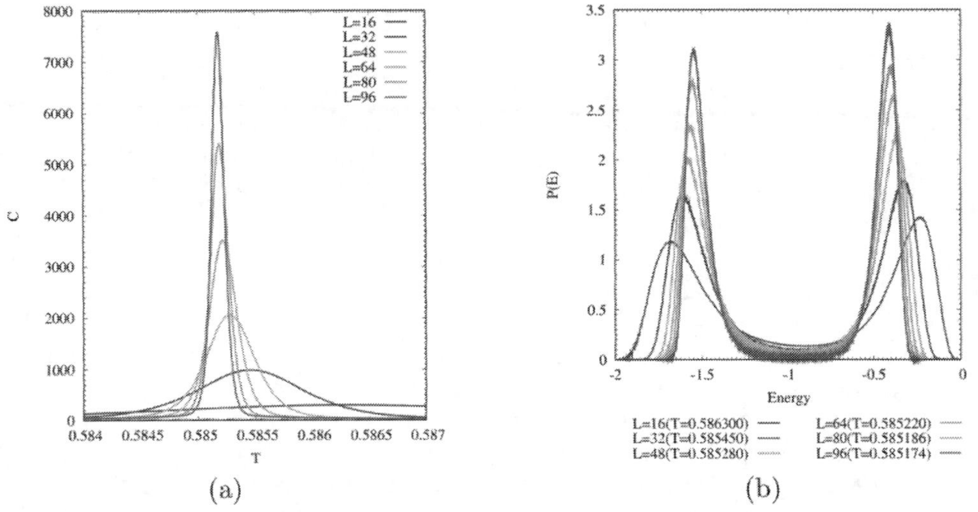


**Figure 4.** (a) Temperature dependency of the specific heat for the (3,27)-state Potts model. (b) Energy histogram for the (3,27)-state Potts model.

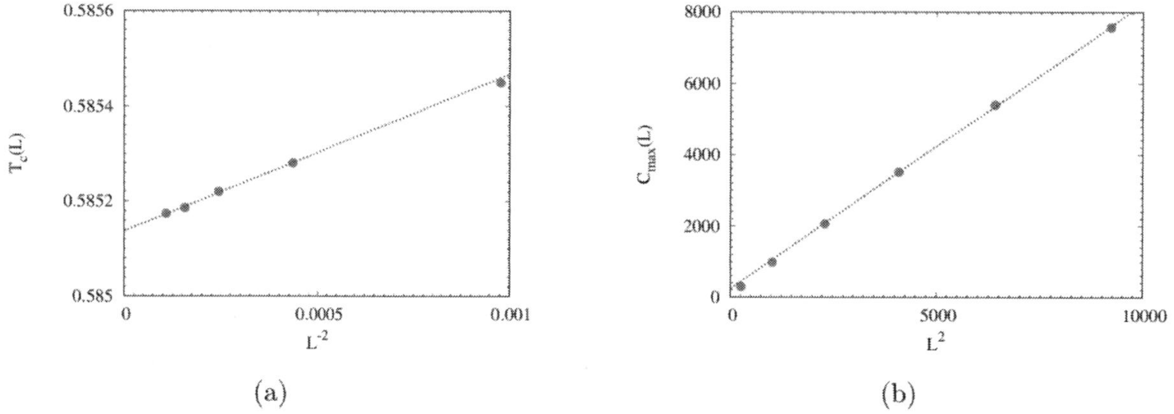


**Figure 5.** (a) $T_c(L)$ as a function of $L^{-2}$. Intercept of the fitting curve indicates the estimated transition temperature for the thermodynamic limit. (b) $C_{\max}(L)$ as a function of $L^2$.

**Table 1.** Transition temperature and latent heat for several parameter set of $(q,r)$-state Potts model. For $r = 0$ case, $T_c$ and $\Delta E$ were obtained exactly[2].

| $(q,r)$ | $T_c$ | $\Delta E$ | Symmetry |
|---|---|---|---|
| (2,0) | 1.13459 | 0 | 2-fold |
| (2,30) | 0.57837(1) | 1.02(2) | 2-fold |
| (2,32) | 0.56857(1) | 1.23(2) | 2-fold |
| (3,0) | 0.994973 | 0 | 3-fold |
| (3,25) | 0.59630(1) | 0.81(2) | 3-fold |
| (3,27) | 0.58513(1) | 1.05(2) | 3-fold |
| (4,0) | 0.910239 | 0 | 4-fold |
| (4,20) | 0.61683(1) | 0.68(2) | 4-fold |
| (4,22) | 0.60396(1) | 0.87(2) | 4-fold |

By the same analysis, we calculate the transition temperature and the latent heat for several parameter set of $(q,r)$ (Table 1). As the number of invisible states increases, the transition temperature decreases and latent heat increases. From Table 1, we conclude the invisible states stimulate a first-order phase transition, which is consistent with the result obtained by the Bragg-Williams approximation.

## 4. Conclusion and Future Perspective

In this paper, we study the nature of the phase transition of the Potts model with invisible states. The Potts model with invisible states is regarded as a straightforward extension of the standard ferromagnetic Potts model, since this model is constructed by just adding the invisible states. The invisible state does not affect the internal energy but contributes to the entropy. It does not change the symmetry which breaks at the transition temperature. From the result obtained by the Bragg-Williams approximation and Monte Carlo simulation, the invisible state drives a first-order phase transition, since the latent heat increases as the number of the invisible states increases. We believe that this model is a fundamental model and it enables us to understand why the first-order phase transition with spontaneous 2,3, and 4-fold symmetry breaking can occur even on two-dimensional lattice.

Recently, the Potts model has been adopted for analysis of problems in information science and technology such as clustering problem[13, 14]. It has been suggested that the effects of thermal and quantum fluctuation are efficient for optimization problems. We expect that the fluctuation which comes from the invisible states is efficient for such problems.

## Acknowledgments

The authors are grateful to Jie Lou, Yoshiki Matsuda, Seiji Miyashita, Takashi Mori, Yohsuke Murase, Masayuki Ohzeki, and Eric Vincent for their valuable comments. S. T. is partly supported by Grant-in-Aid for Young Scientists Start-up (21840021) from the JSPS and the "Open Research Center" Project for Private Universities: matching fund subsidy from MEXT. R.T. is partly supported by Global COE Program "the Physical Sciences Frontier", MEXT, Japan. The present work is financially supported by MEXT Grant-in-Aid for Scientific Research (B) (22340111), and for Scientific Research on Priority Areas "Novel States of Matter Induced by Frustration" (19052004), and by Next Generation Supercomputing Project, Nanoscience Program, MEXT, Japan. The computation in the present work was performed on computers at the Supercomputer Center, Institute for Solid State Physics, University of Tokyo.

## References

[1] Potts R B 1952 *Proc. Cambridge Philos. Soc.* **48** 106
[2] Wu F Y 1982 *Rev. Mod. Phys.* **54** 235
[3] Bretz M 1977 *Phys. Rev. Lett.* **38** 501
[4] Tejwani M J 1980 *Phys. Rev. Lett.* **44** 152
[5] Ahmed M R and Gehring G A 2009 *Phys. Rev. B* **79** 174106
[6] Tamura R and Kawashima N 2008 *J. Phys. Soc. Jpn.* **77** 103002
[7] Stoudenmire E M, Trebst S and Balents L 2009 *Phys. Rev. B* **79** 214436
[8] Okumura S, Kawamura H, Okubo T and Motome Y 2010 *J. Phys. Soc. Jpn.* **79** 114705
[9] Tamura R, Tanaka S and Kawashima N 2010 *Prog. Theor. Phys.* **124** 381
[10] Tanaka S, Tamura R arXiv:1012.4254
[11] Kihara T, Midzuno Y and Shizume T 1954 *J. Phys. Soc. Jpn.* **9** 681
[12] Challa M S S, Landau D P and Binder K *Phys. Rev. B* **34** 1841.
[13] Kurihara K, Tanaka S and Miyashita S 2009 *Proceedings of the 25th Conference on Uncertainty in Artificial Intelligence*
[14] Sato I, Kurihara K, Tanaka S, Nakagawa H, and Miyashita S 2009 *Proceedings of the 25th Conference on Uncertainty in Artificial Intelligence*

Journal of Physics: Conference Series **297** (2011) 012023 doi:10.1088/1742-6596/297/1/012023

# Molecular Dynamics Study of Rotating Nanodroplets: Finite-size Effects and Nonequilibrium Deformation

**Hiroshi Watanabe**[1,2], **Naoki Mitsuda**[3], **Tomoaki Nogawa**[3], **and Nobuyasu Ito**[3]

[1] Institute for Solid State Physics, The University of Tokyo, Kashiwanoha 5-1-5, Kashiwa, Chiba 277-8581, JAPAN

[2] Super Computing Division Information Technology Center University of Tokyo, 2-11-16 Yayoi, Bunkyo, Tokyo 113-8658, Japan

[3] Department of Applied Physics, School of Engineering, The University of Tokyo, Hongo, Bunkyo-ku, Tokyo 113-8656, Japan

E-mail: hwatanabe@issp.u-tokyo.ac.jp

**Abstract.** Noneqiuilibrium dynamics of rotating droplets are studied by molecular dynamics simulations. Small deviations from the theoretical prediction are observed when the size of a droplet is small, and the deviations become smaller as the size of the droplet increases. The characteristic timescale of the deformation is observed, and we find (i) the deformation timescale is almost independent of the rotating velocity with for small frequency and (ii) the deformation timescale becomes shorter as temperature increases. A simple model is proposed to explain the deformation dynamics of droplets.

## 1. Introduction

Droplets are small drops of liquid which are commonly observed in our life. Many things can be considered as ensembles of liquid droplets, such as rain, clouds, and splays, etc. Droplets are not only familiar to us, but their behaviors are also important for engineering, such as spray combustion [1], developments of inkjet printers [2] and electronic sputtering [3], etc. Behaviors of droplets are mainly governed by two kinds of forces, namely, the surface tension and the inertial force. The surface tension works as restoring force trying to keep shape of a droplet, while the inertial force usually tries to deform and destroy it. A simple example of such balance between two forces can be seen in rotating droplets where the centrifugal force plays the role of the inertial force. One of the fundamental studies of such rotating droplets was investigated by Plateau [4]. A water-droplet was put in alcohol with same density in order to mimic a gravity-less system, and then the container was rotated. The droplet was gradually compressed to flat shape as the rotating velocity increased and became unstable when the angular velocity exceeded some critical value. While the experimental technique is quite simple, it is difficult to investigate this phenomena analytically because of the influence from the ambient fluid. Recently, levitated droplets have attracted much researchers' interests. A droplet is levitated by some external force, such as electromagnetic force [5], and then a containerless system is achieved which is convenient for comparison with theoretical predictions. The properties of rotating droplets in

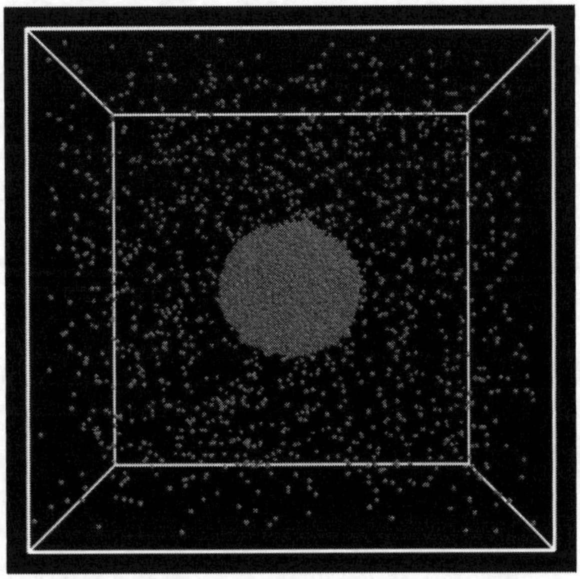

**Figure 1.** A typical snapshot of a static droplet. The white lines denotes the simulation box with the linear size $L = 40$. The total number of particles $N = 16000$ and the temperature $T = 0.6$. The red and gray particles denote a liquid droplet and vapor, respectively.

steady-state are investigated theoretically, and numerical works were followed [6, 7]. Also, the gravityless experiments were performed in spacelab by Wang *et al.* [8]. They reported that while the axisymmetric shapes are well described by the theoretical prediction, the bifurcation point from axisymmetric to nonaxisymmetric shapes locates at a lower rotation velocity than the theoretical prediction. Despite of such past studies, dynamics of the deformation, especially nonlinear motions which cannot be described as simple oscillations, have not been clarified yet. Dynamical aspects of the deformation, such as characteristic timescale of the deformation or how the droplets separates into fragmentations, are very important both for theoretical interests and for applications. Additionally, the validity of the continuum treatments should be tested for small droplets, since a droplet consists of many particles, and consequently, the surface of it has finite thickness, while thickness is usually ignored in the continuum theory. In the present study, we investigate the dynamics of the rotating droplets using molecular dynamics (MD) simulations. Using MD, dynamics of droplets can be studied naturally taking account of its thickness. We first make a brief review of the continuum treatments for rotating droplets in the steady state, and compare it to our results after describing details of numerical methods. Deformation dynamics are observed for different conditions, and a simple model is proposed in order to explain the results.

## 2. Form of Rotating Droplet

The form of rotating droplets are solved exactly by Chandrasekhar [9] when its shape is uniaxial. The steady forms of rotating droplets are determined by the balance between the surface tension and the centrifugal force. Here, we make the brief review of the form of the rotating droplet following the manner by Chandrasekhar.

Consider a droplet rotating around $z$-axis in a system without the gravity. In the steady state, the gradient of the internal pressure of the droplet should be balanced by the centrifugal

Journal of Physics: Conference Series **297** (2011) 012023 doi:10.1088/1742-6596/297/1/012023

force as

$$\rho\omega^2 r = \frac{\mathrm{d}p}{\mathrm{d}r}, \tag{1}$$

where $\rho$ is the density of the droplet, $\omega$ is angular velocity, $p$ is the internal pressure, and $r$ is the distance from the rotating axis, respectively. Equation (1) directly leads to

$$p(r) = p_0 + \frac{1}{2}\rho\omega^2 r^2, \tag{2}$$

with the pressure $p_0$ at the center of the droplet ($r = 0$). At the surface of the droplet, the pressure should satisfy the following relation

$$p = \quad \mathrm{div}\mathbf{n}, \tag{3}$$

with the surface tension  and the normal vector of the droplet surface $\mathbf{n}$. From Eqs. (2) and (3), we have

$$p_0 + \frac{1}{2}\rho\omega^2 r^2 = \quad \mathrm{div}\mathbf{n}. \tag{4}$$

While shape of a rotating droplet varies from the sphere, the shape will be uniaxial when the rotating speed is small. Therefore, the form of the droplet can be expressed as $z = f(r)$, where $z$ is the height from the equatorial plane and $r$ is distance from $z$-axis, namely, $r^2 = x^2 + y^2$, respectively. Then the unit normal vector of the droplet surface $\mathbf{n} = (n_x, n_y, n_z)$ is

$$n_x = -\frac{\phi}{\sqrt{1+\phi^2}}\frac{x}{r}, \tag{5}$$

$$n_y = -\frac{\phi}{\sqrt{1+\phi^2}}\frac{y}{r}, \tag{6}$$

$$n_z = \frac{1}{\sqrt{1+\phi^2}}, \tag{7}$$

where

$$\phi \equiv \frac{\mathrm{d}f}{\mathrm{d}r}. \tag{8}$$

From the above, we easily have,

$$\mathrm{div}\mathbf{n} = \frac{\partial n_x}{\partial x} + \frac{\partial n_y}{\partial y} + \frac{\partial n_z}{\partial z}, \tag{9}$$

$$= -\frac{\phi}{r(1+\phi^2)^{1/2}} - \frac{1}{(1+\phi^2)^{3/2}}\frac{\mathrm{d}\phi}{\mathrm{d}r}, \tag{10}$$

$$= -\frac{1}{r}\frac{\mathrm{d}}{\mathrm{d}r}\frac{r\phi}{\sqrt{1+\phi^2}}. \tag{11}$$

Substituting Eq. (11) to Eq. (4), we obtain

$$p_0 + \frac{1}{2}\rho\omega^2 r^2 = -\frac{\quad}{r}\frac{\mathrm{d}}{\mathrm{d}r}\frac{r\phi}{\sqrt{1+\phi^2}}, \tag{12}$$

which is integrable to be

$$\frac{p_0 r^2}{2} + \frac{\rho\omega^2 r^4}{8} = -\frac{r\phi}{\sqrt{1+\phi^2}}. \tag{13}$$

Journal of Physics: Conference Series **297** (2011) 012023 doi:10.1088/1742-6596/297/1/012023

Note that, the constant of the integration should be zero because of the rotational symmetry around $z$-axis. Let $R_\mathrm{e}$ be the equatorial radius of the rotating droplet. Since $\phi \to -\infty$ when $r \to R_\mathrm{e}$, we have

$$\frac{p_0 R_\mathrm{e}}{2} + \frac{\rho\omega^2 R_\mathrm{e}^3}{8} = 1. \tag{14}$$

We introduce a dimensionless value $\Sigma$ as

$$\Sigma \equiv \frac{\rho\omega^2 R_\mathrm{e}^3}{8} \equiv \left(\frac{\omega}{\omega_0}\right)^2, \tag{15}$$

where $\omega_0$ is the characteristic frequency which is defined as

$$\omega_0 = \sqrt{\frac{8}{\rho R_\mathrm{e}^3}}. \tag{16}$$

Hereafter, we measure the length in the unit of the equatorial radius as $\bar{r} = r/R_\mathrm{e}$. Equation (13) is then reduced to be

$$\frac{\phi}{\sqrt{1+\phi^2}} = -\bar{r}(1 - \Sigma + \Sigma\bar{r}^2), \tag{17}$$

or equivalently,

$$\phi = -\frac{g}{\sqrt{1-g^2}}, \tag{18}$$

where

$$g(\bar{r}) \equiv \bar{r}(1 - \Sigma + \Sigma\bar{r}^2). \tag{19}$$

Finally, we obtain the form of the rotating droplet by the following integration,

$$f(\bar{r}) = \int_0^{\bar{r}} \phi \mathrm{d}\bar{r}. \tag{20}$$

While Eq. (20) can be expressed by the Jacobi elliptic functions, the expression is inconvenient for comparison with numerical results. We, therefore, derive the expression for the ratio of the equatorial radius of the rotating droplet to the radius of the droplet at rest. The volume of the droplet $V$ is expressed in terms of $f(\bar{r})$ by

$$V = 4\pi R_\mathrm{e}^3 \int_0^1 \bar{r} f(\bar{r}) \mathrm{d}\bar{r}. \tag{21}$$

The integration by parts leads to

$$\frac{V}{R_\mathrm{e}^3} = 4\pi \left[\frac{\bar{r}^2}{2} f\right]_0^1 - 2\pi \int_0^1 \bar{r}^2 \frac{\mathrm{d}f}{\mathrm{d}\bar{r}} \mathrm{d}\bar{r}, \tag{22}$$

$$= 2\pi \int_0^1 \frac{\bar{r}^2 g}{\sqrt{1-g^2}} \mathrm{d}\bar{r}, \tag{23}$$

$$\equiv 2\pi h, \tag{24}$$

where $h$ is the function of the reduced rotational frequency $\omega/\omega_0$. Consider the sphere with the radius $R_0$ which volume is the same as the rotating droplet, that is, $V \equiv 4\pi R_0^3/3$. Provided that the volume are not changed by rotation, $R_0$ denotes the radius of the droplet at rest. The radius ratio $R_\mathrm{e}/R_0$ is then expressed as,

$$R_\mathrm{e}/R_0 = \left(\frac{2}{3h}\right)^{1/3}. \tag{25}$$

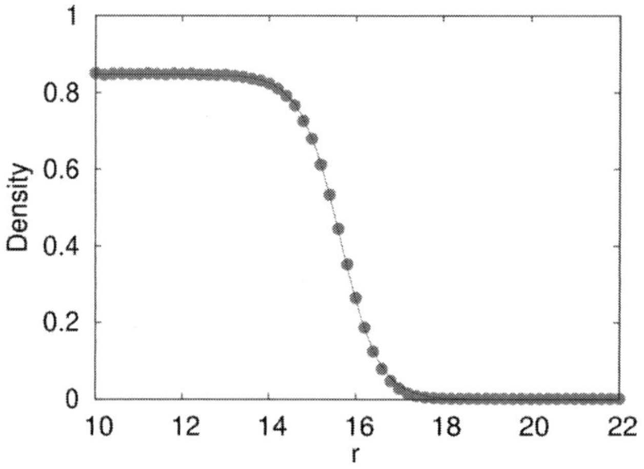


**Figure 2.** The density of the droplet for $N = 16000$. The origin is set at the center of the droplet. The solid line denotes the hyperbolic tangent function of the form $\tanh((r - R_0)/\ )$ with the radius of the droplet $R_0 = 15.7(3)$ and the thickness of the surface $\ = 0.88(5)$. The ratio of the thickness to the radius of the droplet is about 5.6%.

Equation (25) means that deformation behaviors can be scaled for different rotation speed, volume of droplets, surface tension, and so forth. Equation (25) is shown as the solid line in Fig. 5.

A couple of things are worth to be noted. First, the equilibrium form of the rotating droplet can be given by the variational principle with appropriately chosen effective potential function, as described in Ref. [6]. An Euler-Lagrange equation derived by the variational principle leads to the Young-Laplace equation (3). Therefore, the above arguments are completely equivalent to those from the variational principle. Second, one has to be careful with the definition of the characteristic frequency defined in Eq. (15), since some researchers take the radius of the droplet at rest $R_0$ as the scaling length, while we use the equatorial radius of the rotating droplet $R_e$. Accordingly, the definition of $\Sigma$ and $\omega_0$ can be different for researchers.

## 3. Method

In order to study the dynamics of rotating droplets, we perform MD simulations. We use the truncated Lennard-Jones potential of the form

$$V(r) = 4\varepsilon\left[\left(\frac{\sigma}{r}\right)^{12} - \left(\frac{\sigma}{r}\right)^{6} + c_2\left(\frac{r}{\sigma}\right)^{2} + c_0\right], \tag{26}$$

with the well depth $\varepsilon$, the atomic diameter $\sigma$, and the cut-off length $r_c$ [10]. The coefficients $c_0$ and $c_2$ are determined so that $V(r_c) = V'(r_c) = 0$, i.e., the values of potential and the force become continuously zero at the truncation point. We choose the cutoff-length as $r_c = 3.0$. In the following, we measure the physical quantities in the unit of the radius $\sigma$, the well depth $\varepsilon$, the Boltzmann constant $k_B$, and the particle mass $m$. Time step is chosen to be 0.005. The simulation box is the cube with the linear size $L$, and the periodic boundary condition is taken for all directions. Total number of particles $N = 4000, 8000$ and $16000$ are studied. Temperature is controlled by Langevin thermostat [11]. Simulations are partially performed by MDACP which is freely available online [12]

| $N$ | $R_0$ | $\rho$ | | |
|---|---|---|---|---|
| 4000 | 9.61(4) | 0.85(1) | 0.85(6) | 0.62(1) |
| 8000 | 12.4(3) | 0.85(1) | 0.88(5) | 0.67(3) |
| 16000 | 15.7(3) | 0.85(1) | 0.88(5) | 0.72(1) |

**Table 1.** Physical quantities of droplets at rest for $N = 4000, 8000$, and 16000. $R_0$ is the position of Gibbs surface, $\rho$ is the density of the liquid, is the thickness of the surface, and is the surface tension, respectively. While the density and the thickness are almost independent of the size of the droplets, the value of the surface tension increases as the size of the droplet increases.

Particles are initially distributed spherically at the face-centered-cubic lattice, and time evolution is performed under the fixed temperature $T = 0.6$. The density of the total system is $\rho \equiv N/L^3 = 0.25$, which is the liquid-vapor coexistent phase at the temperature. The system, therefore, reaches its equilibrium state after sufficiently long time, and the system contains a single droplet in the saturated vapor. In the present study, $10^6$ steps are spent for the thermalization. An typical snapshot of a droplet is shown in Fig. 1. In order to identify the droplet, we define that two particles within the distance 1.3 belong to the same cluster, and the largest cluster is defined as the droplet which is shown in red particles in the following figures. The density profile is shown in Fig. 2. Since the interface between the liquid and the gas has finite thickness, we have to define the equatorial radius to compare the results with the theory. Here, we use the position of Gibbs surface to define the equatorial radius. Let $r$ be distance from the center of a droplet. The local density $\rho(r)$ is well approximated by the hyperbolic tangent function as,

$$\rho(r) = (\rho_{\mathrm{L}} - \rho_{\mathrm{G}}) \tanh\left((r - R_{\mathrm{G}})/\ \ \right) + \rho_{\mathrm{G}}, \tag{27}$$

with the liquid density $\rho_{\mathrm{L}}$, the gas density $\rho_{\mathrm{G}}$, the position of Gibbs surface $R_{\mathrm{G}}$, and the thickness of the surface . From the fitting using Eq. (27), we determine the values of $\rho_{\mathrm{L}}$, $\rho_{\mathrm{G}}$, and $R_{\mathrm{G}}$. Then we identify the position of Gibbs surface $R_{\mathrm{G}}$ as the equatorial radius $R_{\mathrm{e}}$. We measure the surface tension of the droplets in this equilibrium state following the method proposed by Ikeshoji *et al.* [13]. When the droplet is at rest, the equatorical radius $R_{\mathrm{e}}$ equals $R_0$. The measured physical quantities of the droplet at rest are summarized in Table. 1.

After the equilibrium droplet is obtained, the thermostat is turned off and angular velocity $\omega$ is given to the droplet. As the droplet deforms, the moment of inertia also changes, and consequently, the angular velocity varies. It makes difficult to investigate the relation between the angular velocity and the final form of the droplet precisely. Therefore, we keep the angular velocity fixed throughout the time evolution by the method similar to the velocity scaling scheme [14]. We observe the angular velocity of the droplet, and rescale the velocities of the particles in the droplet so that the total angular velocity of the droplet is kept to be the initial value.

## 4. Results

### *4.1. Static Form of Rotating Droplets*

Two typical time evolutions of droplets are shown in Figs. 3 and 4. The labels "Side" and "Top" in the figures denote that the view from the direction normal and parallel to the rotating axis, respectively. Figure 3 shows the results for the relatively slow rotation with $\omega = 0.025$. The shape of the droplet viewed from the side becomes ellipse, while the shape is kept circle viewed from the top. This implies that the droplet deforms to an oblate spheroid. Figure 4 shows the

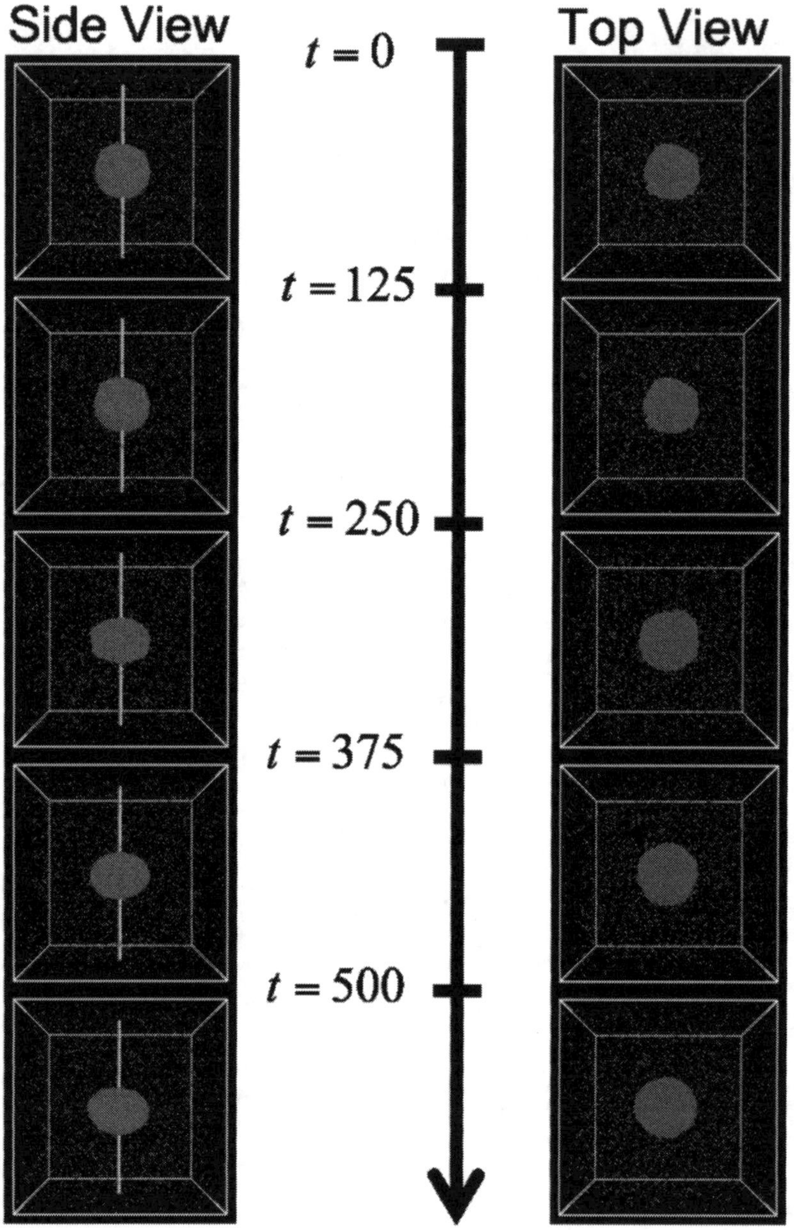


**Figure 3.** Time evolutions of a droplet of $N = 16000$ for $\omega = 0.025$. The yellow line presents the axis of the rotation. The views from the normal direction to the rotation axis are shown in the left, and those from the parallel direction are shown in the right. The form of the droplet changes from the sphere to the uniaxial.

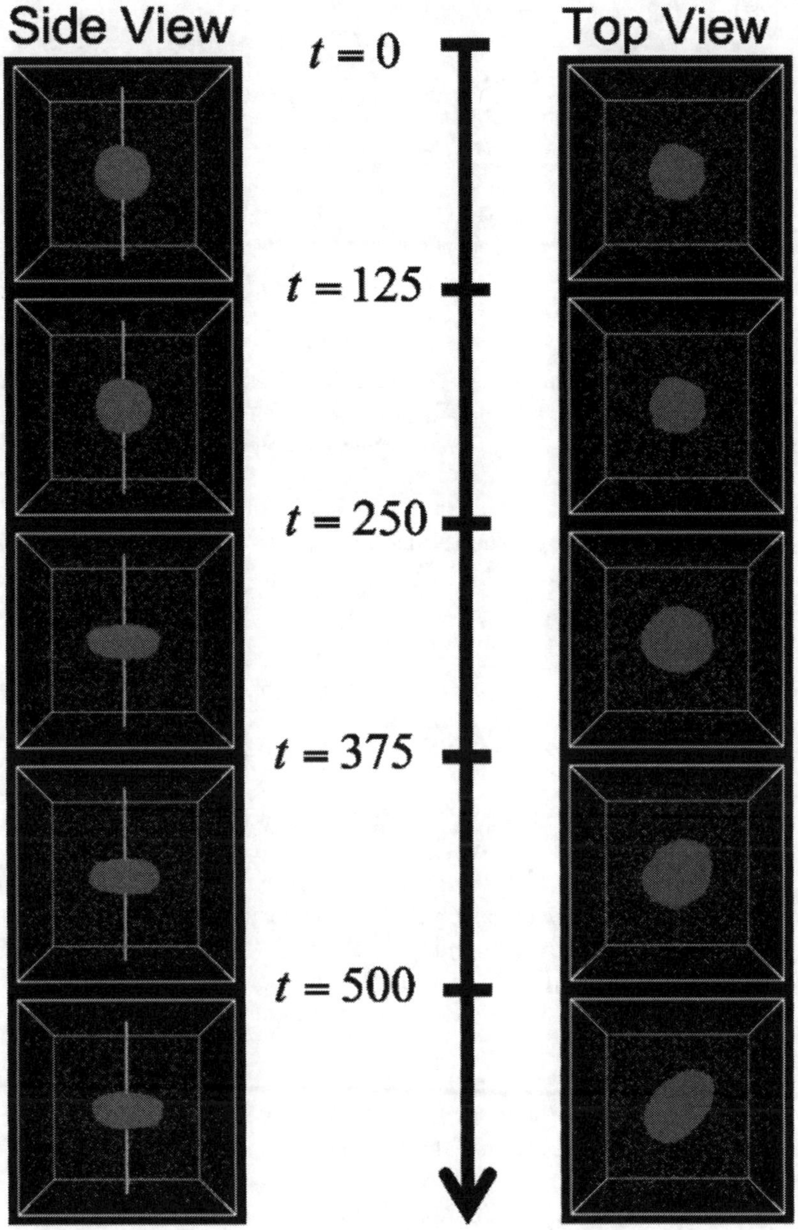


**Figure 4.** Same as Fig. 3 for $\omega = 0.05$. The form of the droplet becomes biaxial.

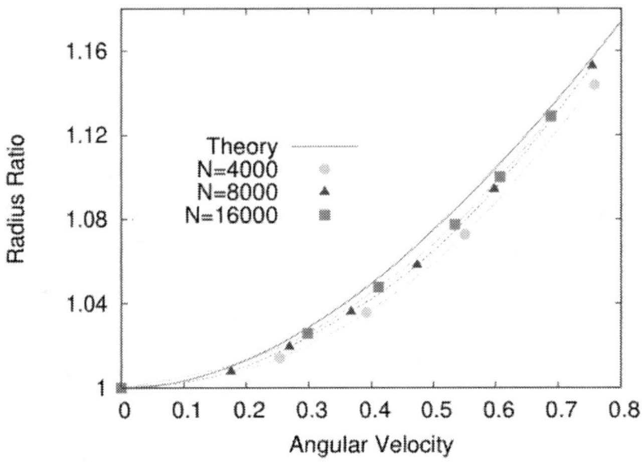


**Figure 5.** The angular velocity dependence of the deformation of rotating droplets. The radius ratio $R_e/R_0$ is shown as a function of the reduced angular velocity $\omega/\omega_0$. The solid line denotes the theoretical prediction in Eq. (25).

results for the larger angular velocity $\omega = 0.05$. The both shapes viewed from side and top become ellipses, *i.e.*, the droplet becomes biaxial shape. We find slight increase of temperature during rotation, but the increase is about $\sim 1\%$. Therefore, the influence of the heating by rotation can be negligible.

For the regime where the final shape of the droplet is uniaxial, we compare the results from MD and the theoretical predictions. We plot the ratio $R_e/R_0$ as a function of the reduced frequency $\omega/\omega_0$ in Fig. 5. While the deviations from the theoretical prediction are observed, the difference between the numerical result and the theory becomes smaller as the size of the droplet increases. Note that, the experimental results such as in Ref. [8] shows good agreements with the theretical prediction, which implies that the size of the treated droplets is large enough so that the influence from the thickness of the surface is negligibly-small.

### *4.2. Deformation Dynamics*

Next, we observe the dynamics of the deformation. While we have used the position of the Gibbs surface for the rotating droplet in steady state, it is difficult to determine it when the droplet undergoes deformation, since the density profile of such cases cannot be determined accurately. Therefore we measure the gyration radius instead of the position of the Gibbs surface in order to measure how a droplet deforms. The gyration radius $R$ is defined as

$$R^2 \equiv \frac{1}{N} \sum r_i^2, \tag{28}$$

where $r_i$ is the distance between particle $i$ to the rotating axis. The time evolutions of the radius ratio $R(t)/R(0)$ for different values of $\omega$ are shown in Fig. 6. Deformation becomes larger for a larger value of $\omega$. The time evolution of the radius is expected to be following exponential form as

$$R(t) = R_{\text{eq}} - [R_{\text{eq}} - R(0)] \exp(-t/\ \ ), \tag{29}$$

Journal of Physics: Conference Series **297** (2011) 012023 doi:10.1088/1742-6596/297/1/012023

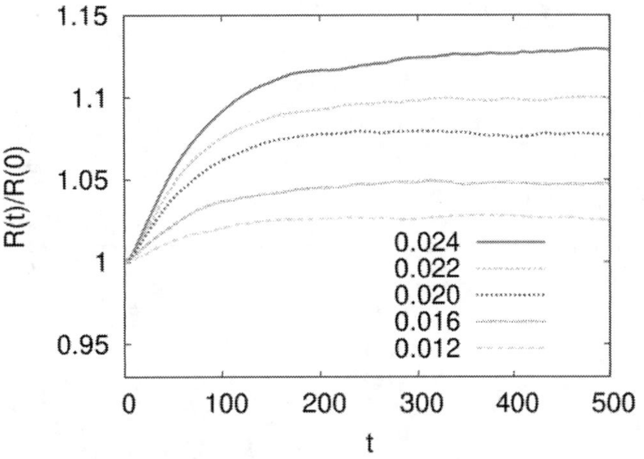


**Figure 6.** Time evolutions of the deformation which is defined by the ratio of the current radius to the initial radius $R(t)/R(0)$ for different values of the angular velocities. Temperature is fixed to be 0.6.

with the final value of the radius $R_{\mathrm{eq}}$, i.e., $R_{\mathrm{eq}} \equiv \lim_{t\to\infty} R(t)$. We define the reduced radius $R^*$ as

$$R^*(t) = \frac{R(t) - R(0)}{R_{\mathrm{eq}} - R(0)}, \tag{30}$$

in order to investigate the characteristic timescale by removing the amplitude of the deformation. The time evolutions of the reduced radii are shown in Fig. 7. The data are collapsed into the single curve without scaling of the horizontal axis. This implies that the deformations for different rotation velocity have the same timescale. In other words, the timescale of the deformation is almost independent of the angular frequency.

We also investigate influence of temperature. Time evolutions of the radii of the droplets for different temperatures are shown in Fig. 8. We estimate relaxation time  defined in Eq. (29), and determine the value of  $= 88.0(3)$, $68.3(6)$, and $61.3(7)$ for temperatures $T = 0.55$, $0.60$, and $0.70$. One can see that the timescale becomes shorter as temperature increases. This means that the droplet deforms quickly for high temperature.

## 5. Summary and Discussion

To summarize, nonequilibrium deformation processes of rotating droplets are investigated by molecular dynamics simulations. First, we study the size-dependence of the deformations. While the deformations of the rotating droplets in steady state are smaller than those of the theoretical predictions, the differences decrease as the size of the droplets increases. The differences are the same order as the ratio of the thickness to the radius. Therefore, the differences come from the fact that the surface of a droplet has finite thickness, and the thickness cannot be ignored for small droplets. This difference was not observed in the past studies such as in Ref. [8], which implies that the treated droplet in the expriments are large enough. Note that, the finite-size effect does not comes from the number of particles consisting the droplet, but the ratio of length of the interface to the radius of the droplet. As the temperature increases, the length of the interface increases and diverges at the critical point. Therefore, the finite-size effect will be observed in the macroscopic experiments in the region near the critical point. For faster speed

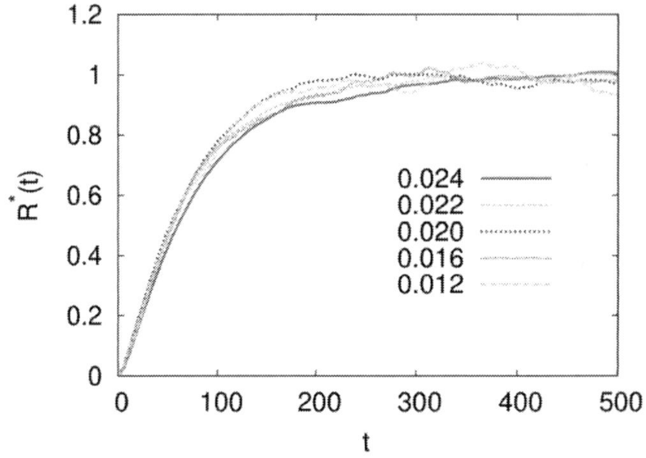


**Figure 7.** Time evolutions of reduced radius $R^*$ for different values of the angular velocity $\omega$. They are well scaled. This implies that they share the identical timescale for their deformations, while their amplitudes of the deformations are different.

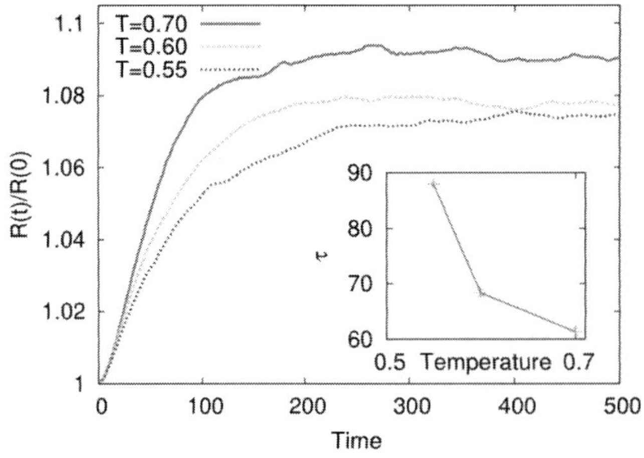


**Figure 8.** Time evolutions of deformations for different temperatures. The ratio of the radius $R(t)/R_0$ is shown, where $R(t)$ is the equatorial radius and $R(0)$ is the initial radius, respectively. (inset) The temperature dependence of the relaxation time . It shows that a droplet in higher temperature deforms quicker.

of rotation, the uniaxial form becomes unstable and biaxial form is observed as reported in the past study using the finite-element method [6].

Nonequilibrium behaviors of the deformation are also observed. We find that the rotation frequency only affects the form of the droplet and does not change the timescale of deformations, when the rotation is small enough so that the final shapes are uniaxial. While the characteristic timescale is independent of the rotation speed, it decreases as temperature increases. In order to understand these nonequilibrium behaviors, we have constructed a following simple model. Consider a spring-mass system in fluid. A particle with mass $M$ is connected to the spring whose spring constant is $k$, and the other end of the spring is fixed. When the particle moves with the angular velocity $\omega$ around the fixed point, then an overdumped equation of motion is written as,

$$\Gamma \frac{\mathrm{d}x}{\mathrm{d}t} = M(x_{\mathrm{n}} + x)\omega^2 - kx, \tag{31}$$

where $x_{\mathrm{n}}$ is the natural length of the spring and $x$ is displacement from it. $\Gamma$ denotes a phenomenological dissipation coefficient which is proportional to viscosity of the fluid. The solution of the equation is

$$x(t) = x_{\mathrm{eq}} - [x_{\mathrm{eq}} - x(0)] \exp(-t/\ ) \tag{32}$$

with the length at the equilibrium

$$x_{\mathrm{eq}} = \left[ \frac{M\omega^2}{k - M\omega^2} \right] x_{\mathrm{n}} \tag{33}$$

and the relaxation time

$$= \frac{\Gamma}{k - M\omega^2}. \tag{34}$$

The solution Eq. (32) corresponds to the behavior Eq. (29). In this model, the spring constant $k$ corresponds to the surface tension , which works as a restoring force (see Appendix). When the rotational frequency is much smaller than the restoring force, i.e., $|\omega^2|\ \ k/M$, then the characteristic timescale is reduced to be,

$$= \frac{\Gamma}{k}. \tag{35}$$

Then the timescale becomes independent of the rotational frequency, which is observed in the numerical simulations. The temperature dependence of the timescale is not trivial, since both restoring force $k$ and the dissipation coefficient $\Gamma$ depend on temperature. Both surface tension and viscosity decreases as temperature increases, and the observed deformation timescale becomes shorter as temperature increases. This implies that the decrease of the viscosity is faster than that of the surface tension, while situations may change for other types of liquid. The dissipation coefficient $\Gamma$ depends not only on temperature, but on size of a droplet. This size-dependence of dissipation is not trivial, since the dissipation caused by deformation is difficult to be solved exactly. These issues should be studied in the future. With larger angular velocity, the deformation of the droplet becomes more complicated. While rotation is almost rigid-body type for small angular velocity, the internal flow of droplets can change the behavior as discussed in Ref. [7]. Non rigid-body rotation will involve dissipation, which causes change of temperature, etc. Dynamics of such behaviors are also worth to be studied by molecular dynamics simulations.

**Acknowledgments**
The authors would like to thank Dr. T. Shimada and Dr. M. Suzuki for fruitful discussion. This work is partially supported by Grants-in Aid for Scientific Research (Contracts No. 19740235), and KAUST GRB (KUK-I1-005-04).

Journal of Physics: Conference Series **297** (2011) 012023 doi:10.1088/1742-6596/297/1/012023

## Appendix A. Restoring Force and Surface Tension

In this appendix, we relate the phenomenological restoring force $k$ in Eq. (32) to the surface tension  and the rotating frequency $\omega$. Consider a rotating droplet consisting $N$ particles. When the droplet rotate slow enough, the form of the droplet is uniaxial and well approximated by an oblate spheroid. The polar radius $R_{\mathrm{p}}$ and the equatorial radius $R$ satisfy the following equation,

$$N = \frac{4\pi}{3}\rho R^2 R_{\mathrm{p}}. \tag{A.1}$$

The surface area $S$ of this droplet is

$$S = 2\pi\left(R^2 + \frac{R_{\mathrm{p}}^2 \tanh^{-1} e}{e}\right), \tag{A.2}$$

with the eccentricity $e \equiv \sqrt{1 - R_{\mathrm{p}}^2/R^2}$. When the rotation velocity is small enough, the eccentricity is very close to unity. Then the approximation $\tanh^{-1} e \sim e$ for $|e| \quad 1$ leads to

$$S = 2\pi(R^2 + R_{\mathrm{p}}^2). \tag{A.3}$$

The moment of inertia $I$ of the droplet is

$$I = \frac{2}{5} m N R^2, \tag{A.4}$$

where $m$ is the mass of the particles. Consequently, the effective potential of the droplet can be written as

$$U = S \quad - \frac{1}{2} I\omega^2 \tag{A.5}$$

$$= \left(2\pi \quad - \frac{mN\omega^2}{5}\right) R^2 + \frac{9N^2}{8\pi\rho^2 R^4}, \tag{A.6}$$

with the rotating frequency $\omega$ and the surface tension  . The contribution from the inner pressure is ignored here, since the observed pressure is almost constant during its rotation in the numerical results. The restoring force for the form of droplets $F(r)$ is written as

$$F(r) = -\frac{\mathrm{d}U}{\mathrm{d}R} = -\left(4\pi \quad - \frac{2mN\omega^2}{5}\right) R + \frac{9N^2}{2\pi\rho^2 R^5}.$$

The mechanical equilibrium radius is given by $F(R_{\mathrm{eq}}) = 0$, which is

$$R_{\mathrm{eq}} = \left[\frac{45N^2}{2\pi\rho^2(20\pi \quad - 2mN^2\omega^2)}\right]^{1/6}. \tag{A.7}$$

The linear approximation around the mechanical equilibrium point gives the effective spring constant of the droplet $K$ as

$$K \equiv -\left.\frac{\mathrm{d}F}{\mathrm{d}R}\right|_{R=R_{\mathrm{eq}}} = \left(24\pi \quad - \frac{12mN\omega^2}{5}\right). \tag{A.8}$$

This $K$ corresponds to the restoring force $k - M\omega^2$ in Eq. (34). Then identifying $mN$ with $M$, we have the characteristic timescale of this system as

$$= \frac{\Gamma}{24\pi \quad - 12M\omega^2/5}. \tag{A.9}$$

When the rotation of the droplet is slow enough, the relaxation time becomes

$$= \frac{\Gamma}{24\pi}, \tag{A.10}$$

which is essentially equivalent to Eq. (35).

## References

[1] Sirignano W A 1983 *Progress in Energy and Combustion Science* **9** 291
[2] Calvert P 2001 *Chemistry of Materials* **13** 3299
[3] Bringa E M, Johnson R E, and Jakas M 1999 *Phys. Rev. B* **60** 15107
[4] Plateau J 1863 *Annual Report of the Board of Regents of the Smithsonian Institution, Washington DC, 1863* (Washington DC: The Smithsonian Institution) 207–285
[5] Hill R J A and Eaves L 2008 *Phys. Rev. Lett.* **101** 234501
[6] Brown R A and Scriven L E1980 *Proc. R. Soc. Lond. A* **371** 331
[7] Watanabe T 2009 *Physics Letters A* **373** 867
[8] Wang T G, Trinh E H, Croonquist A P, and Elleman D D 1986 *Phys. Rev. Lett.* **56** 452
[9] Chandrasekhar S 1965 *Proc. R. Soc. Lond. A.* **286** 1
[10] Stoddard S D and Ford J 1973 *Phys. Rev. A* **8** 1504
[11] Adelman S A and Doll J D 1976 *J. Chem. Phys.* **64** 2375
[12] `http://mdacp.sourceforge.net/`
[13] Ikeshiji T, Hafskjold B, and Furholt H 2003 *Molecular Simulation* **29** 101
[14] Woodcock L V 1971 *Chem. Phys. Lett.* **10** 257

# AUTHOR INDEX